T0140269

Springer Proceedings in Physics

Volume 225

Indexed by Scopus

The series Springer Proceedings in Physics, founded in 1984, is devoted to timely reports of state-of-the-art developments in physics and related sciences. Typically based on material presented at conferences, workshops and similar scientific meetings, volumes published in this series will constitute a comprehensive up-to-date source of reference on a field or subfield of relevance in contemporary physics. Proposals must include the following:

- name, place and date of the scientific meeting
- a link to the committees (local organization, international advisors etc.)
- scientific description of the meeting
- list of invited/plenary speakers
- an estimate of the planned proceedings book parameters (number of pages/ articles, requested number of bulk copies, submission deadline).

More information about this series at http://www.springer.com/series/361

José-Enrique García-Ramos ·
María V. Andrés · José A. Lay Valera ·
Antonio M. Moro · Francisco Pérez-Bernal
Editors

Basic Concepts in Nuclear Physics: Theory, Experiments and Applications

2018 La Rábida International Scientific Meeting on Nuclear Physics

Editors
José-Enrique García-Ramos
Department of Integrated Sciences
Faculty of Experimental Sciences
University of Huelva
Huelva, Spain

María V. Andrés
Department of Atomic, Molecular
and Nuclear Physics, Faculty of Physics
University of Seville
Seville, Spain

José A. Lay Valera
Department of Atomic, Molecular
and Nuclear Physics, Faculty of Physics
University of Seville
Seville, Spain

Antonio M. Moro
Department of Atomic, Molecular
and Nuclear Physics, Faculty of Physics
University of Seville
Seville, Spain

Francisco Pérez-Bernal
Department of Integrated Sciences
Faculty of Experimental Sciences
University of Huelva
Huelva, Spain

ISSN 0930-8989 ISSN 1867-4941 (electronic)
Springer Proceedings in Physics
ISBN 978-3-030-22206-2 ISBN 978-3-030-22204-8 (eBook)
https://doi.org/10.1007/978-3-030-22204-8

This Springer imprint is published by the registered company Springer Nature Switzerland AG
The registered company address is: Gewerbestrasse 11, 6330 Cham, Switzerland

Preface

The name of La Rábida has a special significance for the nuclear physics community. At the beginning of the 80s of the past century, professors from the University of Seville launched a series of Summer Schools on Nuclear Physics, which was first known as "La Rábida Summer Schools" and later as "Hispalensis Summer Schools". The first edition took place in 1982, that is, 36 years ago. Since then, a large fraction of today's world-leading nuclear physicists have participated in one of several editions of the School, either as student or as a keynote speaker or, in some cases, as both. After eight editions and a gap of several years, in 2009 professors of the University of Sevilla and Huelva decided to rekindle this event and organized a new edition, which was named *International Scientific Meeting on Nuclear Physics*, trying to convey the same spirit and zest enjoyed in the pioneer editions.

The 2018 La Rábida International Scientific Meeting on Nuclear Physics was held from June 18 to 22, 2018 in the campus of the International University of Andalucía (UNIA) at La Rábida (Huelva, Spain) and it is the fourth of the new series, with the same structure and general title, i.e., *Basic Concepts in Nuclear Physics: Theory, Experiments and Applications* than the three previous ones. The opening ceremony was presided over by the Director of the International University of Andalucía (UNIA) at La Rábida, Prof. Agustn Galán, with the presence of the Rector of the University of Huelva, Prof. María Antonia Peña, and of the Director of the Meeting, Prof. José-Enrique García-Ramos.

The aim of the meeting, as in the previous editions, has been to provide the participants—mostly graduate students and young postdocs—with a wide and solid education in different aspects of the field of Nuclear Physics. The course was divided into three main topics: theory, experiments, and applications. Six experienced and well-known researchers have participated as keynote speakers in the event; each of them giving four-hour lectures covering some of today's hottest topics in the field of Nuclear Physics. In addition to this, young participants have also presented their own research results through seminar and poster sessions. Most of the lectures and contributions have been collected and published in the present special number of Springer Proceeding in Physics.

The topics presented by the speakers in their lectures covered the whole field of Nuclear Physics, from applications with a significant social impact, as nuclear medicine or medical image processing, to fundamental topics in basic research, theory, and experiment. Here we list the keynote speaker names and affiliation as well as the topics covered in their lectures:

- Prof. Alex Brown. Michigan State University (USA). *Shell model.*
- Prof. Pierre Capel. Johannes Gutenberg-Universität Mainz (Germany). *Nuclear reaction theory.*
- Dr. Tommi Eronen. University of Jyväskylä (Finland). *Experimental techniques for mass measurements.*
- Prof. Juan José Gómez Cadenas. Instituto de Física Corpuscular, IFIC-CSIC (Spain). *Neutrino physics and NEXT experiment.*
- Dr. Alexandre Obertelli. Institut für Kernphysik, Technische Universität Darmstadt (Germany). *Nuclear Reaction Experiments.*
- Prof. Katia Parodi. LMU Munich Physics (Germany). *Medical image processing, treatment planning, PET applications.*

The number of registered Ph.D. students and postdocs was around 60 and they came from very different countries: Algeria, Belgium, Brazil, Bulgaria, China, Colombia, India, Iran, Italy, Germany, Mexico, Nigeria, Poland, Portugal, Spain, and Turkey. In the particular case of Spain, participants came from 9 different Universities and Research Centers, covering basically all institutions where active nuclear physics groups are working.

Grants covering partly lodging and boarding expenses were supplied to 25 participants. This has been possible thanks to the financial support received from the *Consejera de la Presidencia, Administración Local y Memoria Democrática de la Junta de Andaluca,* and the *Cátedra AIQBE (Asociación de Empresas Químicas, Básicas y Energéticas de Huelva) de la Universidad de Huelva.*

The brilliant lectures given by the keynote speakers and the contributions of the young participants have undoubtedly shown the current interest and impact of Nuclear Physics, witnessed by the many bright students working in fundamental research or in the very diverse applications of Nuclear Physics that attended the meeting. Two students were awarded diplomas: Sìlvia Viñals i Onsès with the poster "Electron Capture of ^{8}B into highly excited states of ^{8}Be" and Kajetan Niewczas with the seminar "Modeling nuclear effects for neutrino-nucleus scattering in the few-GeV region", for the most outstanding poster and seminar, respectively.

Nevertheless the event dense academic program, an intense social program was allocated in the evenings. In particular, all participants had the opportunity to enjoy the nearby beach of "El Parador" in Mazagón in two occasions, a special dinner took place in the touristic town of Punta Umbría, and a boat excursion along "la ría de Huelva" that included dinner on board. The participants also visited the replicas of Columbus' "carabelas", located at walking distance from the conference venue.

Last, but not least, a farewell dinner was served at the conference venue, where we enjoyed the show of the rock band "Rosam".

We would like to conclude giving thanks with special gratitude to the students and young postdocs, who have helped us with the daily work and to the staff of the UNIA campus at La Rábida. Organizing this event would have been impossible without their support.

Huelva, Spain
Seville, Spain

José-Enrique García-Ramos
Francisco Pérez-Bernal
María V. Andrés
Antonio M. Moro
José A. Lay Valera

Sponsors

Rábida 2018 Sponsoring Institutions

Rábida 2018 Organizing Institutions

Photographs

un
Universi
INTERNATIONAL SCIENTIFIC
MEETING ON
NUCLEAR PHYSICS
RÁBIDA 18

un
Universidad
Internacional
Andalucía

un
Unive
Intern
de A

Why NEXT? — Advantages of HPXe technology
$T_{1/2}^{-1} \propto a \cdot \epsilon \cdot \sqrt{\frac{Mt}{\Delta E \cdot B}}$

Introduction to Nuclear Reaction Theory
Pierre Capel
JG|U
18–20 June 2018

un
A

Lisbon

Contents

Part I Keynote Speakers

1 **The Nuclear Configuration Interations Method** 3
B. Alex Brown

2 **Introduction to Nuclear-Reaction Theory** 33
Pierre Capel

3 **Nuclear Reaction Experiments** 75
Alexandre Obertelli

Part II Student's Seminars

4 **Measurement of the ^{244}Cm and ^{246}Cm Neutron-Induced Cross Sections at the n_TOF Facility** 117
V. Alcayne, A. Kimura, E. Mendoza, D. Cano-Ott, O. Aberle, S. Amaducci, J. Andrzejewski, L. Audouin, V. Babiano-Suarez, M. Bacak, M. Barbagallo, V. Bécares, F. Bečvář, G. Bellia, E. Berthoumieux, J. Billowes, D. Bosnar, A. S. Brown, M. Busso, M. Caamaño, L. Caballero, M. Calviani, F. Calviño, A. Casanovas, F. Cerutti, Y. H. Chen, E. Chiaveri, N. Colonna, G. P. Cortés, M. A. Cortés-Giraldo, L. Cosentino, S. Cristallo, L. A. Damone, M. Diakaki, M. Dietz, C. Domingo-Pardo, R. Dressler, E. Dupont, I. Durán, Z. Eleme, B. Fernández-Domíngez, A. Ferrari, I. Ferro-Gon calves, P. Finocchiaro, V. Furman, A. Gawlik, S. Gilardoni, T. Glodariu, K. Göbel, E. González-Romero, C. Guerrero, F. Gunsing, S. Heinitz, J. Heyse, D. G. Jenkins, Y. Kadi, F. Käppeler, N. Kivel, M. Kokkoris, Y. Kopatch, M. Krtička, D. Kurtulgil, I. Ladarescu, C. Lederer-Woods, J. Lerendegui-Marco, S. Lo Meo, S.-J. Lonsdale, D. Macina, A. Manna, T. Martínez, A. Masi, C. Massimi, P. F. Mastinu, M. Mastromarco, F. Matteucci, E. Maugeri, A. Mazzone, A. Mengoni, V. Michalopoulou, P. M. Milazzo, F. Mingrone, A. Musumarra, A. Negret, R. Nolte,

F. Ogállar, A. Oprea, N. Patronis, A. Pavlik, J. Perkowski, L. Piersanti, I. Porras, J. Praena, J. M. Quesada, D. Radeck, D. Ramos Doval, T. Rausher, R. Reifarth, D. Rochman, C. Rubbia, M. Sabaté-Gilarte, A. Saxena, P. Schillebeeckx, D. Schumann, A. G. Smith, N. Sosnin, A. Stamatopoulos, G. Tagliente, J. L. Tain, Z. Talip, A. E. Tarifeño-Saldivia, L. Tassan-Got, A. Tsinganis, J. Ulrich, S. Urlass, S. Valenta, G. Vannini, V. Variale, P. Vaz, A. Ventura, V. Vlachoudis, R. Vlastou, A. Wallner, P. J. Woods, T. J. Wright and P. Žugec

5 **First Steps Towards An Understanding of the Relation Between Heavy Ion Double Charge Exchange Nuclear Reactions and Double Beta Decays** 123
Jessica I. Bellone, S. Burrello, Maria Colonna, Horst Lenske and José A. Lay Valera

6 **Study on the Decay of ^{46}Ti*** 127
M. Cicerchia, F. Gramegna, D. Fabris, T. Marchi, M. Cinausero, G. Mantovani, A. Caciolli, G. Collazzuol, D. Mengoni, M. Degerlier, L. Morelli, M. Bruno, M. D'Agostino, C. Frosin, S. Barlini, S. Piantelli, M. Bini, G. Pasquali, P. Ottanelli, G. Casini, G. Pastore, D. Gruyer, A. Camaiani, S. Valdré, N. Gelli, A. Olmi, G. Poggi, I. Lombardo, D. Dell'Aquila, S. Leoni, N. Cieplicka-Orynczak and B. Fornal

7 **Bayesian Reconstruction of Axial Dose Maps Using the Measurements of a Novel Detection System for Verification of Advanced Radiotherapy Treatments** 131
A. D. Domínguez-Muñoz, M. C. Battaglia, J. M. Espino, R. Arráns, M. A. Cortés-Giraldo and M. I. Gallardo

8 **Be-10 Measurements in Atmospheric Filters Using the AMS Technique: The Data Analysis** 133
K. De Los Ríos, C. Méndez-García, S. Padilla, C. Solís, E. Chávez, A. Huerta and L. Acosta

9 **Fission Studies in Inverse Kinematics** 137
M. Feijoo, J. Benlliure, J. L. Rodríguez-Sanchéz and J. Taieb

10 **Iterative Algorithm for Optimal Super Resolution Sampling** 141
P. Galve, A. López-Montes, J. M. Udías and J. López Herraiz

11 **Adiabatic Correction to the Eikonal Approximation** 145
C. Hebborn, D. Baye and Pierre Capel

12 Modeling Neutrino-Nucleus Interactions for Neutrino Oscillation Experiments 149
G. D. Megias, S. Dolan and S. Bolognesi

13 Neutron Radiography at CNA 153
M. A. Millán-Callado, C. Guerrero, B. Fernández, A. M. Franconetti, J. Lerendegui-Marco, M. Macías, T. Rodríguez-González and J. M. Quesada

14 Modeling Nuclear Effects for Neutrino-Nucleus Scattering in the Few-GeV Region 155
K. Niewczas, R. González-Jiménez, N. Jachowicz, A. Nikolakopoulos and J. T. Sobczyk

15 Effect of Outgoing Nucleon Wave Function on Reconstructed Neutrino Energy 157
A. Nikolakopoulos, M. Martini, N. Van Dessel, K. Niewczas, R. González-Jiménez and N. Jachowicz

16 Measurement of the Production Cross Sections of β^+ Emitters for Range Verification in Proton Therapy 159
T. Rodríguez-González, C. Guerrero, M. C. Jiménez-Ramos, J. Lerendegui-Marco, M. A. Millán-Callado, A. Parrado and J. M. Quesada

Part III Student's Posters

17 Calculation of Energy Level and B(E2) Values and G-Factor of Even–Even Isotopes of Sulfur Using the Shell Model Code NuShellX 165
Amin Attarzadeh and Saed Mohammadi

18 Characterization and First Test of an i-TED Prototype at CERN n_TOF 169
V. Babiano-Suarez, L. Caballero, C. Domingo-Pardo, I. Ladarescu, O. Aberle, V. Alcayne, S. Amaducci, J. Andrzejewski, L. Audouin, M. Bacak, M. Barbagallo, V. Bécares, F. Bečvář, G. Bellia, E. Berthoumieux, J. Billowes, D. Bosnar, A. S. Brown, M. Busso, M. Caamaño, M. Calviani, F. Calviño, D. Cano-Ott, A. Casanovas, F. Cerutti, Y. H. Chen, E. Chiaveri, N. Colonna, G. P. Cortés, M. A. Cortés-Giraldo, L. Cosentino, S. Cristallo, L. A. Damone, M. Diakaki, M. Dietz, R. Dressler, E. Dupont, I. Durán, Z. Eleme, B. Fernández-Domíngez, A. Ferrari, I. Ferro-Gon calves, P. Finocchiaro, V. Furman, A. Gawlik, S. Gilardoni, T. Glodariu, K. Göbel, E. González-Romero, C. Guerrero, F. Gunsing, S. Heinitz, J. Heyse, D. G. Jenkins, Y. Kadi, F. Käppeler, A. Kimura, N. Kivel, M. Kokkoris, Y. Kopatch, M. Krtička, D. Kurtulgil,

C. Lederer-Woods, J. Lerendegui-Marco, S. Lo Meo, S.-J. Lonsdale, D. Macina, A. Manna, T. Martínez, A. Masi, C. Massimi, P. F. Mastinu, M. Mastromarco, F. Matteucci, E. Maugeri, A. Mazzone, E. Mendoza, A. Mengoni, V. Michalopoulou, P. M. Milazzo, F. Mingrone, A. Musumarra, A. Negret, R. Nolte, F. Ogállar, A. Oprea, N. Patronis, A. Pavlik, J. Perkowski, L. Piersanti, I. Porras, J. Praena, J. M. Quesada, D. Radeck, D. Ramos Doval, T. Rausher, R. Reifarth, D. Rochman, C. Rubbia, M. Sabaté-Gilarte, A. Saxena, P. Schillebeeckx, D. Schumann, A. G. Smith, N. Sosnin, A. Stamatopoulos, G. Tagliente, J. L. Tain, Z. Talip, A. E. Tarifeño-Saldivia, L. Tassan-Got, A. Tsinganis, J. Ulrich, S. Urlass, S. Valenta, G. Vannini, V. Variale, P. Vaz, A. Ventura, V. Vlachoudis, R. Vlastou, A. Wallner, P. J. Woods, T. J. Wright and P. Žugec

19 Development of a New Radiobiology Beam Line for the Study of Proton RBE at the 18 MeV Proton Cyclotron Facility at CNA 175
A. Baratto-Roldán, M. A. Cortés-Giraldo, M. C. Jiménez-Ramos, M. C. Battaglia, J. García López, M. I. Gallardo and J. M. Espino

20 Role of Competing Transfer Channels on Charge-Exchange Reactions 177
S. Burrello, J. I. Bellone, Maria Colonna, José A. Lay Valera and Horst Lenske

21 Two-Neutron Transfer in the $^{18}O+^{28}Si$ System 181
E. N. Cardozo, J. Lubian, F. Cappuzzello, R. Linares, D. Carbone, M. Cavallaro, J. L. Ferreira, B. Paes, A. Gargano and G. Santagati

22 Study of the Neutron-Rich Region in Vicinity of ^{208}Pb via Multinucleon Transfer Reactions 185
P. Čolović, A. Illana, S. Szilner, J. J. Valiente-Dobón and PRISMA GALILEO MINIBALL Collaborations

23 The QClam-Spectrometer at the S-DALINAC 189
Antonio D'Alessio, Peter von Neumann-Cosel, N. Pietralla, Maxim Singer and V. Werner

24 Kaonic Atoms Measurement at DAΦNE: SIDDHARTA and SIDDHARTA-2 191
L. De Paolis, D. Sirghi, A. Amirkhani, A. Baniahmad, M. Bazzi, G. Bellotti, C. Berucci, D. Bosnar, M. Bragadireanu, M. Cargnelli, C. Curceanu, A. Dawood Butt, R. Del Grande, L. Fabbietti, C. Fiorini, F. Ghio, C. Guaraldo, M. Iliescu, M. Iwasaki, P. Levi Sandri, J. Marton, M. Miliucci, P. Moskal, S. Niedźwiecki,

S. Okada, D. Pietreanu, K. Piscicchia, H. Shi, M. Silarski, F. Sirghi, M. Skurzok, A. Spallone, H. Tatsuno, O. Vazquez Doce, E. Widmann and J. Zmeskal

25 **PIGE Technique Within the EnsarRoot Framework** 197
E. Galiana, D. Galaviz, H. Alvarez-Pol, P. Teubig and P. Cabanelas

26 **Gogny Force Useful for Neutron Star Calculations** 199
C. Gonzalez-Boquera, M. Centelles, X. Viñas and L. M. Robledo

27 **Nuclear Structure and β^- Decay of A = 90 Isobars** 203
Nadjet Laouet and Fatima Benrachi

28 **Real-Time Tomographic Image Reconstruction in PET Using the Pseudoinverse of the System Response Matrix** 207
A. López-Montes, P. Galve, J. M. Udías and J. López Herraiz

29 **Photoacoustic Dose Monitoring in Radiosurgery** 211
O. M. Giza, D. Sánchez-Parcerisa, J. Camacho, V. Sánchez-Tembleque, S. Avery and J. M. Udías

30 **Structure of Light Nuclei Studied with ^{7}Li+^{6,7}Li Reactions** 215
D. Nurkić, M. Uroić, M. Milin, A. Di Pietro, P. Figuera, M. Fisichella, M. Lattuada, I. Martel, Đ. Miljanić, M. G. Pellegriti, L. Prepolec, A. M. Sánchez-Benítez, V. Scuderi, N. Soić, E. Strano and D. Torresi

31 **ICH15: A Linac Accelerator for Proton Therapy and Radioisotope Production Using IH/CH Cavities** 217
A. K. Orduz, I. Martel, A. C. C. Villari, J. Sánchez-Segovia, C. Bontoui, F. Manchado de Sola, R. Berjillos, J. Pérez, J. López-Morillas, J. Díaz, A. Jurado, A. M. López-Antequera, J. Vazquez, J. L. Aguado-Casas, T. Pérez, A. Pinto, D. Ablanedo, E. Hidalgo and M. Trueba

32 **Near Coulomb Barrier Scattering of ^{15}C on ^{208}Pb** 221
J. D. Ovejas

33 **Preliminary Data of ^{10}Be/^{9}Be Ratios in Aerosol Filters in Mexico City** 225
S. Padilla, C. G. Méndez, C. Solís, L. Acosta, M. Rodríguez-Ceja and E. Chávez

34 **Measurement of Signal-to-Noise Ratio in Straw Tube Detectors for PANDA Forward Tracker** 229
Narendra Rathod, Jerzy Smyrski and Akshay Malige

35 Investigation of the Pygmy Dipole Resonance Across the Shell Closure in Chromium Isotopes 233
P. C. Ries, T. Beck, J. Beller, M. Bhike, U. Gayer, J. Isaak, B. Löher, Krishichayan, L. Mertes, H. Pai, N. Pietralla, V. Yu. Ponomarev, C. Romig, D. Savran, M. Schilling, W. Tornow, V. Werner and M. Zweidinger

36 Modification of UO_2 Fuel Thermal Conductivity Model at High Burnup Structure 235
B. Roostaii, H. Kazeminejad and S. Khakshournia

37 Shell Model Calculations for Nuclei with Two Valence Nucleons Around the Doubly Magic ^{78}Ni Core 239
Hanane Saifi and Fatima Benrachi

38 Investigation of the Mechanism of Proton Induced Spallation Reactions 243
U. Singh, I. Ciepał, B. Kamys, P. Lasko, J. Łukasik, A. Magiera, P. Pawłowski, K. Pysz, Z. Rudy and S. K. Sharma

39 Challenging the Calorimeter CALIFA for FAIR Using High Energetic Photons 245
P. Teubig, P. Remmels, P. Klenze, H. Alvarez-Pol, E. Alves, J. M. Boillos, P. Cabanelas, R. C. da Silva, D. Cortina-Gil, J. Cruz, D. Ferreira, M. Fonseca, D. Galaviz, E. Galiana, R. Gernhäuser, D. González, A. Henriques, A. P. Jesus, H. Luís, J. Machado, L. Peralta, J. Rocha, A. M. Sánchez-Benítez, H. Silva and P. Velho

40 Research and Development of a Position-Sensitive Scintillator Detector for γ- and X-Ray Imaging and Spectroscopy 247
Zh. Toneva, V. Bozhilov, G. Georgiev, S. Ivanov, D. Ivanova, V. Kozhuharov, S. Lalkovski and G. Vankova-Kirilova

41 The Path Towards Low Dose CT: The Case of Breast CBCT 251
A. Villa-Abaunza, P. Ibáñez García, J. López Herraiz and J. M. Udías

42 Electron Capture of ^{8}B into Highly Excited States of ^{8}Be 255
S. Viñals

43 Examining the Helium Cluster Decays of the ^{12}Be Excited States by Triton Transfer to the ^{9}Li Beam 257
N. Vukman, N. Soić, P. Čolović, M. Uroić, M. Freer, T. Davinson, A. Di Pietro, M. Alcorta, D. Connolly, A. Lennarz, C. Ruiz, A. Shotter, M. Williams and A. Psaltis

44 Fast-Timing Lifetime Measurement of ^{174,176,178,180}Hf 259
J. Wiederhold, V. Werner, R. Kern, N. Pietralla, D. Bucurescu, R. Carroll, N. Cooper, T. Daniel, D. Filipescu, N. Florea, R.-B. Gerst, D. Ghita, L. Gurgi, J. Jolie, R. Ilieva, R. Lica, N. Marginean, R. Marginean, C. Mihai, I. O. Mitu, F. Naqvi, C. Nita, M. Rudigier, S. Stegemann, S. Pascu and P. H. Regan

45 Searching for Halo Nuclear Excited States Using Sub-Coulomb Transfer Reactions 261
J. Yang, Pierre Capel and Alexandre Obertelli

Contributors

O. Aberle European Organization for Nuclear Research (CERN), Meyrin, Switzerland

D. Ablanedo ATI Sistemas, Bergondo (A Coruña), Spain

L. Acosta Laboratorio Nacional de Espectrometría de Masas Con Acelerador (LEMA), Instituto de Física, Universidad Nacional Autónoma de México (UNAM), Circuito de la Investigación Científica Ciudad Universitaria, Mexico, D.F, Mexico

J. L. Aguado-Casas Science and Technology Research Centre, University of Huelva, Huelva, Spain

V. Alcayne Centro de Investigaciones Energéticas Medioambientales y Tecnológicas (CIEMAT), Madrid, Spain

M. Alcorta TRIUMF, Vancouver, BC, Canada

B. Alex Brown Department of Physics and the National Superconducting Cyclotron Laboratory, Michigan State University, East Lansing, MI, USA

H. Alvarez-Pol Departamento de Física de Partículas, Instituto Gallego de Física de Altas Energías (IGFAE), Universidad de Santiago de Compostela, Santiago de Compostela, Spain

E. Alves IPFN, IST-UL, Lisbon, Portugal

S. Amaducci INFN Laboratori Nazionali del Sud, Catania, Italy

A. Amirkhani Informazione e Bioingegneria and INFN Sezione di Milano, Dipartimento di Elettronica, Politecnico di Milano, Milan, Italy

J. Andrzejewski University of Lodz, Lodz, Poland

R. Arráns Hospital Universitario Virgen Macarena (HUVM), Seville, Spain

Amin Attarzadeh Institute for Higher Education ACECR, Khouzestan, Iran

L. Audouin IPN, CNRS-IN2P3, University of Paris-Sud, Université Paris-Saclay, Orsay, France

S. Avery Department of Radiation Oncology, Hospital of the University of Pennsylvania, Philadelphia, PA, USA

V. Babiano-Suarez Instituto de Física Corpuscular, CSIC - Universidad de Valencia, Valencia, Spain

M. Bacak European Organization for Nuclear Research (CERN), Meyrin, Switzerland;
Technische Universität Wien, Vienna, Austria

A. Baniahmad Informazione e Bioingegneria and INFN Sezione di Milano, Dipartimento di Elettronica, Politecnico di Milano, Milan, Italy

A. Baratto-Roldán Centro Nacional de Aceleradores, Seville, Spain

M. Barbagallo European Organization for Nuclear Research (CERN), Meyrin, Switzerland;
Istituto Nazionale di Fisica Nucleare, Bari, Italy

S. Barlini INFN Sezione di Firenze e Dipartimento di Fisica e Astronomia, Università Di Firenze, Florence, Italy

M. C. Battaglia Departamento de Física Atómica, Molecular y Nuclear, Universidad de Sevilla, Seville, Spain

D. Baye Physique Nucléaire et Physique Quantique (CP 229), Université libre de Bruxelles (ULB), Brussels, Belgium

M. Bazzi INFN, Laboratori Nazionali di Frascati, Frascati (Roma), Italy

V. Bécares Centro de Investigaciones Energéticas Medioambientales y Tecnológicas (CIEMAT), Madrid, Spain

T. Beck Institut für Kernphysik, Darmstadt, Germany

F. Bečvář Charles University, Prague, Czech Republic

J. Beller Institut für Kernphysik, Darmstadt, Germany

G. Bellia INFN Laboratori Nazionali del Sud, Dipartimento di Fisica e Astronomia, Università di Catania, Catania, Italy

Jessica I. Bellone INFN - LNS, Catania, Italy

G. Bellotti Informazione e Bioingegneria and INFN Sezione di Milano, Dipartimento di Elettronica, Politecnico di Milano, Milan, Italy

J. Benlliure Department of Particle Physics, IGFAE, University of Santiago de Compostela, Santiago de Compostela, Spain

Fatima Benrachi Laboratoire de Physique Mathematique et de Physique Subatomique (LPMS), Frères Mentouri Constantine-1 University, Constantine, Algeria

R. Berjillos Science and Technology Research Centre, University of Huelva, Huelva, Spain;
TTI Norte, La Rinconada (Sevilla), Spain

E. Berthoumieux CEA Saclay, Irfu, Université Paris-Saclay, Gif-sur-Yvette, France

C. Berucci INFN, Laboratori Nazionali di Frascati, Frascati (Roma), Italy;
Stefan-Meyer-Institut für Subatomare Physik, Vienna, Austria

M. Bhike Triangle Universities Nuclear Laboratory, Durham, NC, USA

J. Billowes University of Manchester, Manchester, UK

M. Bini INFN Sezione di Firenze e Dipartimento di Fisica e Astronomia, Università Di Firenze, Florence, Italy

J. M. Boillos IGFAE, Universidade de Santiago de Compostela, Santiago de Compostela, Spain

S. Bolognesi DPhP, IRFU, CEA Saclay, Gif-sur-Yvette, France

C. Bontoui Hospital Juan Ramón Jiménez, Servicio Andaluz de Salud, Science and Technology Research Centre, University of Huelva, Huelva, Spain

D. Bosnar Department of Physics, Faculty of Science, University of Zagreb, Zagreb, Croatia

V. Bozhilov Faculty of Physics, University of Sofia “St. Kliment Ohridski”, Sofia, Bulgaria

M. Bragadireanu Horia Hulubei National Institute of Physics and Nuclear Engineering (IFIN-HH), Magurele, Romania

A. S. Brown University of York, York, UK

M. Bruno INFN Sezione di Bologna e Dipartimento di Fisica e Astronomia, Università Di Bologna, Bologna, Italy

D. Bucurescu “Horia Hulubei” NIPNE, Bucharest-Magurele, Romania

S. Burrello INFN - LNS, Catania, Italy;
Departamento de FAMN, Universidad de Sevilla, Seville, Spain

M. Busso Dipartimento di Fisica e Geologia, Istituto Nazionale di Fisica Nazionale, Università di Perugia, Perugia, Italy

M. Caamaño University of Santiago de Compostela, Santiago de Compostela, Spain

L. Caballero Instituto de Física Corpuscular, CSIC - Universidad de Valencia, Valencia, Spain

P. Cabanelas Departamento de Física de Partículas, Instituto Gallego de Física de Altas Energías (IGFAE), Universidad de Santiago de Compostela, Santiago de Compostela, Spain

A. Caciolli INFN Sezione di Padova, Dipartimento di Fisica e Astronomia dell'Università di Padova, Padua, Italy

M. Calviani European Organization for Nuclear Research (CERN), Meyrin, Switzerland

F. Calviño Universitat Politècnica de Catalunya, Barcelona, Spain;
University of Santiago de Compostela, Santiago de Compostela, Spain

J. Camacho ITEFI, Spanish National Research Council (CSIC), Madrid, Spain

A. Camaiani INFN Sezione di Firenze e Dipartimento di Fisica e Astronomia, Università Di Firenze, Florence, Italy

D. Cano-Ott Centro de Investigaciones Energéticas Medioambientales y Tecnológicas (CIEMAT), Madrid, Spain

Pierre Capel Physique Nucléaire et Physique Quantique (CP 229), Université libre de Bruxelles (ULB), Brussels, Belgium;
Institut für Kernphysik, Johannes Gutenberg-Universität Mainz, Mainz, Germany

F. Cappuzzello Laboratori Nazionali del Sud, Istituto Nazionale di Fisica Nucleare, Catania, Italy

D. Carbone Laboratori Nazionali del Sud, Istituto Nazionale di Fisica Nucleare, Catania, Italy

E. N. Cardozo Universidade Federal Fluminense, Niteroi, Brazil

M. Cargnelli Stefan-Meyer-Institut für Subatomare Physik, Vienna, Austria

R. Carroll Department of Physics, University of Surrey, Guildford, Surrey, UK

A. Casanovas Universitat Politècnica de Catalunya, Barcelona, Spain;
University of Santiago de Compostela, Santiago de Compostela, Spain

G. Casini INFN Sezione di Firenze e Dipartimento di Fisica e Astronomia, Università Di Firenze, Florence, Italy

M. Cavallaro Laboratori Nazionali del Sud, Istituto Nazionale di Fisica Nucleare, Catania, Italy

M. Centelles Departament de Física Quàntica i Astrofísica and Facultat de Física, Institut de Ciències del Cosmos (ICCUB), Universitat de Barcelona, Barcelona, Spain

F. Cerutti European Organization for Nuclear Research (CERN), Meyrin, Switzerland

E. Chávez Laboratorio Nacional de Espectrometría de Masas Con Acelerador (LEMA), Instituto de Física, Universidad Nacional Autónoma de México (UNAM), Circuito de la Investigación Científica Ciudad Universitaria, Mexico, D.F, Mexico

Y. H. Chen IPN, CNRS-IN2P3, University of Paris-Sud, Université Paris-Saclay, Orsay, France

E. Chiaveri European Organization for Nuclear Research (CERN), Meyrin, Switzerland;
University of Manchester, Manchester, UK

M. Cicerchia INFN Laboratori Nazionali di Legnaro, Legnaro (PD), Italy;
Dipartimento di Fisica e Astronomia dell'Università di Padova, Padua, Italy

I. Ciepał H. Niewodniczański Institute of Nuclear Physics PAN, Cracow, Poland

N. Cieplicka-Oryńczak INFN Sezione di Milano e Dipartimento di Fisica, Università Di Milano, Milan, Italy;
Institute of Nuclear Physics, Polish Academy of Sciences Krakow, Krakow, Poland

M. Cinausero INFN Laboratori Nazionali di Legnaro, Legnaro (PD), Italy

PRISMA GALILEO MINIBALL Collaborations Ruder Bošković Institute, Zagreb, Croatia

G. Collazzuol INFN Sezione di Padova, Dipartimento di Fisica e Astronomia dell'Università di Padova, Padua, Italy

Maria Colonna INFN - LNS, Catania, Italy

N. Colonna Istituto Nazionale di Fisica Nucleare, Bari, Italy

P. Čolović Ruđer Bošković Institute, Zagreb, Croatia

D. Connolly TRIUMF, Vancouver, BC, Canada

N. Cooper Wright Nuclear Structure Laboratory, Yale University, New Haven, CT, USA

G. P. Cortés Universitat Politècnica de Catalunya, Barcelona, Spain

M. A. Cortés-Giraldo Instituto de Física Corpuscular, CSIC - Universidad de Valencia, Valencia, Spain;
Departamento de Física Atómica, Molecular y Nuclear, Universidad de Sevilla, Seville, Spain

D. Cortina-Gil IGFAE, Universidad de Santiago de Compostela, Santiago de Compostela, Spain

L. Cosentino INFN Laboratori Nazionali del Sud, Catania, Italy

S. Cristallo Istituto Nazionale di Fisica Nazionale, Perugia, Italy;
Istituto Nazionale di Astrofisica - Osservatorio Astronomico di Teramo, Teramo, Italy

J. Cruz LIBPhys-UNL, Lisbon, Portugal

C. Curceanu INFN, Laboratori Nazionali di Frascati, Frascati (Roma), Italy

M. D'Agostino INFN Sezione di Bologna e Dipartimento di Fisica e Astronomia, Università Di Bologna, Bologna, Italy

Antonio D'Alessio Institut für Kernphysik, TU Darmstadt, Darmstadt, Germany

R. C. da Silva IPFN, IST-UL, Lisbon, Portugal

L. A. Damone Dipartimento di Fisica, Istituto Nazionale di Fisica Nucleare, Università degli Studi di Bari, Bari, Italy

T. Daniel Department of Physics, University of Surrey, Guildford, Surrey, UK

T. Davinson University of Edinburgh, Edinburgh, UK

A. Dawood Butt Informazione e Bioingegneria and INFN Sezione di Milano, Dipartimento di Elettronica, Politecnico di Milano, Milan, Italy

K. De Los Ríos Instituto de Física, UNAM, Mexico City, Mexico

L. De Paolis INFN, Laboratori Nazionali di Frascati, Frascati (Roma), Italy;
Department of Physics, Faculty of Science MM.FF.NN., University of Rome 2 (Tor Vergata), Rome, Italy

M. Degerlier Science and Art Faculty, Physics Department, Nevsehir Haci Bektas Veli University, Nevsehir, Turkey

R. Del Grande INFN, Laboratori Nazionali di Frascati, Frascati (Roma), Italy

D. Dell'Aquila INFN Sezione di Napoli e Dipartimento di Fisica, Università Federico II Napoli, Naples, Italy;
Institut de Physique Nuclèaire (IPN) Université Paris-Sud 11, Orsay, Île-de-France, France

A. Di Pietro INFN-Laboratori Nazionali del Sud, Catania, Italy

M. Diakaki National Technical University of Athens, Athens, Greece

J. Díaz E.T.S.I. Informática, University of Granada, Granada, Spain

M. Dietz School of Physics and Astronomy, University of Edinburgh, Edinburgh, UK

S. Dolan Laboratoire Leprince-Ringuet, IN2P3-CNRS, Palaiseau, France

C. Domingo-Pardo Instituto de Física Corpuscular, CSIC - Universidad de Valencia, Valencia, Spain

A. D. Domínguez-Muñoz Departamento de Física Atómica, Molecular y Nuclear, Universidad de Sevilla, Seville, Spain

R. Dressler Paul Scherrer Institut (PSI), Villigen, Switzerland

E. Dupont CEA Saclay, Irfu, Université Paris-Saclay, Gif-sur-Yvette, France

I. Durán University of Santiago de Compostela, Santiago de Compostela, Spain

Z. Eleme University of Ioannina, Ioannina, Greece

J. M. Espino Departamento de Física Atómica, Molecular y Nuclear, Centro Nacional de Aceleradores, Universidad de Sevilla, Seville, Spain

L. Fabbietti Excellence Cluster Universe, Technische Universiät München, Garching, Germany

D. Fabris INFN Sezione di Padova, Padua, Italy

M. Feijoo Department of Particle Physics, IGFAE, University of Santiago de Compostela, Santiago de Compostela, Spain

B. Fernández Centro Nacional de Aceleradores, Universidad de Sevilla, CSIC, Junta de Andalucía, Seville, Spain

B. Fernández-Domíngez University of Santiago de Compostela, Santiago de Compostela, Spain

A. Ferrari European Organization for Nuclear Research (CERN), Meyrin, Switzerland

D. Ferreira LIP/FCUL, Lisbon, Portugal

J. L. Ferreira Universidade Federal Fluminense, Niteroi, Brazil

I. Ferro-Gon calves Instituto Superior Técnico, Lisbon, Portugal

P. Figuera INFN-Laboratori Nazionali del Sud, Catania, Italy

D. Filipescu “Horia Hulubei” NIPNE, Bucharest-Magurele, Romania

P. Finocchiaro INFN Laboratori Nazionali del Sud, Catania, Italy

C. Fiorini Informazione e Bioingegneria and INFN Sezione di Milano, Dipartimento di Elettronica, Politecnico di Milano, Milan, Italy

M. Fisichella INFN-Laboratori Nazionali del Sud, Catania, Italy

N. Florea “Horia Hulubei” NIPNE, Bucharest-Magurele, Romania

M. Fonseca LIBPhys-UNL, Lisbon, Portugal

B. Fornal Institute of Nuclear Physics, Polish Academy of Sciences Krakow, Krakow, Poland

A. M. Franconetti Centro Nacional de Aceleradores, Universidad de Sevilla, CSIC, Junta de Andalucía, Seville, Spain

M. Freer University of Birningham, Birningham, UK

C. Frosin INFN Sezione di Bologna e Dipartimento di Fisica e Astronomia, Università Di Bologna, Bologna, Italy

V. Furman Joint Institute for Nuclear Research (JINR), Dubna, Russia

D. Galaviz LIP/FCUL, Lisbon, Portugal

E. Galiana LIP/FCUL, Lisbon, Portugal;
Departamento de Física de Partículas, Instituto Gallego de Física de Altas Energías (IGFAE), Universidad de Santiago de Compostela, Santiago de Compostela, Spain

M. I. Gallardo Departamento de Física Atómica, Molecular y Nuclear, Universidad de Sevilla, Seville, Spain

P. Galve Grupo de Física Nuclear and Iparcos, Facultad de Ciencias Físicas, Universidad Complutense de Madrid, Madrid, Spain

J. García López Departamento de Física Atómica, Molecular y Nuclear, Centro Nacional de Aceleradores, Universidad de Sevilla, Seville, Spain

A. Gargano Istituto Nazionale di Fisica Nucleare, Sezione di Napoli, Catania, Italy

A. Gawlik University of Lodz, Lodz, Poland

U. Gayer Institut für Kernphysik, Darmstadt, Germany

N. Gelli INFN Sezione di Firenze e Dipartimento di Fisica e Astronomia, Università Di Firenze, Florence, Italy

G. Georgiev Faculty of Physics, University of Sofia "St. Kliment Ohridski", Sofia, Bulgaria

R. Gernhäuser TUM, Munich, Germany

R.-B. Gerst Institut für Kernphysik, Universität zu Köln, Cologne, Germany

F. Ghio INFN Sezione di Roma and Istituto Superiore di Sanità, Roma, Italy

D. Ghita "Horia Hulubei" NIPNE, Bucharest-Magurele, Romania

S. Gilardoni European Organization for Nuclear Research (CERN), Meyrin, Switzerland

O. M. Giza Grupo de Física Nuclear and Iparcos, Facultad de Ciencias Físicas, Instituto de Investigación Sanitaria del Hospital Clínico San Carlos (IdISSC), Universidad Complutense de Madrid, Madrid, Spain

T. Glodariu Horia Hulubei National Institute of Physics and Nuclear Engineering (IFIN-HH), Bucharest, Romania

K. Göbel Goethe University Frankfurt, Frankfurt, Germany

D. González IGFAE, Universidad de Santiago de Compostela, Santiago de Compostela, Spain

C. Gonzalez-Boquera Departament de Física Quàntica i Astrofísica and Institut de Ciències del Cosmos (ICCUB), Facultat de Física, Universitat de Barcelona, Barcelona, Spain

R. González-Jiménez Universidad Complutense de Madrid, CEI Moncloa, Madrid, Spain

E. González-Romero Centro de Investigaciones Energéticas Medioambientales y Tecnológicas (CIEMAT), Madrid, Spain

F. Gramegna INFN Laboratori Nazionali di Legnaro, Legnaro (PD), Italy

D. Gruyer Grand Accélérateur National d'Ions Lourds, Caen, France

C. Guaraldo INFN, Laboratori Nazionali di Frascati, Frascati (Roma), Italy

C. Guerrero Departamento de Fisica Atomica, Molecular y Nuclear, Facultad de Física, Centro Nacional de Aceleradores, Universidad de Sevilla, Seville, Spain;
Instituto de Física Corpuscular, CSIC - Universidad de Valencia, Valencia, Spain

F. Gunsing CEA Saclay, Irfu, Université Paris-Saclay, Gif-sur-Yvette, France

L. Gurgi Department of Physics, University of Surrey, Guildford, Surrey, UK

C. Hebborn Physique Nucléaire et Physique Quantique (CP 229), Université libre de Bruxelles (ULB), Brussels, Belgium

S. Heinitz Paul Scherrer Institut (PSI), Villigen, Switzerland

A. Henriques LIP/FCUL, Lisbon, Portugal

J. Heyse European Commission, Joint Research Centre, Geel, Geel, Belgium

E. Hidalgo ATI Sistemas, Bergondo (A Coruña), Spain

A. Huerta Instituto de Física, UNAM, Mexico City, Mexico

P. Ibáñez García Grupo de Física Nuclear and Iparcos, Facultad de Ciencias Físicas, Universidad Complutense de Madrid, Madrid, Spain

M. Iliescu INFN, Laboratori Nazionali di Frascati, Frascati (Roma), Italy

R. Ilieva Wright Nuclear Structure Laboratory, Yale University, New Haven, CT, USA;
Department of Physics, University of Surrey, Guildford, Surrey, UK

A. Illana INFN, Laboratori Nazionali di Legnaro, Legnaro, Italy

J. Isaak Institut für Kernphysik, Darmstadt, Germany

S. Ivanov Faculty of Physics, University of Sofia "St. Kliment Ohridski", Sofia, Bulgaria

D. Ivanova Military Medical Academy, Sofia, Bulgaria

M. Iwasaki RIKEN, Tokyo, Japan

N. Jachowicz Ghent University, Gent, Belgium

D. G. Jenkins University of York, York, UK

A. P. Jesus LIBPhys-UNL, Lisbon, Portugal

M. C. Jiménez-Ramos Centro Nacional de Aceleradores, Seville, Spain

J. Jolie Institut für Kernphysik, Universität zu Köln, Cologne, Germany

A. Jurado E.T.S.I. Informática, University of Granada, Granada, Spain

Y. Kadi European Organization for Nuclear Research (CERN), Meyrin, Switzerland

B. Kamys M. Smoluchowski Institute of Physics, Jagiellonian University, Cracow, Poland

F. Käppeler Karlsruhe Institute of Technology, Campus North, IKP, Karlsruhe, Germany

H. Kazeminejad Nuclear Science and Technology Research Institute (NSTRI), Tehran, Iran

R. Kern Institut für Kernphysik, TU Darmstadt, Darmstadt, Germany

S. Khakshournia Nuclear Science and Technology Research Institute (NSTRI), Tehran, Iran

A. Kimura Japan Atomic Energy Agency (JAEA), Tokai-mura, Japan

N. Kivel Paul Scherrer Institut (PSI), Villigen, Switzerland

P. Klenze TUM, Munich, Germany

M. Kokkoris National Technical University of Athens, Athens, Greece

Y. Kopatch Joint Institute for Nuclear Research (JINR), Dubna, Russia

V. Kozhuharov Faculty of Physics, University of Sofia "St. Kliment Ohridski", Sofia, Bulgaria

Krishichayan Triangle Universities Nuclear Laboratory, Durham, NC, USA

M. Krtička Charles University, Prague, Czech Republic

D. Kurtulgil Goethe University Frankfurt, Frankfurt, Germany

I. Ladarescu Instituto de Física Corpuscular, CSIC - Universidad de Valencia, Valencia, Spain

S. Lalkovski Faculty of Physics, University of Sofia “St. Kliment Ohridski”, Sofia, Bulgaria

Nadjet Laouet Laboratoire de Physique Mathematique et de Physique Subatomique (LPMS), Frères Mentouri Constantine-1 University, Constantine, Algeria

P. Lasko H. Niewodniczański Institute of Nuclear Physics PAN, Cracow, Poland

M. Lattuada INFN-Laboratori Nazionali del Sud, Catania, Italy

C. Lederer-Woods School of Physics and Astronomy, University of Edinburgh, Edinburgh, UK

A. Lennarz TRIUMF, Vancouver, BC, Canada

Horst Lenske Institut für Theoretische Physik, Justus-Liebig-Universitat Giessen, Giessen, Germany

S. Leoni INFN Sezione di Milano e Dipartimento di Fisica, Università Di Milano, Milan, Italy

J. Lerendegui-Marco Instituto de Física Corpuscular, CSIC - Universidad de Valencia, Valencia, Spain;
Departamento de Física Atómica, Molecular y Nuclear, Facultad de Física, Universidad de Sevilla, Seville, Spain

P. Levi Sandri INFN, Laboratori Nazionali di Frascati, Frascati (Roma), Italy

R. Lica “Horia Hulubei” NIPNE, Bucharest-Magurele, Romania

R. Linares Universidade Federal Fluminense, Niteroi, Brazil

S. Lo Meo Agenzia nazionale per le nuove tecnologie, l’energia e lo sviluppo economico sostenibile (ENEA), Bologna, Italy;
Istituto Nazionale di Fisica Nucleare, Sezione di Bologna, Bologna, Italy

B. Löher Institut für Kernphysik, Darmstadt, Germany

I. Lombardo INFN Sezione di Napoli e Dipartimento di Fisica, Università Federico II Napoli, Naples, Italy

S.-J. Lonsdale School of Physics and Astronomy, University of Edinburgh, Edinburgh, UK

J. López Herraiz Grupo de Física Nuclear and Iparcos, Facultad de Ciencias Físicas, Universidad Complutense de Madrid, Madrid, Spain

A. M. López-Antequera E.T.S.I. Informática, University of Granada, Granada, Spain

A. López-Montes Grupo de Física Nuclear and Iparcos, Facultad de Ciencias Físicas, Universidad Complutense de Madrid, Madrid, Spain

J. López-Morillas TTI Norte, La Rinconada (Sevilla), Spain

J. Lubian Universidade Federal Fluminense, Niteroi, Brazil

H. Luís IPFN, IST-UL, Lisbon, Portugal

J. Łukasik H. Niewodniczański Institute of Nuclear Physics PAN, Cracow, Poland

J. Machado LIBPhys-UNL, Lisbon, Portugal

M. Macías Department FAMN, Facultad de Física, Centro Nacional de Aceleradores, Universidad de Sevilla, CSIC, Junta de Andalucía, Seville, Spain

D. Macina European Organization for Nuclear Research (CERN), Meyrin, Switzerland

A. Magiera M. Smoluchowski Institute of Physics, Jagiellonian University, Cracow, Poland

Akshay Malige The Marian Smoluchowski Institute of Physics, Jagiellonian University, Kraków, Poland

F. Manchado de Sola Hospital Juan Ramón Jiménez, Servicio Andaluz de Salud, Science and Technology Research Centre, University of Huelva, Huelva, Spain

A. Manna Istituto Nazionale di Fisica Nucleare, Sezione di Bologna, Bologna, Italy;
Dipartimento di Fisica e Astronomia, Università di Bologna, Bologna, Italy

G. Mantovani INFN Laboratori Nazionali di Legnaro, Legnaro (PD), Italy;
Dipartimento di Fisica e Astronomia dell'Università di Padova, Padua, Italy

T. Marchi INFN Laboratori Nazionali di Legnaro, Legnaro (PD), Italy

N. Marginean "Horia Hulubei" NIPNE, Bucharest-Magurele, Romania

R. Marginean "Horia Hulubei" NIPNE, Bucharest-Magurele, Romania

I. Martel Departamento de Fisica Aplicada, Science and Technology Research Centre, Universidad de Huelva, Huelva, Spain

T. Martínez Centro de Investigaciones Energéticas Medioambientales y Tecnológicas (CIEMAT), Madrid, Spain

M. Martini ESNT, CEA, IRFU, SPN, Université Paris-Saclay, Gif-sur-Yvette, France

J. Marton Stefan-Meyer-Institut für Subatomare Physik, Vienna, Austria

A. Masi European Organization for Nuclear Research (CERN), Meyrin, Switzerland

C. Massimi Istituto Nazionale di Fisica Nucleare, Sezione di Bologna, Bologna, Italy;
Dipartimento di Fisica e Astronomia, Università di Bologna, Bologna, Italy

P. F. Mastinu Istituto Nazionale di Fisica Nucleare, Sezione di Legnaro, Legnaro, Italy

M. Mastromarco European Organization for Nuclear Research (CERN), Meyrin, Switzerland

F. Matteucci Dipartimento di Fisica, Istituto Nazionale di Fisica Nazionale, Università di Trieste, Trieste, Italy

E. Maugeri Paul Scherrer Institut (PSI), Villigen, Switzerland

A. Mazzone Consiglio Nazionale delle Ricerche, Istituto Nazionale di Fisica Nucleare, Bari, Italy

G. D. Megias Departamento de Física Atómica, Molecular y Nuclear, Universidad de Sevilla, Sevilla, Spain

C. G. Méndez Cátedra CONACYT, Laboratorio Nacional de Espectrometría de Masas Con Acelerador (LEMA), Instituto de Física, Universidad Nacional Autónoma de México (UNAM), Circuito de la Investigación Científica Ciudad Universitaria, Mexico, D.F, Mexico

C. Méndez-García CONACyT, Instituto de Física, UNAM, Mexico City, Mexico

E. Mendoza Centro de Investigaciones Energéticas Medioambientales y Tecnológicas (CIEMAT), Madrid, Spain

A. Mengoni Agenzia nazionale per le nuove tecnologie, l'energia e lo sviluppo economico sostenibile (ENEA), Bologna, Italy;
Istituto Nazionale di Fisica Nucleare, Sezione di Bologna, Bologna, Italy

D. Mengoni Dipartimento di Fisica e Astronomia dell'Università di Padova, Padua, Italy;
INFN Sezione di Padova, Padua, Italy

L. Mertes Institut für Kernphysik, Darmstadt, Germany

V. Michalopoulou National Technical University of Athens, Athens, Greece

C. Mihai "Horia Hulubei" NIPNE, Bucharest-Magurele, Romania

P. M. Milazzo Istituto Nazionale di Fisica Nazionale, Trieste, Italy

M. Milin Department of Physics, Faculty of Science, University of Zagreb, Zagreb, Croatia

M. Miliucci INFN, Laboratori Nazionali di Frascati, Frascati (Roma), Italy

Đ. Miljanić Ruđer Bošković Institute, Zagreb, Croatia

M. A. Millán-Callado Departamento de Física Atómica, Molecular y Nuclear, Facultad de Física, Centro Nacional de Aceleradores, Universidad de Sevilla, CSIC, Junta de Andalucía, Seville, Spain

F. Mingrone European Organization for Nuclear Research (CERN), Meyrin, Switzerland

I. O. Mitu "Horia Hulubei" NIPNE, Bucharest-Magurele, Romania

Saed Mohammadi Department of Physics, Payame Noor University (PNU), Tehran, Iran

L. Morelli INFN Sezione di Bologna e Dipartimento di Fisica e Astronomia, Università Di Bologna, Bologna, Italy

P. Moskal The M. Smoluchowski Institute of Physics, Jagiellonian University, Kraków, Poland

A. Musumarra INFN Laboratori Nazionali del Sud, Catania, Italy;
Dipartimento di Fisica e Astronomia, Università di Catania, Catania, Italy

F. Naqvi Wright Nuclear Structure Laboratory, Yale University, New Haven, CT, USA

A. Negret Horia Hulubei National Institute of Physics and Nuclear Engineering (IFIN-HH), Bucharest, Romania

S. Niedźwiecki The M. Smoluchowski Institute of Physics, Jagiellonian University, Kraków, Poland

K. Niewczas Ghent University, Gent, Belgium;
University of Wrocław, Wrocław, Poland

A. Nikolakopoulos Ghent University, Gent, Belgium

C. Nita "Horia Hulubei" NIPNE, Bucharest-Magurele, Romania

R. Nolte Physikalisch-Technische Bundesanstalt (PTB), Braunschweig, Germany

D. Nurkić Department of Physics, Faculty of Science, University of Zagreb, Zagreb, Croatia

Alexandre Obertelli Institut für Kernphysik, Technische Universität Darmstadt, Darmstadt, Germany

F. Ogállar University of Granada, Granada, Spain

S. Okada RIKEN, Tokyo, Japan

A. Olmi INFN Sezione di Firenze e Dipartimento di Fisica e Astronomia, Università Di Firenze, Florence, Italy

A. Oprea Horia Hulubei National Institute of Physics and Nuclear Engineering (IFIN-HH), Bucharest, Romania

A. K. Orduz Science and Technology Research Centre, University of Huelva, Huelva, Spain

P. Ottanelli INFN Sezione di Firenze e Dipartimento di Fisica e Astronomia, Università Di Firenze, Florence, Italy

J. D. Ovejas IEM - CSIC, Madrid, Spain

S. Padilla Laboratorio Nacional de Espectrometría de Masas Con Acelerador (LEMA), Instituto de Física, Universidad Nacional Autónoma de México (UNAM), Circuito de la Investigación Científica Ciudad Universitaria, Mexico, D.F, Mexico

B. Paes Universidade Federal Fluminense, Niteroi, Brazil

H. Pai Saha Institute of Nuclear Physics, Kolkata, West Bengal, India

A. Parrado Centro Nacional de Aceleradores, Seville, Spain

S. Pascu "Horia Hulubei" NIPNE, Bucharest-Magurele, Romania

G. Pasquali INFN Sezione di Firenze e Dipartimento di Fisica e Astronomia, Università Di Firenze, Florence, Italy

G. Pastore INFN Sezione di Firenze e Dipartimento di Fisica e Astronomia, Università Di Firenze, Florence, Italy

N. Patronis University of Ioannina, Ioannina, Greece

A. Pavlik Faculty of Physics, University of Vienna, Vienna, Austria

P. Pawłowski H. Niewodniczański Institute of Nuclear Physics PAN, Cracow, Poland

M. G. Pellegriti INFN-Laboratori Nazionali del Sud, Catania, Italy

L. Peralta LIP/FCUL, Lisbon, Portugal

J. Pérez TTI Norte, La Rinconada (Sevilla), Spain

T. Pérez Science and Technology Research Centre, University of Huelva, Huelva, Spain

J. Perkowski University of Lodz, Lodz, Poland

S. Piantelli INFN Sezione di Firenze e Dipartimento di Fisica e Astronomia, Università Di Firenze, Florence, Italy

L. Piersanti Istituto Nazionale di Fisica Nazionale, Perugia, Italy;
Istituto Nazionale di Astrofisica - Osservatorio Astronomico di Teramo, Teramo, Italy

N. Pietralla Institut für Kernphysik, TU Darmstadt, Darmstadt, Germany

D. Pietreanu INFN, Laboratori Nazionali di Frascati, Frascati (Roma), Italy; Horia Hulubei National Institute of Physics and Nuclear Engineering (IFIN-HH), Magurele, Romania

A. Pinto Science and Technology Research Centre, University of Huelva, Huelva, Spain

K. Piscicchia INFN, Laboratori Nazionali di Frascati, Frascati (Roma), Italy; Museo Storico della Fisica e Centro Studi e Ricerche Enrico Fermi, Rome, Italy

G. Poggi INFN Sezione di Firenze e Dipartimento di Fisica e Astronomia, Università Di Firenze, Florence, Italy

V. Yu. Ponomarev Institut für Kernphysik, Darmstadt, Germany

I. Porras University of Granada, Granada, Spain

J. Praena University of Granada, Granada, Spain

L. Prepolec Ruđer Bošković Institute, Zagreb, Croatia

A. Psaltis McMaster University, Hamilton, ON, Canada

K. Pysz H. Niewodniczański Institute of Nuclear Physics PAN, Cracow, Poland

J. M. Quesada Departamento de Física Atómica, Molecular y Nuclear, Facultad de Física, Universidad de Sevilla, Seville, Spain;
Instituto de Física Corpuscular, CSIC - Universidad de Valencia, Valencia, Spain

D. Radeck Physikalisch-Technische Bundesanstalt (PTB), Braunschweig, Germany

D. Ramos Doval IPN, CNRS-IN2P3, University of Paris-Sud, Université Paris-Saclay, Orsay, France;
Instituto de Física Corpuscular, CSIC - Universidad de Valencia, Valencia, Spain

Narendra Rathod The Marian Smoluchowski Institute of Physics, Jagiellonian University, Kraków, Poland

T. Rausher Department of Physics, University of Basel, Basel, Switzerland; School of Physics, Astronomy and Mathematics, University of Hertfordshire, Hertfordshire, UK

P. H. Regan Department of Physics, University of Surrey, Guildford, Surrey, UK; National Physical Laboratory, Teddington, Middlesex, UK

R. Reifarth Goethe University Frankfurt, Frankfurt, Germany

P. Remmels TUM, Munich, Germany

P. C. Ries Institut für Kernphysik, Darmstadt, Germany

L. M. Robledo Departamento de Física Teórica, Facultad de Física, Universidad Autónoma de Madrid, Madrid, Spain;
Center for Computational Imulation and Center for Computational Simulation, Universidad Politécnica de Madrid, Madrid, Spain

J. Rocha IPFN, IST-UL, Lisbon, Portugal

D. Rochman Paul Scherrer Institut (PSI), Villigen, Switzerland

M. Rodríguez-Ceja Laboratorio Nacional de Espectrometría de Masas Con Acelerador (LEMA), Instituto de Física, Universidad Nacional Autónoma de México (UNAM), Circuito de la Investigación Científica Ciudad Universitaria, Mexico, D.F, Mexico

T. Rodríguez-González Department FAMN, Facultad de Física, Centro Nacional de Aceleradores, Universidad de Sevilla, CSIC, Junta de Andalucía, Seville, Spain

J. L. Rodríguez-Sanchéz Department of Particle Physics, IGFAE, University of Santiago de Compostela, Santiago de Compostela, Spain

C. Romig Institut für Kernphysik, Darmstadt, Germany

B. Roostaii Nuclear Science and Technology Research Institute (NSTRI), Tehran, Iran

C. Rubbia Japan Atomic Energy Agency (JAEA), Tokai-mura, Japan;
European Organization for Nuclear Research (CERN), Meyrin, Switzerland

M. Rudigier Institut für Kernphysik, Universität zu Köln, Cologne, Germany;
University of Surrey, Guildford, Surrey, UK

Z. Rudy M. Smoluchowski Institute of Physics, Jagiellonian University, Cracow, Poland

C. Ruiz TRIUMF, Vancouver, BC, Canada

M. Sabaté-Gilarte European Organization for Nuclear Research (CERN), Meyrin, Switzerland;
Instituto de Física Corpuscular, CSIC - Universidad de Valencia, Valencia, Spain

Hanane Saifi Laboratoire de Physique Mathématique et Subatomique, Université Frères Mentouri, Constantine, Algeria

A. M. Sánchez-Benítez Departamento de Fisica Aplicada, Universidad de Huelva, Huelva, Spain;
Departamento Ciencias Integradas, Facultad de Ciencias Experimentales, Universidad de Huelva, Huelva, Spain

D. Sánchez-Parcerisa Grupo de Física Nuclear and Iparcos, Facultad de Ciencias Físicas, Instituto de Investigación Sanitaria del Hospital Clínico San Carlos (IdISSC), Universidad Complutense de Madrid, Madrid, Spain

J. Sánchez-Segovia Hospital Juan Ramón Jiménez, Servicio Andaluz de Salud, Science and Technology Research Centre, University of Huelva, Huelva, Spain

V. Sánchez-Tembleque Grupo de Física Nuclear and Iparcos, Facultad de Ciencias Físicas, Instituto de Investigación Sanitaria del Hospital Clínico San Carlos (IdISSC), Universidad Complutense de Madrid, Madrid, Spain

G. Santagati Universidade Federal Fluminense, Niteroi, Brazil;
Laboratori Nazionali del Sud, Istituto Nazionale di Fisica Nucleare, Catania, Italy

D. Savran GSI Helmholtzzentrum für Schwerionenforschung, Darmstadt, Germany

A. Saxena Bhabha Atomic Research Centre (BARC), Mumbai, India

P. Schillebeeckx European Commission, Joint Research Centre, Geel, Geel, Belgium

M. Schilling Institut für Kernphysik, Darmstadt, Germany

D. Schumann Paul Scherrer Institut (PSI), Villigen, Switzerland

V. Scuderi INFN-Laboratori Nazionali del Sud, Catania, Italy

S. K. Sharma M. Smoluchowski Institute of Physics, Jagiellonian University, Cracow, Poland

H. Shi Institute for High Energy Physics of Austrian Academy of Science, HEPHY - Institut für Hochenergiephysik der ÖAW, Vienna, Austria

A. Shotter TRIUMF, Vancouver, BC, Canada

M. Silarski The M. Smoluchowski Institute of Physics, Jagiellonian University, Kraków, Poland

H. Silva LIBPhys-UNL, Lisbon, Portugal

Maxim Singer Institut für Kernphysik, TU Darmstadt, Darmstadt, Germany

U. Singh M. Smoluchowski Institute of Physics, Jagiellonian University, Cracow, Poland

D. Sirghi INFN, Laboratori Nazionali di Frascati, Frascati (Roma), Italy;
Horia Hulubei National Institute of Physics and Nuclear Engineering (IFIN-HH), Magurele, Romania

F. Sirghi INFN, Laboratori Nazionali di Frascati, Frascati (Roma), Italy;
Department of Physics, Faculty of Science MM.FF.NN., University of Rome 2 (Tor Vergata), Rome, Italy

M. Skurzok The M. Smoluchowski Institute of Physics, Jagiellonian University, Kraków, Poland

A. G. Smith University of Manchester, Manchester, UK

Jerzy Smyrski The Marian Smoluchowski Institute of Physics, Jagiellonian University, Kraków, Poland

J. T. Sobczyk University of Wrocław, Wrocław, Poland

N. Soić Ruđer Bošković Institute, Zagreb, Croatia

C. Solís Laboratorio Nacional de Espectrometría de Masas Con Acelerador (LEMA), Instituto de Física, Universidad Nacional Autónoma de México (UNAM), Circuito de la Investigación Científica Ciudad Universitaria, Mexico, D.F, Mexico

N. Sosnin University of Manchester, Manchester, UK

A. Spallone INFN, Laboratori Nazionali di Frascati, Frascati (Roma), Italy

A. Stamatopoulos National Technical University of Athens, Athens, Greece

S. Stegemann Institut für Kernphysik, Universität zu Köln, Cologne, Germany

E. Strano INFN-Laboratori Nazionali del Sud, Catania, Italy

S. Szilner Ruder Bošković Institute, Zagreb, Croatia

G. Tagliente Istituto Nazionale di Fisica Nucleare, Bari, Italy

J. Taieb CEA DAM Bruyères-le-Châtel, Arpajon, France

J. L. Tain Instituto de Física Corpuscular, CSIC - Universidad de Valencia, Valencia, Spain

Z. Talip Paul Scherrer Institut (PSI), Villigen, Switzerland

A. E. Tarifeño-Saldivia Universitat Politècnica de Catalunya, Barcelona, Spain

L. Tassan-Got European Organization for Nuclear Research (CERN), Meyrin, Switzerland;
IPN, CNRS-IN2P3, Université Paris-Sud, Université Paris-Saclay, Orsay, France;
National Technical University of Athens, Athens, Greece

H. Tatsuno Lund Univeristy, Lund, Sweden

P. Teubig LIP/FCUL, Lisbon, Portugal

Zh. Toneva Faculty of Physics, University of Sofia "St. Kliment Ohridski", Sofia, Bulgaria

W. Tornow Triangle Universities Nuclear Laboratory, Durham, NC, USA

D. Torresi INFN-Laboratori Nazionali del Sud, Catania, Italy

M. Trueba ATI Sistemas, Bergondo (A Coruña), Spain

A. Tsinganis European Organization for Nuclear Research (CERN), Meyrin, Switzerland

J. M. Udías Grupo de Física Nuclear and Iparcos, Facultad de Ciencias Físicas, Universidad Complutense de Madrid, Madrid, Spain

J. Ulrich Paul Scherrer Institut (PSI), Villigen, Switzerland

S. Urlass European Organization for Nuclear Research (CERN), Meyrin, Switzerland;
Helmholtz-Zentrum Dresden-Rossendorf, Dresden, Germany

M. Uroić Ruđer Bošković Institute, Zagreb, Croatia

S. Valdré INFN Sezione di Firenze e Dipartimento di Fisica e Astronomia, Università Di Firenze, Florence, Italy

S. Valenta Charles University, Prague, Czech Republic

José A. Lay Valera Departamento de FAMN, Universidad de Sevilla, Seville, Spain

J. J. Valiente-Dobón INFN, Laboratori Nazionali di Legnaro, Legnaro, Italy

N. Van Dessel Ghent University, Gent, Belgium

G. Vankova-Kirilova Faculty of Physics, University of Sofia "St. Kliment Ohridski", Sofia, Bulgaria

G. Vannini Istituto Nazionale di Fisica Nucleare, Sezione di Bologna, Bologna, Italy;
Dipartimento di Fisica e Astronomia, Università di Bologna, Bologna, Italy

V. Variale Istituto Nazionale di Fisica Nucleare, Bari, Italy

P. Vaz Instituto Superior Técnico, Lisbon, Portugal

J. Vazquez FAYSOL S.A.L., Palos de La Frontera (Huelva), Spain

O. Vazquez Doce INFN, Laboratori Nazionali di Frascati, Frascati (Roma), Italy;
Excellence Cluster Universe, Technische Universiät München, Garching, Germany

P. Velho LIP/FCUL, Lisbon, Portugal

A. Ventura Istituto Nazionale di Fisica Nucleare, Sezione di Bologna, Bologna, Italy

A. Villa-Abaunza Grupo de Física Nuclear and Iparcos, Facultad de Ciencias Físicas, Universidad Complutense de Madrid, Madrid, Spain

A. C. C. Villari FRIB, East Lansing, MI, USA

S. Viñals IEM - CSIC, Madrid, Spain

X. Viñas Departament de Física Quàntica i Astrofísica and Institut de Ciències del Cosmos (ICCUB), Facultat de Física, Universitat de Barcelona, Barcelona, Spain

V. Vlachoudis European Organization for Nuclear Research (CERN), Meyrin, Switzerland

R. Vlastou National Technical University of Athens, Athens, Greece

Peter von Neumann-Cosel Institut für Kernphysik, TU Darmstadt, Darmstadt, Germany

N. Vukman Ruđer Bošković Institute, Zagreb, Croatia

A. Wallner Australian National University, Canberra, Australia

V. Werner Institut für Kernphysik, TU Darmstadt, Darmstadt, Germany;
Wright Nuclear Structure Laboratory, Yale University, New Haven, CT, USA

E. Widmann Stefan-Meyer-Institut für Subatomare Physik, Vienna, Austria

J. Wiederhold Institut für Kernphysik, TU Darmstadt, Darmstadt, Germany

M. Williams TRIUMF, Vancouver, BC, Canada

P. J. Woods School of Physics and Astronomy, University of Edinburgh, Edinburgh, UK

T. J. Wright University of Manchester, Manchester, UK

J. Yang Physique Nucléaire et Physique Quantique (CP 229), Université libre de Bruxelles, Brussels, Belgium;
Afdeling Kern- en Stralingsfysica, KU Leuven, Leuven, Belgium

J. Zmeskal INFN, Laboratori Nazionali di Frascati, Frascati (Roma), Italy;
Stefan-Meyer-Institut für Subatomare Physik, Vienna, Austria

P. Žugec University of Zagreb, Zagreb, Croatia

M. Zweidinger Institut für Kernphysik, Darmstadt, Germany

Participants

Guler Aggez
guler_aggez@hotmail.com
Istanbul University, Istanbul (Turkey)
Istanbul University, Science Faculty, Physics Department,
34134 Vezneciler, Fatih-Istanbul, Turkey

Victor Alcayne Aicua
victor.alcayne@ciemat.es
Ciemat, Madrid (Spain)
CIEMAT Avenida Complutense 40, E17.P0.12b, 28040 Madrid, Spain

M.ª Victoria Andrés Martín
m-v-andres@us.es
Universidad de Sevilla (Spain)
Departamento de Física Atómica, Molecular y Nuclear.
Facultad de Física. Universidad de Sevilla. Apartado 1065, E-41080 Sevilla, Spain

Amin Attarzadeh
Attarzadeh_amin@yahoo.com
ACECR, Tehran (Iran)
No. 54, Rastin Building, Movahedin gharbi St., Kianpars, Ahvaz,
61 5561 5738, Khuzestan province, Iran

Víctor Babiano Suárez
vbabiano@ific.uv.es
IFIC-CSIC, Valencia (Spain)
Avda. Campanar, 11, 20, 46009 Valencia, Spain

Anna Baratto Roldan
abaratto@us.es
CNA/Universidad de Sevilla (Spain)
C/Thomas Alva Edison, 7, 41092, Sevilla, Spain

Jessica Ilaria Bellone
bellone@lns.infn.it
University of Catania/INFN-LNS (Italy)
Largo Madonna del Sorriso, n. 6, Aci Catena, 95022, CT, Italia

Fatima Benrachi
fatima.benrachi@umc.edu.dz
University of Frères Mentouri (Algeria)
LPMS Laboratory, Universite Freres Mentour, Route Ain El Bey, Constantine, Algeria

Alex Brown
brown@nscl.msu.edu
Michigan State University (USA)
Michigan State University, Cyclotron Laboratory, 640 S. Shaw Lane, East Lansing, MI 48824-1321, USA

S. Burrello
burrello@lns.infn.it
LNS-INFN (Italy)
INFN—Laboratori Nazionali del sud, Via S. Sofia, 62 95125 Catania, Italy

Pierre Capel
pierre.capel@ulb.ac.be
Johannes Gutenberg-Universität Mainz (Germany)
Institut für Kernphysik, Johannes Gutenberg-Universitüt Mainz, Johann-Joachim-Becher Weg 45 D-55099 Mainz, Germany

Magda Cicerchia
cicerchia@lnl.infn.it
Universitá di Padova/LNL-INFN (Italy)
UniPd, Dipartimento di Fisica e Astronomia. via Marzolo, 8. I-35131 Padova, Italy

Petra Colovic
petra.colovic@irb.hr
Rudjer Boskovic Institute (Croatia)
Ruder Boskovic Institute, Bijenicka Cesta 54, 10 000 Zagreb, Croatia

Antonio D'alessio
adalessio@ikp.tu-darmstadt.de
TU Darmstadt (Germany)
Institut für Kernphysik Antonio D'Alessio Schlossgartenstraße 9 64289 Darmstadt, Germany

Luca De Paolis
luca.depaolis@lnf.infn.it
University of Rome 2/LNF—INFN (Italy)
Marco Valerio Corvo Street, N 121, CAP 00174, Rome, Italy

Javier Díaz Ovejas
j.diaz@csic.es
IEM-CSIC, Madrid (Spain)
IEM-CSIC, Serrano 113bis, ES-28006, Madrid (Madrid), Spain

Katerine Viviana Díaz Hernández
kvdiazh@unal.edu.co
Universidad de Sevilla (Spain)
Departamento de Física Atómica, Molecular y Nuclear, Facultad de Física, Universidad de Sevilla.
Apartado 1065, E-41080 Sevilla, Spain

Antonio Damián Domínguez Muñoz
adominguez18@us.es
Universidad de Sevilla (Spain)
Departamento de Física Atómica, Molecular y Nuclear.
Facultad de Física. Universidad de Sevilla.
Apartado 1065, E-41080 Sevilla, Spain

Tommi Eronen
tommi.eronen@jyu.fi
University of Jyväskylä (Finland).
University of Jyväskylä, Department of Physics, P.O. Box 35 (YFL),
FI-40014, Finland

Manuel Feijoo Rodríguez
manuelfeijoo.rodriguez@usc.es
Universidade de Santiago de Compostela (Spain)
Department of Particle Physics,
University of Santiago de Compostela,
Monte da Condesa Building, s/n, Campus Vida,
Santiago de Compostela, 15782, Spain

Axel Frotscher
afrotscher@ikp.tu-darmstadt.de
Institute for Nuclear Physics Darmstadt (Germany)
Flachsbachweg 48, 64285 Darmstadt, Germany

Karen Patricia Gaitán De Los Ríos
krndelosrios@gmail.com
Universidad Nacional Autónoma de México (México)
Instituto de Física, Circuito de la Investigación
Científica Ciudad Universitaria
CP 04510 Ciudad de México, México

Elisabet Galiana Baldó
elisabet.galiana@usc.es
Universidade de Lisboa (Portugal)
Avenue Xativa 13, 03820 Cocentaina, Spain

Pablo Galve
pgalve@nuclear.fis.ucm.es
Universidad Complutense de Madrid (Spain).
Avenida Complutense, s/n, Facultad de Ciencias Físicas,
Departamento de Estructura de la Materia,
Física Térmica y Electrónica, Madrid, 28040, Spain

José-Enrique García-Ramos
enrique.ramos@dfaie.uhu.es
Universidad de Huelva (Spain)
Departamento de Física Aplica, Facultad de Ciencias Experimentales,
Universidad de Huelva, Avda de las Fuerzas Armadas s/n, 21071, Spain

Juan José Gómez Cadenas
jjgomezcadenas@gmail.com
IFIC-CSIC (Spain).
Parque Científico, C/Catedrático José Beltrán, 2. E-46980 Paterna, Spain

Raul Gonzalez Jimenez
raugonjim@gmail.com
Universidad Complutense de Madrid (Spain)
Manuela Malasaña 4, 2nd derecha, 28004 Madrid, Spain

Claùdia Gonzalez-Boquera
cgonzalezboquera@ub.edu
Universidad de Barcelona (Spain)
Dept. Física Quàntica i Astrofísica. Facultat de Física Universitat de Barcelona,
Martí i Franquès, 1 E-08028 Barcelona, Spain

Chloë Hebborn
chebborn@ulb.ac.be
Université libre de Bruxelles (Belgium)
251, chaussée de Haecht 1030 Bruxelles, Belgium

Yamil Khalouf
yamil.khalouf@dci.uhu.es
Universidad de Huelva (Spain)
Departamento de Física Aplica, Facultad de Ciencias Experimentales,
Universidad de Huelva, Avda de las Fuerzas Armadas s/n, 21071, Spain

José A. Lay Valera
lay@us.es
Universidad de Sevilla (Spain)
Departamento de Física Atómica, Molecular y Nuclear.
Facultad de Física. Universidad de Sevilla.
Apartado 1065, E-41080 Sevilla, Spain

A. López-Montes
alelopez@ucm.es
Universidad Complutense de Madrid (Spain)
c/Papagayo N. 33 2A. 28025 Madrid, Spain

Manuel Lozano
lozano@us.es
Universidad de Sevilla (Spain)
Departamento de Física Atómica, Molecular y Nuclear.
Facultad de Física. Universidad de Sevilla.
Apartado 1065, E-41080 Sevilla, Spain

Giorgia Mantovani
giorgia.mantovani@lnl.infn.it
Universitá degli Studi di Padova/LNL-INFN (Italy)
Piazza G. Matteotti, 40, 44021 Codigoro (Ferrara), Italy

Guillermo D. Megias
megias@us.es
CEA-Irfu (France)/Universidad de Sevilla (Spain)
Departamento de Física Atómica, Molecular y Nuclear.
Facultad de Física. Universidad de Sevilla.
Apartado 1065, E-41080 Sevilla, Spain

M.ª Ángeles Millán Callado
mmillan5@us.es
Universidad de Sevilla (Spain)
Departamento de Física Atómica, Molecular y Nuclear.
Facultad de Física. Universidad de Sevilla.
Apartado 1065, E-41080 Sevilla, Spain

Olivia Miszczynska Giza
oliviami@ucm.es
Universidad Complutense de Madrid (Spain)
Travesía del Roble 2K 2A. 19139 Yebes (Guadalajara), Spain

Antonio M. Moro Muñoz
moro@us.es
Universidad de Sevilla (Spain)
Departamento de Física Atómica, Molecular y Nuclear.
Facultad de Física. Universidad de Sevilla.
Apartado 1065, E-41080 Sevilla, Spain

Kajetan Niewczas
kajetan.niewczas@ugent.be
Ghent University (Belgium)
Department of Physics and Astronomy, Ghent University,
Proeftuinstraat 86 N3, 9000 Gent, Belgium

Alexis Nikolakopoulos
alexis.nikolakopoulos@ugent.be
Ghent University (Belgium)
Proeftuinstraat 86, 9000 Gent, Belgium

Erica Nunes Cardozo
ericanunes@fisica.if.uff.br
Universidade Federal Fluminense, Rio de J. (Brazil)
Travessa Adalgisa, 51—Patronato, 24.435-320,
São Gonçalo (Rio de Janeiro), Brazil

Deni Nurkic
dnurkic@phy.hr
University of Zagreb (Croatia)
University of Zagreb, Faculty of Science, Department of Physics,
Bijenièka cesta 32, 10000 Zagreb, Croatia

Alexandre Obertelli
aobertelli@ikp.tu-darmstadt.de
Institut für Kernphysik, TU Darmstadt (Germany).
Schlossgartenstraße 9 64289 Darmstadt, Germany

Chidera Opara
chiopa@alum.us.es
Universidad de Sevilla (Spain)
Departamento de Física Atómica, Molecular y Nuclear.
Facultad de Física. Universidad de Sevilla.
Apartado 1065, E-41080 Sevilla, Spain

Angie Orduz
angie.orduz@dfa.uhu.es
Universidad de Huelva (Spain)
Departamento de Física Aplica, Facultad de Ciencias Experimentales,
Universidad de Huelva, Avda de las Fuerzas Armadas s/n, 21071, Spain

Santiago Padilla Domínguez
spadilla@fisica.unam.mx
UNAM (Mèxico)
Laboratorio Nacional de Espectrometría de Masas (LEMA).
Instituto de Física. UNAM, Circuito de la Investigación Científica,
Ciudad Universitaria CP 04510 Ciudad de Mèxico, Mèxico

Katia Parodi
katia.parodi@physik.uni-muenchen.de
LMU Munich Physics (Germany).
Ludwig-Maximilians-Universität München, Fakultät für Physik,
Lehrstuhl für Medizinische Physik,
Am Coulombwall 1 D-85748 Garching, Germany

Wiktor Parol
wiktor.parol@ifj.edu.pl
Polish Academy of Sciences, Warsaw (Poland)
ul. Radzikowskiego 152, Krakow, Poland

Francisco Pérez-Bernal
francisco.perez@dfaie.uhu.es
Universidad de Huelva (Spain)
Departamento de Física Aplica, Facultad de Ciencias Experimentales,
Universidad de Huelva, Avda de las Fuerzas Armadas s/n, 21071, Spain

Narendra Rathod
nsrathore.rajput@gmail.com
Jagiellonian University (Poland)
Dom Studenski Uniwersytetu Jagiellonskiego
ul. Bydgoska 19, room- no-D-416, 30-056 Kraków, 12 36-36-100, Poland

Philipp Ries
philipp-ries@gmx.de
Institut für Kernphysik, TU Darmstadt (Germany)
Schlossgartenstrasse 9 64289 Darmstadt, Germany

Mª Teresa Rodríguez González
sarete6@hotmail.com
Universidad de Sevilla (Spain)
Urb. Macarena Tres Huertas, bloque 17, 6oA. 41009, Sevilla, Spain

Bahareh Roostaii
broostaii@aeoi.org.ir
NSTRI, Tehran (Iran)
North Kargar Ave. Tehran, Iran

William Saenz
wdsaenza@unal.edu.co
Universidad de Sevilla (Spain)
Departamento de Física Atómica, Molecular y Nuclear.
Facultad de Física. Universidad de Sevilla.
Apartado 1065, E-41080 Sevilla, Spain

Hanane Saifi
hanane.saifi25@gmail.com
University of Constantine 1, Constantine (Algeria)
N 156 Lot Belkarfa Ain Smara Constantine, 25150, Algeria

Hadi Shamoradifar
h.shamoradifar@gmail.com
Payam-e Noor University, Tehran (Iran)
Payam-e Noor University, Physics Department, Mashhad, Iran

Udai Singh
udai.singh@doctoral.uj.edu.pl
Jagiellonian University, Krakow (Poland)
D 326 Bydgoska 19, 30-056 Kraków, Poland

Natalia Targosz-Sleczka
natalia.targosz@usz.edu.pl
University of Szczecin (Poland)
Wielkopolska 15, 70-451 Szczecin, Poland

Pamela Teubig
pteubig@fc.ul.pt
LIP, Lisbon (Portugal)
Rua dos Lirios 14, 2760-055 Caxias, Portugal

Zhulieta Toneva
zh.h.toneva@phys.uni-sofia.bg
Sofia University (Bulgaria)
Faculty of Physics University of Sofia "St. Kliment Ohridski",
5 Bld. James Bourcher 1164 Sofia, Bulgaria

Mina Torabi
minatorabi90@gmail.com
Khajeh Nasir University of Technology, Tehran (Iran)
Tehran, Mirdamad Blvd, No. 470, Iran

Michelangelo Traina
michelangelotraina@yahoo.com
Universidad de Sevilla (Spain)
Departamento de Física Atómica, Molecular y Nuclear.
Facultad de Física. Universidad de Sevilla.
Apartado 1065, E-41080 Sevilla, Spain

Amaia Villa Abaunza
amavil01@ucm.es
Universidad Complutense de Madrid (Spain)
Calle Berezikoetxe 7, 8A 48960 Galdakao (Bizkaia), Spain

Sílvia Viñals I Onsès
s.vinals@csic.es
CSIC-IEM, Madrid (Spain)
C/Serrano 113bis, ES-28006, Madrid, Spain

Nikola Vukman
nvukman@irb.hr
Ruder Boskovic Institute, Zagreb (Croatia)
Ruder Boskovic Institute Bijenicka cesta 54, 10 000 Zagreb, Croatia

Johannes Christof Wiederhold
jwiederhold@ikp.tu-darmstadt.de
TU Darmstadt (Germany)
Institut für Kernphysik—TU Darmstadt Schlossgartenstraße 9 64289 Darmstadt, Germany

Jiecheng Yang
jiecyang@ulb.ac.be
Université Libre de Bruxelles/KU Leuven (Belgium)
Afdeling Kern-en Stralingsfysica, KU Leuven Celestijnenlaan 200d, bus 2418, 3001, Leuven, Belgium

Part I
Keynote Speakers

Chapter 1
The Nuclear Configuration Interations Method

B. Alex Brown

Abstract This contribution consists of several connected parts: a brief introduction to nuclear physics and nuclear theory; the evidence for magic numbers in nuclear properties, and their interpretation in terms of the single-particle nuclear shell model; the construction of many-body wave functions for configuration mixing calculations; model space truncations of the many-body basis; types of configuration interaction codes and the Lanczos method; types of effective Hamiltonians for specific model spaces; and an introduction to the NuShellX code for calculating, wavefunctions, energies and observables for a given model space and Hamiltonian.

1.1 Introduction

The study of nuclear physics is part of much broader intellectual endeavor to understand the fundamental building blocks of matter and how they combine, interact and evolve to form the universe. At the most microscopic level, the quarks and leptons are the building blocks of the Standard Model of electro-weak and strong interactions. But at this fundamental level there are many questions and unknowns that nuclear physics will help to understand: what are the properties of neutrinos and what is their mass?, what is dark matter?, why is the universe made of matter and not anti-matter?, is there an even more fundamental way of understanding the number or particles, their mass and their weak-strong integration mixing angles?

Quarks and gluons interact to form the proton and neutron. The proton has never been observed to decay, and the neutron beta decays into the proton emitting an electron and a antineutrino. Nucleons interact via the exchanges of mesons to form nuclei. The extra binding energy of this interaction means that many nuclei are stable. The heaviest nuclei are unstable due to the repulsive Coulomb interaction between protons. The complex interplay of motion between these nucleons in nuclei leads

B. Alex Brown (✉)
Department of Physics and the National Superconducting Cyclotron Laboratory,
Michigan State University, East Lansing, MI 42284-1321, USA
e-mail: brown@nscl.msu.edu

J.-E. García-Ramos et al. (eds.), *Basic Concepts in Nuclear Physics: Theory, Experiments and Applications*, Springer Proceedings in Physics 225,
https://doi.org/10.1007/978-3-030-22204-8_1

to a wide variety of mesoscopic structures. At one level this structure gives rise to collective motion involving all of the nucleons such as in the rotations and vibrations. At another level this structure is dominated by the motion of one nucleon relative to the others—the foundation of the nuclear shell model. All nuclei involve an interplay between these extremes which is the subject of nuclear theory.

The nuclei are at the center of atoms. The interaction between the electrons in the atom and nucleus provides an essential experimental tool for studying nuclear properties. Nuclei and atoms constitute matter. Particles, nuclei and matter interact and evolve during the evolution of the universe. How this process proceeds depends upon the properties of nuclei and their extrapolation to the extremes of nucleon number (neutron stars) and temperature.

The theoretical models used for nuclear structure fall into the three broad categories. This division is determined by the limitations of each model. Ab initio models that include all degrees of freedom for the nucleons in the nucleus are limited to light nuclei. Density Functional models are based primarily on single Slater determinants as an approximation for the ground state and low-lying excited states of spherical or deformed nuclei. Configuration Interaction models include many Slater determinants (up to the order of one billion) usually in a spherical basis. This limits the applications to a subset of the states in nuclei with up to about 100 nucleons and to nuclei near the closed shells (up to ^{132}Sn and ^{208}Pb). There are many variations within these models including those based upon limiting the number of configurations based on symmetries. There are also many additions to these models including, for example, the quasi-particle random-phase approximation model for particle-hole states. At present, many interrelated models are needed to describe everything. My lectures concentrate on the practical applications of the configuration interaction model.

1.2 The Nuclear Shell Model

In the shell model, the quantum mechanical problem for the motion of one nucleon in a nucleus is similar to that for the motion of an electron in the hydrogen atom, except that overall scale is determined by the size of the nucleus (10^{-15} m) rather than the size of the atom (10^{-10} m). Another important difference between the atomic and nuclear potentials is that the dependence of the potential on the relative orientation of the intrinsic nucleon (electron) spin and its orbital angular momentum is much stronger and opposite in sign for the nucleon compared to that for the electron.

The single-particle potential has eigenstates that are characterized by their single-particle energies and their quantum numbers. The properties of a nucleus with a given number of protons and neutrons are determined by the filling of the lowest energy single-particle levels allowed by the Pauli exclusion principle which must be obeyed in a system of identical Fermions (the nucleons in this case). The Pauli exclusion principle allows only one proton or neutron to occupy a state with a given set of quantum numbers. The average nuclear potential arises from the short-ranged

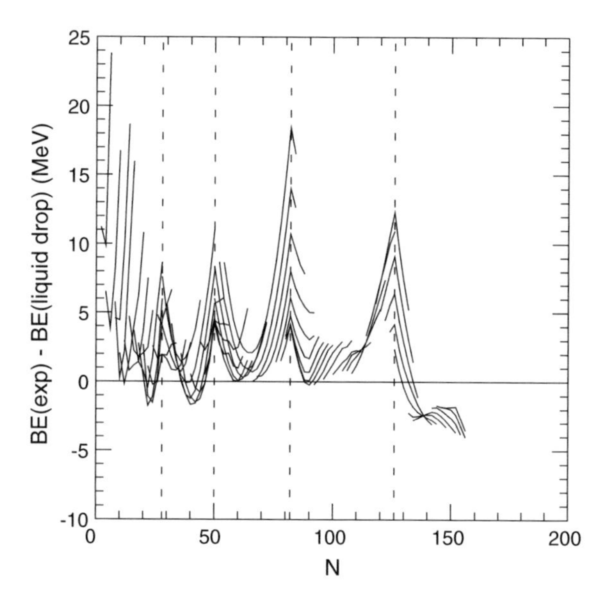

Fig. 1.1 The difference between the experimental and liquid-drop binding energies as a function of N for even-even nuclei. The peaks occur at the magic numbers 28, 50, 82 and 126

attractive nucleon-nucleon interaction and is determined by the shape of the nuclear density distribution.

Evidence for the validity of the nuclear shell model comes from the observation of shell effects in experimental observables such as binding energy, size, spin, and level density. In particular, the nuclear binding energy is not a smooth function of proton and neutron number, but exhibits small fluctuations. The liquid-drop model binding energy is a smooth function of proton and neutron number. When the liquid-drop values for the binding energies are subtracted from the experimental values, the differences show peaks at the magic numbers: $N_{\mathrm{m}} = 28$, 50, 82 and 126 as shown in Figs. 1.1 and 1.2.

The peaks indicate that the nuclei with these magic numbers are more tightly bound than average. Those nuclei that are magic with respect to both neutron and proton numbers are referred to as doubly-magic; an example is the nucleus ^{208}Pb with $N = 126$ and $Z = 82$. Although not so obvious from the binding energy data $N_{\mathrm{m}} = 2$, 6, 8, 14, 16, 20 and 32 are also magic numbers for some nuclei. At these magic numbers the shell gaps are relatively large. This is also reflected in the energy of the first excited 2^+ in even-even nuclei as shown in Fig. 1.3 where the energies range from over 4 MeV for the doubly magic nuclei down to less than 100 keV for the deformed nuclei. Many of these shell gaps are weakened in neutron-rich nuclei to the extent that the energy of particle-hole configurations can come lower in energy than those for the filled shell, leading to "islands of inversion" [1].

The calculated single-particle energy levels appropriate for the neutrons in ^{208}Pb are shown in Fig. 1.4. The potential arises from the average interaction of one neutron with the 207 other nucleons. Since the nuclear force is short-ranged, the shape of

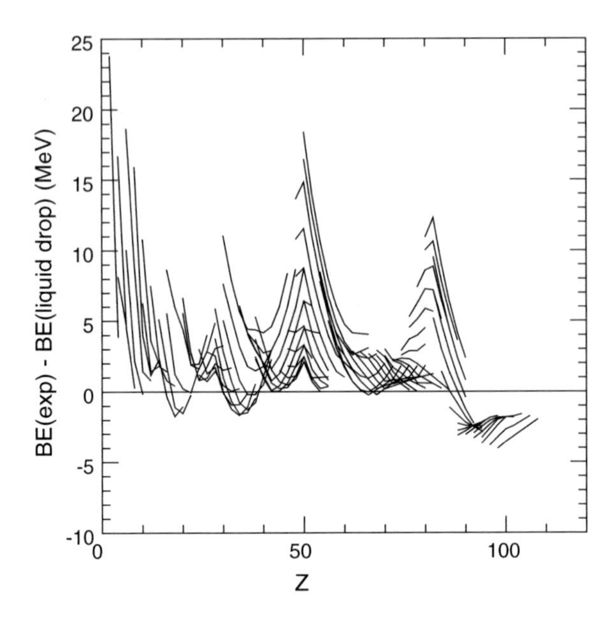

Fig. 1.2 The difference between the experimental and liquid-drop binding energies as a function of Z for even-even nuclei. The peaks occur at the magic numbers 28, 50, 82 and 126

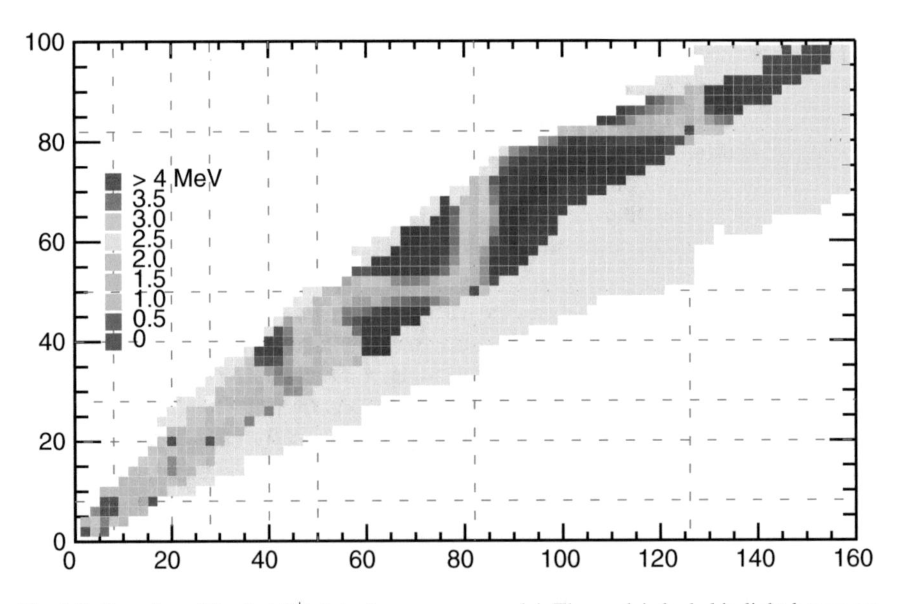

Fig. 1.3 Energies of the first 2^+ states in even-even nuclei. The nuclei shaded in light-brown are those predicted to exist out to the drip lines but not yet observed

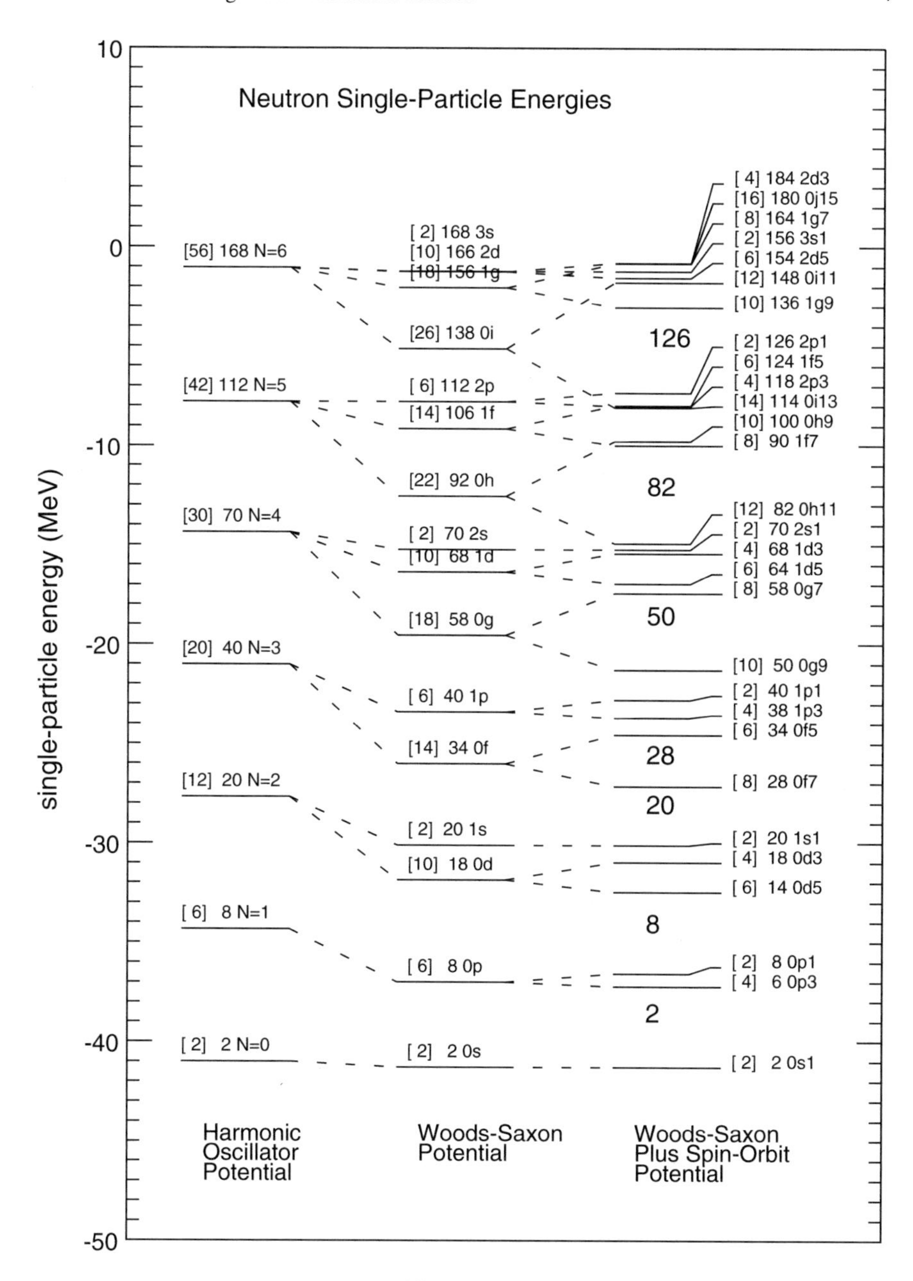

Fig. 1.4 Neutron single-particle states in ^{208}Pb with three potential models, harmonic oscillator (left), Woods-Saxon without spin-orbit (middle) and Woods-Saxon with spin orbit (right). The numbers in square brackets are the maximum number of neutrons in that each level can contain, the following number is a running sum of the total. In addition the harmonic oscillator is labeled by the major quantum number $N = 2n + \ell$, the Woods Saxon is labeled by n, ℓ and the Woods-Saxon with spin-orbit is labeled by n, ℓ, $2j$

potential is similar to the nucleon density in ^{208}Pb which is experimentally known to be close to the Fermi or Woods-Saxon shape of

$$V(r) = \frac{V_o}{[1 + \exp(r - R)/a]}, \tag{1.1}$$

where $R \approx 1.2A^{1/3}$ fm and $a \approx 0.60$ fm. The single-particle energy levels for a potential of approximately this shape and with a central depth of about $V_o = -50$ MeV are shown in the middle of Fig. 1.4. The number of neutrons that are allowed by the Pauli principle to occupy one of these levels, the occupation number, is given by the number in square brackets. In addition, each level is labeled by its cumulative occupation number (the total number of neutrons needed to fill up to the given level) and its $n\ell$ value. n is the radial quantum number (the number of times the radial wave function changes sign) and ℓ is the angular momentum quantum number represented in the spectroscopic notations s, p, d, f, g, h, i and j for ℓ = 0, 1, 2, 3, 4, 5, 6 and 7, respectively. Each ℓ value can have $2\ell + 1$ m states and each m state can contain a proton or neutron with spin up and spin down ($s_z = \pm 1/2$). The occupation number given by the Pauli principle is thus $N_o = 2(2\ell + 1)$.

The Woods-Saxon results are compared to the levels of an harmonic oscillator potential on the left-hand side of Fig. 1.4. Note that the ℓ degeneracy present in the oscillator is broken in the Woods-Saxon potential with levels of larger ℓ coming lower in energy.

The relative spacing of the neutron and proton levels for all nuclei are qualitatively similar to those shown in Fig. 1.4. (The overall spacing between levels goes approximately as $A^{-1/3}$.) According to the Pauli principle, as neutrons are added to nuclei they go into the lowest energy level not already occupied. When a nucleon is added to a nucleus in which the neutron number is equal to one of the cumulative occupation numbers, the neutron must be placed into a relatively higher-energy (more loosely bound) state. Thus the nuclei with the highest relative binding energy are those for which the proton or neutron number is equal to one of the cumulative occupation numbers. A magic number occurs when there is a relatively large energy gap above one of the cumulative numbers. The magic numbers are thus related to the bunching of energy levels. The Woods-Saxon potential gives the correct magic numbers for $N_{\rm m}$ = 2, 8 and 20 but is incorrect for the higher values.

In 1949 the key role of the spin-orbit splitting in the one-body potential was proposed by Mayer [2] and Haxel et al. [3]. This one-body potential model is the starting point for the nuclear shell model. A short history of Mayer's contributions is given in [4]. The spin-orbit potential has the form, $V_{\rm so}(r)\boldsymbol{\ell} \cdot \mathbf{s}$, where $\boldsymbol{\ell}$ is the orbital angular momentum and $\mathbf{s}$ is the intrinsic spin angular momentum of the nucleon. The radial shape conventionally takes the form

$$V_{\rm so}(r) = V_{\rm so} \frac{1}{r} \frac{{\rm d}f_{\rm so}(r)}{{\rm d}r}, \tag{1.2}$$

with

$$f_{\rm so}(r) = \frac{1}{1 + [\exp(r - R_{\rm so})/a_{\rm so}]}. \tag{1.3}$$

The form of (1.2) is based on the the relativistic Thomas interaction energy in atomic physics [5].

With the $\ell \cdot \mathbf{s}$ term, m and s_z are no longer good quantum numbers. The orbital and spin angular momentum must be coupled to a definite total angular momentum, $\mathbf{j} = \ell + \mathbf{s}$. Eigenstates of the spin-orbit potential are determined by the total angular momentum quantum number $j = \ell \pm 1/2$ (except $j = 1/2$ for $\ell = 0$) and the quantum number m_j associated with the z component of j. The expectation value of $\boldsymbol{\ell} \cdot \mathbf{s}$ can be obtained from

$$-\ell \cdot \mathbf{s} \mid \psi_j\rangle = -\frac{1}{2}\left(j^2 - \ell^2 - s^2\right) \mid \psi_j\rangle = -\frac{1}{2}[j(j+1) - \ell(\ell+1) - s(s+1)] \mid \psi_j\rangle \tag{1.4}$$

(the minus sign takes into account the observed sign of the $\langle V_{so}(r)\rangle$) which gives

$$\langle \psi_{j=\ell+1/2} \mid -\ell \cdot \mathbf{s} \mid \psi_{j=\ell+1/2}\rangle = -\frac{\ell}{2} \tag{1.5}$$

and

$$\langle \psi_{j=\ell-1/2} \mid -\ell \cdot \mathbf{s} \mid \psi_{j=\ell-1/2}\rangle = +\frac{(\ell+1)}{2} \tag{1.6}$$

Each j has $2j + 1$ m_j values and hence each j orbit can contain up $N_o = 2j + 1$ protons or neutrons. The energy levels obtained when the spin-orbit potential is added to the Woods-Saxon potential are shown on the right-hand side of Fig. 1.4. The dashed lines that connect to the middle of Fig. 1.4 indicate the effect of the spin-orbit potential in splitting the states of a given ℓ value. The overall strength of the spin-orbit potential has been determined empirically. Each level is labeled by the occupation number (in square brackets), the cumulative occupation number, and the values for n, ℓ, $2j$ ($2j$ is twice the angular momentum quantum number j). The values of the neutron number for which there are large gaps in the cumulative occupation number now reproduce all of the observed magic numbers (as emphasized by the numbers shown in the energy gaps on the right-hand side).

The average nuclear potential can be calculated microscopically from the nucleon–nucleon interaction by using Hartree–Fock theory together with the Brueckner theory for taking into account the repulsion at very short distances between the nucleons. The strength of the spin-orbit potential for nucleons is much larger and opposite in sign to spin-orbit potential for electrons atoms. The radial part of the spin-orbit potential, $V_{\rm so}(r)$, is largest at the nuclear surface and is often taken to be proportional to the derivative of the Woods-Saxon form.

The essential physics behind the shell model is that the many-nucleon collisions that might be expected are greatly suppressed in the nuclear ground states and low-lying nuclear levels because the nucleons would be scattered into states which are forbidden by the Pauli principle. At higher excitation energy the number of allowed states becomes much greater and the nuclear properties indeed become complex and chaotic.

The shell model in its simplest form is able to successfully predict the properties of nuclei which are one nucleon removed or added to the one of the magic number. The shell model can also be extended to include the more complex configurations that arise for the nuclei with nucleon numbers that are in between the magic numbers. For many applications these complex configurations can be taken into account exactly by the diagonalization of a Hamiltonian matrix. In other cases approximations must be used; these include the use of a deformed intrinsic single-particle potential, and the use of group theory to classify the configurations. Current theoretical investigations using the shell model focus on these complex configurations.

We will refer to the closed-shell configurations for the oscillator magic numbers as LS closed shells since they can be described in a basis with good ℓ and s quantum numbers. A equation for the LS sequence of magic numbers, 2, 8, 20, 40, 70, 112, ..., is

$$N = (n)(n + 1)(n + 2)/3, \tag{1.7}$$

where $n = 1, 2, 3, \ldots$. We will refer to the closed-shell configurations for cases where only one of the spin-orbit partners is filled as JJ closed shells since only the total angular momentum j is a good quantum number. A equation for the JJ sequence of magic numbers, 2, 6, 14, 28, 50, 82, 126, ..., is

$$N = (n^3 + 5n)/3, \tag{1.8}$$

where $n = 1, 2, 3, \ldots$. In addition, in neutron-rich nuclei far from stability magic numbers associated with the filling of specific $n\ell_j$ states appear, $N = 16$ ($1s_{1/2}$), $N = 32$ ($1p_{3/2}$) and $N = 34$ ($1p_{1/2}$). Other magic numbers can appear due to shell gaps in a deformed basis.

1.3 Many-Body Wavefunctions in the m-Scheme

For a spherical potential the single-particle wavefunctions are labeled by their radial, orbital angular momentum, and total angular momentum quantum numbers, n_r, ℓ and j, respectively. This set of quantum numbers will be denoted by $k \equiv (n_r, \ell, j)$. Each j value has $(2m + 1)$ m-states, and the associated single-particle wavefunctions will be labeled by $\alpha \equiv (km)$. It is useful to think of k as unique numerical sequence of numbers associated with the complete set of single-particle states. A particular choice for this labeling which is often used [6] is $k = 1, 2, 3, 4, 5, 6, 7, 8, 9, 10, \ldots$, for the sequence $0s_{1/2}, 0p_{3/2}, 0p_{1/2}, 0d_{5/2}, 0d_{3/2}, 1s_{1/2}, 0f_{7/2}, 0f_{5/2}, 1p_{3/2}, 1p_{1/2}$, ... The k value for a given $n\ell j$ can be computed from:

$$k = \frac{1}{2}[(2n + \ell)(2n + \ell + 3) - 2j + 3]. \tag{1.9}$$

The formalism developed thus far is basically all that is needed for an M-scheme calculation [7]. In the M-scheme one starts with a set of basis states Φ for a given M value

$$M = \sum_{\alpha} m_{\alpha}, \tag{1.10}$$

where the sum is over the m values for the occupied states. In general there are an infinite number of basis states Φ, but a for a given situation one truncates the number of states based upon those that are lowest in unperturbed energy. Since the Hamiltonian is diagonal in M one need only consider the subset of basis states Φ with a single value of M in the construction of the many-particle wavefunctions Ψ. The many-body matrix elements of the relevant one- and two-body operators can then be calculated with the techniques of second quantization. Many computer codes have been written to do this.

A basis state with a given value of M does not in general have a definite (good) value of the total angular momentum J. However, since the Hamiltonian is spherically symmetric, the Hamiltonian is also diagonal in J. Thus, the eigenvalues of H, which are linear combinations of the Φ basis, will automatically have good a J value with $J \geq M$, as long as the basis contains the complete set of states that are connected by the $\hat{J}^2$ operator. The J value can be determined by calculating the expectation value $\langle \Psi \mid \hat{J}^2 \mid \Psi \rangle$.

We can calculate the total number of states for a given J, the J-dimension $D(J)$, from the M-scheme dimensions $d(M)$. This is based upon the fact that for each J state there must be $(2J + 1)$ M-states. Thus we find:

$$D(J) = d(M = J) - d(M = J + 1). \tag{1.11}$$

The meaning of (1.11) is that the number of extra $M = J$ states compared to the number of $M = J + 1$ states must be the number of states with angular momentum J. Since $d(-M) = d(M)$ one only has to consider $M \geq 0$. Equation 1.11 will be illustrated with with some examples. The results for these examples only depend upon the (j, m) quantum numbers, and thus the (n_r, ℓ) values are not given explicitly.

Table 1.1 gives all possible $M \geq 0$-states for the $[(j = 5/2)^2]$ configuration for identical nucleons (the $M < 0$ states are obtained from $M > 0$ with a sign change in the m). Due to the Pauli principle we cannot have $M = 5$, and thus $J = 5$ is not allowed. The M-scheme dimensions for this case are given in Table 1.2. $M_{\rm max} = 4$ means that the highest J value is $J = 4$ and this will account for nine of the M states, M= 4, 3, 2, 1, 0, -1, -2, -3 and -4. $J = 3$ is not allowed since there is only one $M = 3$ state which must go with $J = 4$. The extra $M = 2$ state means that there is a state with $J = 2$ which accounts for five more M states M= 2, 1, 0, -1, and -2. $J = 1$ is not allowed since all of the $M = 1$ states are now used, and the extra $M = 0$

Table 1.1 $M \geq 0$ values for the $[(j = 5/2)^2]$ configuration for identical nucleons. The x's under columns headed by $2m$ indicate that the state is occupied, and the total M value is given on the right-hand side

$2m$	5	3	1	−1	−3	−5	
							M
	x	x					4
	x		x				3
	x			x			2
	x				x		1
	x					x	0
		x	x				2
		x		x			1
		x			x		0
			x	x			0

Table 1.2 Table of dimensions for $[(j = 5/2)^2]$

M	$d(M)$	J	$D(J)$
4	1	4	1
3	1	3	0
2	2	2	1
1	2	1	0
0	3	0	1

states means that there is one state with $J = 0$. Thus, the allowed J values are 0, 2 and 4.

A second example is given for the $[(j = 5/2)]^3$ configuration given in Table 1.3. From the multiplicity of the M values one can deduce that only $J = 3/2, 5/2$ and $9/2$ are allowed.

In general, the maximum J-value allowed for a $[k^n]$ configuration is given by the sum of the n largest possible m values,

$$J_{\rm max} = \sum_{i=1}^{n} m_i^{\rm max}. \tag{1.12}$$

In the $[(j = 5/2)^3]$ example, $J_{\rm max} = 5/2 + 3/2 + 1/2 = 9/2$. Lawson [8] discusses other rules which can be deduced from these counting procedures. In particular, the k^n state with $J = J_{\rm max} - 1$ is not allowed by the Pauli principle.

For combinations of more than one orbital, all possible J values are allowed that satisfy the triangle condition for the total J or each orbital. For example, for the $[(j = 5/2)^2(j = 1/2)]$ configuration all possible J values allowed by the triangle

Table 1.3 M values ($M \geq 0$) for the $[(j = 5/2)^3]$ configuration for identical nucleons. The x's under columns headed by $2m$ indicate that the state is occupied, and the total M value is given on the right-hand side

$2m$	5	3	1	-1	-3	-5	
							$2M$
	x	x	x				9
	x	x		x			7
	x	x			x		5
	x	x				x	3
	x		x	x			5
	x		x		x		3
	x		x			x	1
	x			x	x		1
		x	x	x			3
		x	x		x		1

condition of coupling $j = 1/2$ to the $J = 0$, 2 and 4 states allowed for two particles in the $j = 5/2$ state are allowed: $J = 9/2$, 7/2, 5/2, 3/2 and 1/2.

If there is only one M value allowed, the wavefunctions of the J state is given by that of the single M state. If there is more that one M value allowed, then one must diagonalize the Hamiltonian in a space which has the dimension $d(M)$. The states of good J will be linear combinations of the M states. For a given J value of interest, one usually chooses $M = J$, since the number of M states is minimized in this case. However, one could also choose any $M \leq \mid J \mid$ value.

Alternatively, the linear combination of M-states with good J can be calculated using angular-momentum projection methods. This is the method used by the code OXBASH to construct a matrix with dimension $D(J)$ corresponding to the states of good J in terms of the M-scheme basis.

A partition is defined as a specific distribution of the nucleons into the allowed (active) set of k states. The examples discussed above include the three partitions allowed for putting three protons (or three neutrons) into the $j = 5/2$ and $j = 1/2$ states. Written in the form $[(j = 5/2)^a(j = 1/2)^b]$, we have (a, b) = (3, 0), (2, 1) and (1, 2).

The first step in a shell model calculation is to specify the number of particles, the number of active orbits, and then to make a list of the complete set of partitions. Then for each partition we calculate the number of states for each J value. The total dimension is obtained by summing this over all partitions. The calculation may proceed with the full set of partitions within a given model space (a full-space calculation) or it may be restricted to some subset of the partitions (a truncated calculation).

Compilations of $D(J)$ for j^n configurations for some values of j are available in the literature [9], and computer codes are available for the general case [6, 10, 11].

1.4 Model Spaces

Shell-model configuration mixing is carried out within a model space. The model space is a truncation of the infinite set of orbitals into a finite set. All operators in this model space must be renormalized to account for the orbitals left out.

For light nuclei up to about $A = 12$ we can consider a no-core basis where we start with the $0s_{1/2}$ orbital ($N = 0$) and go up to some maximum value $N_{\max}$ in the oscillator basis [12, 13]. Typically up to about $N_{\max} = 8$, but can go higher for few nucleon systems or lower above $A = 12$. The Hamiltonian in the no-core space is

$$H = T + G, \quad (1.13)$$

where T is the kinetic energy operator, and G is the G-matrix type interaction that would be obtained from the V_{lowk} [14] or similarity renormalization group (SRG) [15] methods. Some calculations also include a three-body interaction.

For heavier nuclei calculations can only be carried out exactly in a much smaller model space based upon a closed-shell, some valence orbitals in the model space, and followed by an infinite set of empty orbitals. All operators in this model space must be renormalised to account for this truncation. This is usually done in two steps, first a G matrix type interaction is obtained that takes into account the short-ranged repulsion, and then this is renormalised with pertubation theory to account for the mixing of configurations beyond the model space. The choice of model space is determined by the computational limitations together with guidance from experiment as to which orbitals contribute to the observables for a given mass region. The model space is also guided by the observed shell gaps as indicated for example by the energies of the 2^+ states in even-even nuclei shown in Fig. 1.3. Some model spaces for $A = 4-100$ are shown in Fig. 1.5 and model spaces for A up to 300 are shown in Fig. 1.6.

In the NuShellX Hamiltonian library [15], the names of some model spaces for heavy nuclei are labeled by the number of orbitals that are between the standard magic numbers;

$$k = 4 \ (0f_{5/2}, 1p_{3/2}, 1p_{1/2}, 0g_{9/2}) \ \text{for} \ 28-50;$$

$$k = 5 \ (0g_{5/2}, 1d_{5/2}, 1d_{3/2}, 2s_{1/2}, 0h_{11/2}) \ \text{for} \ 50-82;$$

$$k = 6 \ (0h_{9/2}, 1f_{5/2}, 1f_{5/2}, 2p_{3/2}, 2p_{1/2}, 0i_{13/2}) \ \text{for} \ 82-126;$$

etc. The model space names in proton-neutron formalism where isospin is not necessarily conserved are labeled jjk_pk_npn. For example, the model space called $jj45pn$ is for protons in the group of four above and neutrons in the group of five above. The model space in isospin formalism where total isospin is an explicit quantum number is labeled by jjk_pk_n (without the pn on the end).

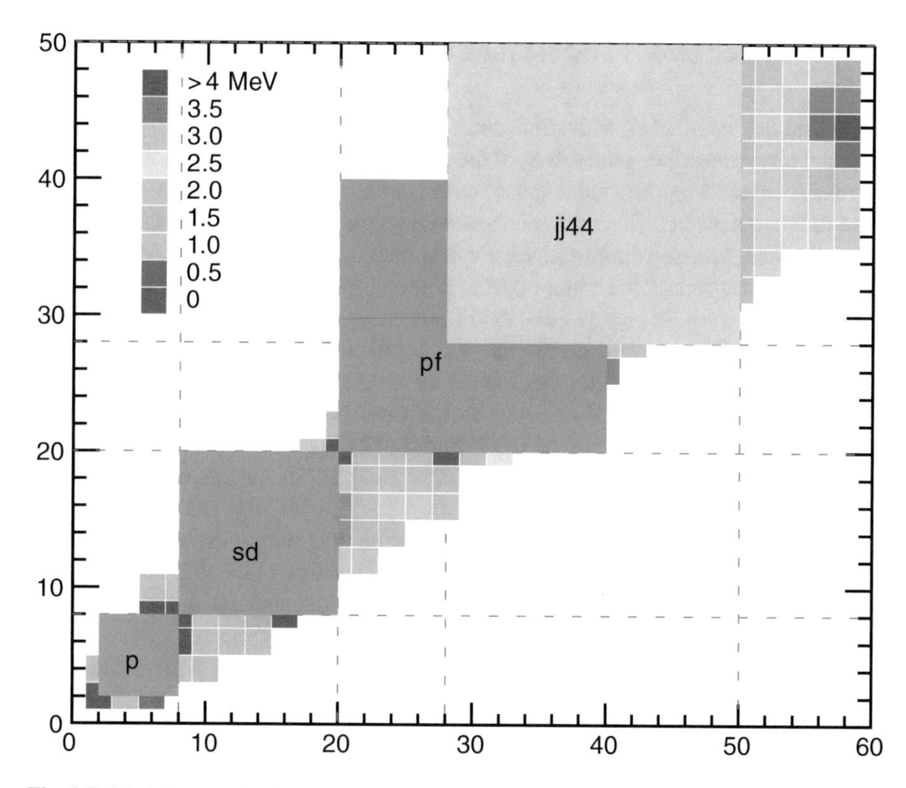

Fig. 1.5 Model spaces for light nuclei

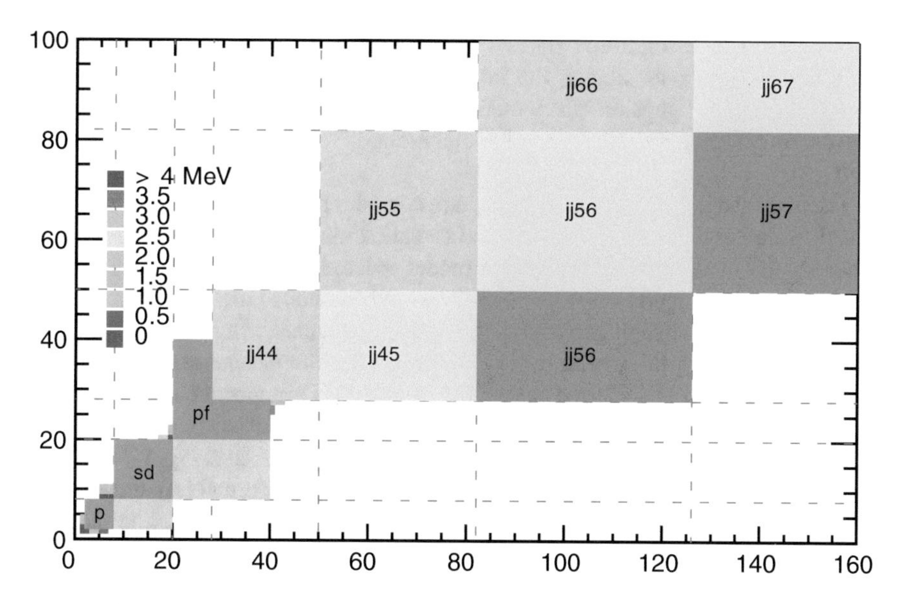

Fig. 1.6 Model spaces for heavy nuclei

1.5 Intruder States and Islands of Inversion

A shell-model calculation with some choice of model space can be considered successful if it can describe some subset of the observed energy levels and transitions for the nuclei covered by the model space, with Hamiltonians and operators which are close to those expected from the renormalized properties of free nucleons. Intruder states are those whose configurations are dominated by nucleons excited across the shell gap into or out of the model space. If the single-particle energy gap is E_{gap}, then the single-particle energy contribution to the excitation energy of a configuration with n particles excited across the gap is just $n\,E_{\mathrm{gap}}$. However, there is also a correlation energy contribution that lowers the energy of the intruder states relative to nE_{gap}. For example, the $N = 8$ gap energy near ^{16}O is about 11.5 MeV. But the experimental excitation energy of the intruder 1/2^{+} state in ^{19}F that involves one nucleon excited across the $N = 8$ gap is only 100 keV. In the sd model space the 1/2^{+} ground state of ^{19}F has the configuration $(sd)^3$, while the 1/2^{-} intruder state has the configuration $(p)^{-1}(sd)^4$. The extra correlation energy associated with the $(sd)^4$ "alpha-particle-like" configuration lowers the energy of the 1/2^{-} state relative to E_{gap}. Another example in ^{11}Be where the 1/2^{+} intruder state (with one neutron in the sd shell) comes 320 keV below the p shell 1/2^{-} state.

Two-particle two-hole $(2p - 2h)$ intruder states have the same parity as those in the model space and can mix with them. For example, the structure of low-lying states in ^{18}O involves a complex mixture of $(sd)^2$ and $(p)^{-2}(sd)^4$ configurations [16]. Part of the correlation energy associated with $2p - 2h$ intruder states is the gain in pairing energy in the $2p$ and $2h$ configurations. These $2p - 2h$ intruder states are seen in all nuclei near the closed shells.

In light nuclei the intruder states are associated with excitations across the LS shell gaps. For heavier nuclei, the intruder states are associated with excitations across the JJ shell gaps. In heavy nuclei the intruder states can also come very low in excitation energy, as in the case of nuclei around ^{186}Pb [17] related to the $Z = 82$ shell gap.

The history of the shell model has been to consider progressively larger and larger model spaces, so that states which would be called intruders in a small model space become fully incorporated into a larger model space. For example, the model space for ^{18}O can be enlarged to include both the p and sd shells [18].

A decrease in the shell gap can lead to situations where whole regions of nuclei have intruder states that come lower in energy than the normal state. The phrase "island of inversion" was used in [1, 19] to describe a region of nuclei centered around ^{31}Na where the configuration of the ground states consists of intruder np–nh excitations across the $N = 20$ shell gap. This region is illustrated in Fig. 1.7 by the red circle just above the arrow for $N = 20$. As an example, the normal (A) and $2p - 2h$ (B) neutron configurations for ^{32}Mg are shown in upper part of Fig. 1.7. The energy of the $2p$–$2h$ state is lowered by the additional pairing correlation and by a change in the proton–neutron interaction which gives a larger quadrupole deformation energy [19].

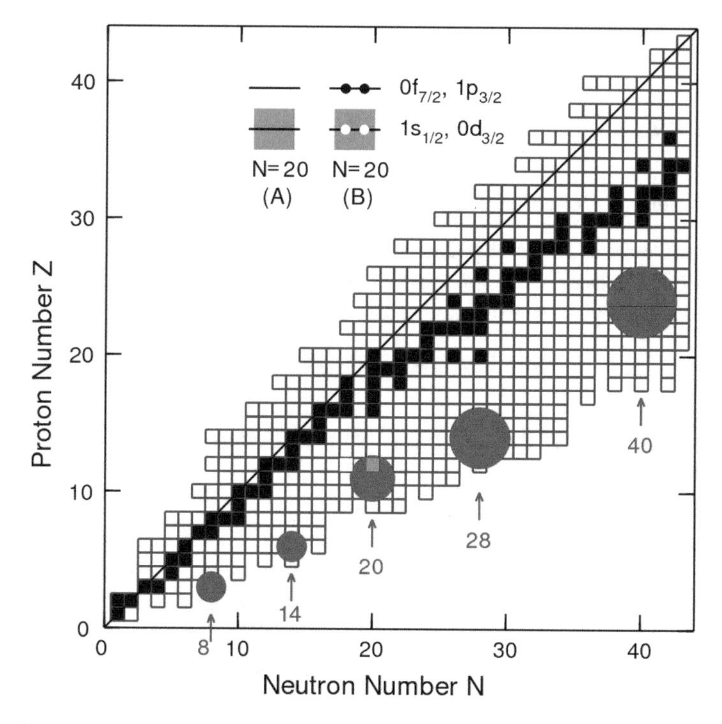

Fig. 1.7 Islands of inversion. The red circles are regions of neutron-rich nuclei where the indicated magic numbers are no longer valid. The green inset shows an example for a configuration involving the $N = 20$ magic number. The energy of the cross-shell configuration (two-particles and two-holes in this case) is lower than that of the closed-shell configuration due to the addition the energy coming from pairing and deformed correlations in the open-shell configuration

The island-of-inversion near ^{31}Na is part of an archipelago of islands related to the breaking of LS magic numbers, starting with an islet near ^{11}Li ($N = 8$) and extending to a larger island centered around ^{64}Cr ($N = 40$); the red circles in Fig. 1.7. It has been predicted that there is an island-of-inversion related to excitations of neutrons across (JJ) $N = 50$ gap in the region of ^{76}Fe [20].

The main reason why these islands-of-inversion are found in neutron-rich nuclei is that E_{gap} tends to decrease as one approaches the neutron drip line. This is a generic feature of loosely bound levels in a finite potential well. More specifically, the single-particle energies with low ℓ values decrease in energy relative to those with higher ℓ values [21]. For example, in the region of $N = 8$, the $1s_{1/2}$ orbital from the sd shell is lowered in energy relative to $0p_{1/2}$ orbital as one goes from ^{16}O to ^{10}He with an associated decrease in the $N = 8$ shell gap [22]. In the region of ^{32}Mg, the lowering of the $1p_{3/2}$ energy relative to that of $0d_{3/2}$ is important [23]. The proton-neutron tensor interaction [24, 25] also plays a role in reducing the shell gap in neutron-rich nuclei [23]. For example, for $N = 20$, as $0d_{5/2}$ protons are removed starting at $Z = 14$ the single-particle energy of the $0d_{3/2}$ orbital moves up the energy of the $0f_{7/2}$ orbital moves down, reducing the $N = 20$ energy gap [23] near $Z = 8$.

1.6 Configuration-Interaction Codes

1.6.1 Types of Basis States

For configuration-interaction model calculations we need to;

(i) select a basis in terms of partitions, and J states within each partition,
(ii) calculate the matrix elements of the Hamiltonian in this basis,
(iii) find the eigenvalues and eigenvectors of this matrix.

The matrix is symmetric and Hermitian. There are several ways to specify the basis:

1. The $M - pn$ scheme makes a basis that has fixed total M and T_z values. Diagonalization will give all states with $J \geq M$ and $T \geq T_z$. The J values are obtained by comparing $\langle \psi_f \mid \hat{J}^2 \mid \psi_i \rangle$ to expected results of $J(J+1)\delta_{fi}$. Isospin is obtained by $\langle \psi_f \mid \hat{T}^2 \mid \psi_i \rangle$. If the Hamiltonian conserves isospin, then this will be $T(T+1)\delta_{fi}$. If the Hamiltonian does not conserve isospin then the results will deviate from this and there will be non-zero off-diagonal terms.
2. The $J - pn$ scheme makes a basis that has fixed J and T_z value. The J basis states are linear combinations of M states obtained by applying the J-projection operator to a selected set of M basis states. Diagonalization will give all states $T \geq T_z$.
3. The $J - T$ scheme makes a basis that has fixed J and T. The $J - T$ basis states are obtained by applying the $J - T$ projection operator to a selected set of the M, T_z basis states.

The matrix dimension depends on the type of basis. For example, for ^{20}Ne in the sd shell model space there are:

- $M - pn$: 640 states with $M = 0$ and $T_z = 0$,
- $J - pn$: 46 states with $J = 0$ and $T_z = 0$,
- $J - T$: 21 states with $J = 0$ and $T = 0$.

If we only want states with $J = 0$ and $T = 0$, the matrix will be smallest for the J–T basis. But the construction of the matrix is easiest in the M–pn basis. If the Hamiltonian does not conserve isospin then we must use pn basis. Various CI codes have been written to be most efficient for a given type of basis.

One of the first M–pn code was written by Whitehead et al. [26]. Other codes that use this basis are ANTOINE [27], MFDn [28] and BIGSTICK [29].

The code Oxbash [30] uses the J–T or the J–pn basis. Since the entire matrix is stored, the maximum basis dimension is on the order of 100,000.

The code NuShellX [31] only uses the J–pn basis. For states with both protons and neutrons, dimensions up to about 10^8 can be considered. Details of this code are given in the next section.

1.6.2 The Lanczos Method

We are usually interested in only the lowest n states, where typically $n = 10$. But the dimension d of the matrix H is much larger. All CI codes use a version of what is called the Lanczos method. The Lanczos method is based on the power-iteration method for finding the eigenfunction for the eigenvalue λ that has the largest absolute value. For nuclear Hamiltonians this is the most negative λ corresponding to the ground state for a given matrix, λ_g with wavefunction $\mid v_g\rangle$. We start with any vector $\mid v_0\rangle$ that contains some of $\mid v_g\rangle$, for example a vector generated by random numbers. (One might also start with the unit vector $(1, 1, 1, \ldots)$. But in some cases there might be some eigenfunctions which are orthogonal to this unit vector.) Then multiply this by the matrix and normalize

$$\mid v_1\rangle = H \mid v_0\rangle, \quad \mid v_1\rangle / \mid \langle v_1 \mid v_1\rangle \mid \rightarrow \mid v_1\rangle. \tag{1.14}$$

Continue this for k times.

$$\mid v_k\rangle = H \mid v_{k-1}\rangle \tag{1.15}$$

The $\mid v_k\rangle$ will converge to $\mid v_g\rangle$ for $k \ll d$. The reason for this is that $\mid v_0\rangle$ contains a random linear combination of all of the eigenfunctions, $\mid u_i\rangle$

$$\mid v_0\rangle = \sum_{i=1,d} \alpha_i \mid u_i\rangle, \tag{1.16}$$

with

$$H \mid u_i\rangle = \lambda_i \mid u_i\rangle. \tag{1.17}$$

Operating with $(H^k/\lambda_g)^k$ gives

$$(H/\lambda_g)^k \mid u_0\rangle = \sum_{i=1,d} \alpha_i (\lambda_i/\lambda_g)^k \mid u_i\rangle. \tag{1.18}$$

All of the terms with become small except for the one with $i = g$ leaving the (unnormalized) wavefunction $\alpha_g \mid v_g\rangle$.

The Lanczos method starts with power method and then also generates orthogonal excited states that converge into series of excited state wavefunctions [32]. The procedure is as follows:

1. Start with some any initial vector $\mid v_i\rangle$, for example, one obtained with random numbers in the range of -1 to 1 and normalized.
2. Multiply by the Hamiltonian matrix to get $\mid w_i\rangle = H \mid v_i\rangle$.
3. Take the overlap, $\alpha_i = \langle v_i \mid w_i\rangle$.
4. Orthogonalize against the initial vector, $\mid w_i\rangle - \alpha_i \mid v_i\rangle \rightarrow \mid w_i\rangle$.
5. If $i\rangle 1$ orthogonalize against the prior vector, $\mid w_i\rangle - \beta_{i-1} \mid v_{i-1}\rangle \rightarrow \mid w_i\rangle$.
6. Find the norm to get $\beta_i = \sqrt{\langle w_i \mid w_i\rangle}$.

7. Normalize to get the next Lanczos vector $| v_{i+1}\rangle = | w_i\rangle/\beta_i$.
8. Repeat this process this process n times.

Due to round-off errors it is better to replace (1.4) with an orthogonalization against all previous vectors. This procedure results in an n dimensional symmetric tridiagonal matrix with the α_i as the diagonal and β_i as the off diagonal, e.g.

$$H \mid v_1\rangle = \alpha_1 \mid v_1\rangle + \beta_1 \mid v_2\rangle$$

$$H \mid v_2\rangle = \beta_1 \mid v_1\rangle + \alpha_2 \mid v_2\rangle + \beta_2 \mid v_3\rangle$$

$$H \mid v_3\rangle = \beta_2 \mid v_2\rangle + \alpha_3 \mid v_3\rangle + \beta_3 \mid v_4\rangle$$

continued up to

$$H \mid v_n\rangle = \beta_{n-1} \mid v_{n-1}\rangle + \alpha_n \mid v_n\rangle.$$

The eigenvalues and eigenvectors of this tridiagonal matrix converge to the exact results. Typically 10 states converged to 1 keV accuracy requires about $n = 100$ Lanczos iterations. An example of the convergence in energy for the first 30 states in ^{48}Cr is shown in Fig. 1.8.

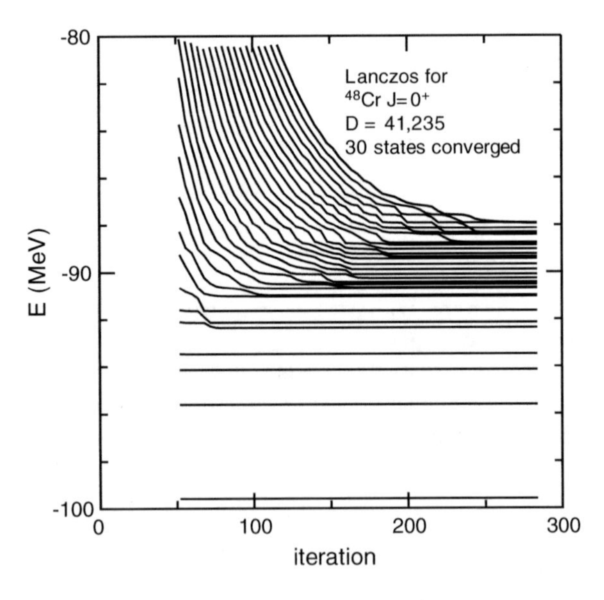

Fig. 1.8 Energies for first 30 states with $J = 0$ in ^{48}Cr as a function of the number of Lanczos iterations

1.7 Effective Hamiltonians

The Hamiltonian has the form

$$H = H_0 + H_1 + H_2. \tag{1.19}$$

The first term is the binding energy of the closed shell. The second term is a one-body operator whose matrix elements are given by the single-particle energies (SPE) multiplied by the orbital occupancy. It implicitly contains both the kinetic energy and the monopole interaction of the valence nucleon with those in the closed core. One often uses experimental information for the single-particle energies, or when this is not known, some approximate calculations such as results from a Woods-Saxon potential. The two-body part represented, H_2, could be obtained in a variety of ways. For a qualitative understanding one could use schematic forms such as the delta-function interaction. More microscopic forms start with the potentials that reproduce nucleon-nucleon scattering data and then are renormalized for the chosen model space. Empirical Hamiltonians are based on fitting the two-body matrix elements (TBME) to a selection of experimental data. These empirical Hamiltonians implicitly contain the corrections due the addition of three-body interactions and to the limitations of the many-body approaches.

It is conventional to determine a Hamiltonian for one mass value and then to use the same Hamiltonian over a wide mass region. Sometimes a smooth scaling of H_2 is used to account for an implicit mass scaling. For example, the USD interaction in the sd-shell model space assumes the form $\text{TBME}^{A} = \text{TBME}^{A=18}(A/18)^{-0.3}$, where the power was determined empirically [33]. The scaling is a result of the change of the (implicit) radial wave function as a function of mass (A). If one takes the simplest oscillator dependence of $\hbar\omega = 41A^{-1/3}$, then the TBME calculated with a delta function vary as $A^{-1/2}$ and the TBME for the long-range Coulomb interaction varies as $A^{-1/6}$. Nuclear interactions of intermediate range would vary somewhere between these two extremes. The smooth scaling is an approximation. As we approach the drip line one might expect to observe deviations from experiment due to the larger radial extent of the valence wave functions.

Usually the core, H_0 and one-body, H_1 (the SPE) terms are not changed as a function of mass. But there is no reason they should not change. The effective values for H_2 implicitly compensate for possible changes in H_0 and H_1.

A common element of most configuration-mixing calculations in light nuclei is that the interactions used conserve isospin and are specified in terms of TBME which have good J and T. The effects of the Coulomb interaction are usually treated by adding a Z-dependent constant. It can be deduced from the energy shift between analogue states. In practice, this means that a shell-model result for a given isotopic chain (fixed Z) gives the relative binding energy and excitation energy as a function of neutron number. To obtain the absolute binding energy, a constant must be added to connect it to a given experimental value which can be taken as that of the $T = 0$ ground state for even Z and that of the $T = 1/2$ ground state for odd Z. The levels in mirror nuclei are identical in this approximation. The experimental excitation energies in mirror nuclei are typically shifted by 100 keV or less relative to the ground state energy. This is small compared to the accuracy of most configuration-mixing calculations (200 keV or more). There are larger shifts in light nuclei associated with the Thomas-Ehrman shift of low ℓ orbitals, and these should be treated more carefully in terms of the structure change between mirror nuclei. Configuration mixing in light nuclei can be carried out in a proton-neutron basis [34], but the main interest is in the special problem of overlaps in Fermi beta decay [35] and in isospin forbidden processes [36].

1.7.1 Types of Hamiltonians

One usually begins with a set of TBME obtained from a realistic Hamiltonian based on the renormalized G matrix. But quality of the predictions and applications for a given model space and mass region can be greatly improved if the values of the TBME are constrained to reproduce some selection of energy data. There a several approaches to doing this that are outlined below.

ETBME: For very small model spaces the wavefunctions for a large number of states arise from a relatively few TBME. One of the best examples is the $0f_{7/2}$ shell [37] where this one orbital is rather isolated. For the $0f_{7/2}$ shell there are only eight TBME and one SPE which can be used to obtain the energies and wave functions of several hundred levels in the $A = 40$–56 mass region [38] (a more complete model space for these nuclei is possible now). In small model spaces such as the p-shell, essentially all of the TBME can be obtained empirically, and these will be referred to as effective-TBME (ETBME) hamiltonians.

ETBME+G: In larger model spaces the number of TBME is large and they cannot all be well determined by existing data. The data are sensitive to particular linear combinations of SPE and TBME. The well determined linear combinations can be obtained from the single-values decomposition (SVD) discussed in the next section. The other linear combinations are left at their G matrix values. Early examples of this are [39, 40]. The SVD method was used in the *sd* model space to obtain the USD Hamiltonians [33, 41], and and in the *pf* shell to obtain the GXPF pf-shell hamiltonians [42].

G+MON: In larger model spaces the most important part of the Hamiltonian are the monopole combinations of TBME, because these determine the evolution of the SPE a function of mass. Effective interactions where some or all of the monopole are used to fit some selected set of energy data will be called G+MON. Examples of G+MON Hamiltonians are the KB3 [43] and KB3G [44] in the pf shell.

POT: Empirical hamiltonians can also be based upon two-body potential models in which the strengths of the various channels (central, spin-orbit and tensor) are obtained from a fit to energy data. These will be called potential "POT" models for the shell-model hamiltonian. The simplest form of the potential models are the delta functions and MSDI mentioned above. A more elaborate form is the modified-surface-one-boson-exchange-potential (MSOBEP) [45]. Other types of fits related to the potential models can be based upon the relative matrix elements [46] and Talmi integrals [45, 47].

For heavy nuclei one usually finds Hamiltonians specifically designed for use only for a relatively small region of nuclei, such as those around ^{132}Sn or ^{208}Pb.

These empirical hamiltonians provide a way to generate realistic wave functions (from which one calculates observables) and to extrapolate the known properties of nuclei to the unknown. When one observes something which does not agree with these extrapolations, it is usually an indication of "new physics" involving degrees of freedom which are not in the assumed model space. The observation of such "new physics" in neutron-rich nuclei is the essential aspect of the current and proposed radioactive-beam experimentations.

1.7.2 Hamiltonians for Specific Model Spaces

The names of some commonly used effective hamiltonians and the model-space they are associated with are given in Table 1.1. These hamiltonians have been determined by a least-squares fit to binding energy and excitation energy data. This was accomplished by varying the full set of two-body matrix elements (ETBME), by varying well determined linear combinations of two-body matrix elements and keeping the rest fixed at some G matrix values (ETBME+G), by adding monopole corrections to the G matrix (G+MON), and by varying potential parameters (POT).

1.7.3 Hamiltonians in the $jj44$ Model Space

Historically, the first $jj44x$ type of Hamiltonian for the $jj44$ model space was called $jj44pna$ in the NuShellX library for the $jj44pn$ model space [59]. This Hamiltonian contains one set of two-body matrix elements (TBME) with $T = 1$ for neutrons that are constrained to reproduce the binding energies and excitation energies for the nickel isotopes ($Z = 28$) with $N = 33$–44, and another set of TBME with $T = 1$ for protons that are constrained to reproduce the binding energies and excitation energies for isotones with $N = 50$ and $Z = 32$–50. The $jj44pna$ Hamiltonian does not contain proton–neutron TBME and cannot be used away from $Z = 28$ or $N = 50$. For $jj44pna$ the neutron and proton TBME are different. As a consequence of this, the 8^+ seniority isomers obtained in ^{94}Ru and ^{96}Pd are not present in the analogous nuclei ^{72,74}Ni due to a crossing of some states dominated by seniority two and four [59] (Table 1.4).

The $jj44b$ and JUN45 Hamiltonians are for the $jj44$ model space. Both of these contain an assumed mass dependence of $(A/58)^{-0.3}$. The TBME for $jj44b$

Table 1.4 Table of model spaces and interactions names

Model space	A	Interaction	Type	References
0p (p)	$A = 5$–16	(6–16)TBME	ETBME	[48]
		PJT	ETBME	[49]
		PJP	POT	[49]
	$A = 8$–16	(8-16)POT	POT	[48]
		(8–16)TBME	ETBME	[48]
	$A = 10$–16	PWBT	ETBME	[18]
1s0d (sd)	$A = 16$–40	USD	ETBME+G	[33]
		SDPOTA	POT	[45]
		USDA	ETBME+G	[41]
		USDB	ETBME+G	[41]
1p0f (pf)	$A = 40$–50	FPMG	G+MON	[50]
		FPD6	POT	[51]
		KB3	G+MON	[43]
	$A = 40$–60	KB3G	G+MON	[44]
		GXPF1	ETBME+G	[42]
		GXPF1A	ETBME+G	[52]
sd-pf	Near $N = 20$	WBMB		[19, 53]
		RCNP		[54]
		Utsuno et al.		[55]
		Dean et al.		[56]
	Near $N = 28$ $Z > 14$	SDPFU	G+MON	[57]
	Near $N = 28$ $Z \leq 14$	SDPFUSI	G+MON	[sdpfu]
jj44	$A = 70$–100	SDPFU	G+MON	[57]
		SDPFUSI	G+MON	[57]
		jun45	ETBME+G	[58]
		jj44pna	ETBME+G	[59]
		jj44b	ETBME+G	(see text)

are based on those obtained with the renormalized Bonn-C potential. The singlular-value decomposition (SVD) method was used to constrain 30 linear combinations of the 133 TBME to 77 binding energies and 470 excitation energies in nuclei with $Z = 28$–30 ($N = 28 = 50$), and $N = 48$–50 ($Z = 28$–50). For a given Z, the binding energies are corrected by an overall shift obtained from the Coulomb part of a Skyrme energy-density functional calculation. The rms deviation between the theoretical and experimental energies was about 240 keV. When the $jj44b$ Hamiltonian is used in the proton-neutron model space $jj44pn$ it is called $jj44bpn$; the results with $jj44b$ and $jj44bpn$ are the same. Starting in 2007, the $jj44b$ Hamiltonian has been used for comparison to data in many publications [where it is sometimes called $jj4b$ and sometimes cited as B. A. Brown and A. F. Lisetskiy (private communication)] [60–70].

The $T = 1$ TBME for $jj44b$ are approximately an average of those for protons and neutrons in the the $jj44pna$ Hamiltonian. When the $jj44b$ Hamiltonian is applied to $Z = 28$ or $N = 50$ it does work as well as the $jj44pna$ Hamiltonian. Starting with $jj44b$ another Hamiltonian called $jj44c$ was obtained by leaving out energy data above $Z = 38$. This is a better Hamiltonian to use for $Z = 28$–30. These $jj44$ are described in the appendix of [71].

A method similar to that used to obtain the $jj44b$ Hamiltonian was used by Honma et al. to obtain the JUN45 Hamiltonian [58]. For the JUN45 Hamiltonian, 45 SVD linear combinations were determined by a fit to about 69 binding energies and 330 excitation energies for nuclei in the range $N = 30$–32 and $Z = 46$–50, as shown in Fig. 1 of [58]. The rms deviation was 185 keV. These data included the ground state and first three excited states in ^{76}Ge.

1.8 The NuShellX Code

NuShellX is a set of computer codes written by Bill Rae. NuShellX@MSU is a set of wrapper codes written by Brown [31] that use data files for model spaces and Hamiltonians to generate input for NuShellX. The wrapper codes also convert the NuShellX output into figures and tables for energy levels, gamma decay and beta decay.

Any paper that is written using the NuShellX@MSU code should contain a sentence of the type—"The calculations were carried out in the x model space with the y Hamiltonian (give the reference for the interaction) using the code NuShellX@MSU [31]". If the model space, model-space truncation or Hamiltonian is changed from the original reference it must be discussed and justified in the text. The NuShellX library contains some Hamiltonians without references or documentation. These are only for tests and examples and cannot be used for publications.

All calculations with NuShellX are in the proton-neutron formalism. When single-particle state (*.sp) files and Hamiltonian (*.int) files in isospin formalism are used, they are converted to proton-neutron formalism. If the file iso.nux exists, the expectation value of the T^2 operator will be calculated with the results given in the file *.ovl. If isospin is conserved only the diagonal matrix elements of the *.ovl file are non-zero and have the value $T(T+1)$. If isospin is not conserved then the *.ovl file will have off diagonal matrix elements that can be used to calculate isospin mixing.

Detailed information on the installation and use of NuShellX is given in the file help.pdf contained in the NuShellX package. Truncations can be made only in the proton and/or neutron orbital occupations. They cannot be made in both at the same time. For this reason it is not possible to make $\hbar\omega$-type truncations for model spaces such as s–p–sd–pf as use in [18]. The model space sd–pf can only be used in the case where active protons are restricted to the sd shell (the proton pf shell is empty), the neutron sd shell is full, and the active neutrons are in the pf shell.

The NuShellX code uses a proton-neutron basis and a technique similar to that used for the code NATHAN [72]. The first step is to make a list of partitions with a given set of orbital occupation restrictions for protons (neutrons). Then for each partition the $d(M)$ M states are constructed and stored in terms a string of binary numbers (0 of 1) that give the state occupancy. Then the $D(J)$ J basis states are obtained as linear combinations of these M states using an angular momentum projection operator.

The two-body part of the Hamiltonian is written as a sum of three terms:

$$H = H_{nn} + H_{pp} + H_{pn} \tag{1.20}$$

for the sum of the neutron–neutron (nn), proton–proton (pp) and proton–neutron (pn) interactions. The pp (nn) Hamiltonian matrix elements are calculated for each partition and its connection to all other partitions that differ in not more that two protons (neutrons) being moved between different orbitals.

The second quantized form for H_{pn} is

$$H_{pn} = \sum_{pnp'n',J_o} \langle pnJ_o \mid V \mid p'n'J_o \rangle \left\{[a_p^+ a_n^+]^{J_o} \otimes [\tilde{a}_{p'}\tilde{a}_{n'}]^{J_o}\right\}^{(0)} . \tag{1.21}$$

We can recouple the operators to:

$$\left\{[a_p^+ a_n^+]^{J_o} \otimes [\tilde{a}_{p'}\tilde{a}_{n'}]^{J_o}\right\}^{(0)} = -\sum_{\lambda} \sqrt{(2\lambda+1)(2J_o+1)}\,(-1)^{j_n+j_{p'}-\lambda-J_o}$$
$$\times \begin{Bmatrix} j_p & j_n & J_o \\ j_{n'} & j_{p'} & \lambda \end{Bmatrix} \times \left\{[a_p^+ \tilde{a}_{p'}]^{\lambda} \otimes [a_n^+ \tilde{a}_{n'}]^{\lambda}\right\}^{(0)} , \tag{1.22}$$

where, for example, p stands for the single-particle wavefunction (n_p, ℓ_p, j_p). H_{pn} can thus be written in the particle-hole form:

$$H_{pn} = \sum_{pp'nn'\lambda} F_{\lambda}(pp'nn')\ \{[a_p^+ \tilde{a}_{p'}]^{\lambda} \otimes [a_p^+ \tilde{a}_{n'}]^{\lambda}\}^{(0)}, \tag{1.23}$$

where

$$F_{\lambda}(pp'nn') = -\sum_{J_o} \sqrt{(2\lambda+1)(2J+1)}\,(-1)^{n+p'-\lambda-J_o}$$
$$\times \begin{Bmatrix} p & n & J_o \\ n' & p' & \lambda \end{Bmatrix} \langle pnJ_o \mid V \mid p'n'J_o \rangle. \tag{1.24}$$

The NuShellX basis states have the form:

$$\mid B_i, J\rangle = \mid [(J_{p_i}, \alpha_{p_i}) \otimes (J_{n_i}, \alpha_{n_i})]J\rangle \equiv \mid [P_i \otimes N_i]J\rangle, \tag{1.25}$$

where, P_i stands for labels (J_{p_i}, α_{p_i}), where J_{p_i} is the proton angular momentum and α_{p_i} are all of the other quantum numbers needed to specify the complete basis.

These proton (neutron) basis states can be subdivided into partitions of the protons (neutrons) among the the orbitals p. For the Lanczos multiplications we need the matrix elements of H_{pp}

$$\langle B_f, J \mid H_{pp} \mid B_i, J\rangle = \delta_{N_f,N_i}\delta_{J_{P_f},J_{P_i}}\langle P_f \mid H_{pp} \mid P_i\rangle, \tag{1.26}$$

H_{nn}

$$\langle B_f, J \mid H_{nn} \mid B_i, J\rangle = \delta_{P_f,P_i}\delta_{J_{n_f},J_{n_i}}\langle N_f \mid H_{nn} \mid N_i\rangle, \tag{1.27}$$

and H_{pn}

$$\begin{aligned}\langle B_f, J \mid H_{pn} \mid B_i, J\rangle &= \sum_{pp'nn',\lambda} F_\lambda(pp'nn')\left\langle B_f, J \mid \{[a_p^+\tilde{a}_{p'}]^\lambda \otimes [a_p^+\tilde{a}_{n'}]^\lambda\}^{(0)} \mid B_i, J\right\rangle. \\ &= \sum_{pp'nn'\lambda} F_\lambda(pp'nn')\ \Gamma_\lambda(J_{p_f}, J_{p_i}, J_{n_f}, J_{n_i}, J) \\ &\times \mathrm{RDM}(P_f, P_i, p, p', \lambda)\ \ \mathrm{RDM}(N_f, N_i, n, n', \lambda),\end{aligned} \tag{1.28}$$

where

$$\Gamma_\lambda(J_{p_f}, J_{p_i}, J_{n_f}, J_{n_i}, J) = \begin{Bmatrix} J_{p_f} & J_{p_i} & \lambda \\ J_{n_f} & J_{n_i} & \lambda \\ J & J & 0 \end{Bmatrix} \tag{1.29}$$

and RDM are the reduced density matrices:

$$\mathrm{RDM}(P_f, P_i, p, p', \lambda) = \langle P_f || [a_p^+\tilde{a}_{p'}]^\lambda || P_i\rangle, \tag{1.30}$$

and

$$\mathrm{RDM}(N_f, N_i, n, n', \lambda) = \langle N_f || [a_n^+\tilde{a}_{n'}]^\lambda || N_i\rangle.$$

These RDM are precalculated and stored.

Starting with an initial vector

$$\mid \omega_i, J\rangle = \sum_{P_i,N_i} A(P_i, N_i, J)\ \ \mid [P_i \otimes N_i]J\rangle. \tag{1.31}$$

Multiplying H_{pn} on $\mid \omega_i, J\rangle$ gives:

$$H_{pn} \mid \omega_i, J\rangle = \mid \omega_f, J\rangle = \sum_{P_f,N_f} A'(P_f, N_f, J)\ \ \mid [P_f \otimes N_f]J\rangle, \tag{1.32}$$

with

$$A'(P_f, N_f, J) = \delta_{N_f,N_i}\delta_{J_{p_f},J_{p_i}}\langle P_f \mid H_{pp} \mid P_i\rangle + \delta_{P_f,P_i}\delta_{J_{n_f},J_{n_i}}\langle N_f \mid H_{nn} \mid N_i\rangle,$$
$$+ \sum_{\lambda pp'nn'} F_\lambda(pp'nn') \sum_{P_i} \mathrm{RDM}(P_f, P_i, p, p', \lambda)$$
$$\times \sum_{N_i} \mathrm{RDM}(N_f, N_i, n, n', \lambda)\ \Gamma_\lambda(J_{p_f}, J_{p_i}, J_{n_f}, J_{n_i}, J)\ A(P_i, N_i, J). \tag{1.33}$$

The Hamiltonian matrix elements are calculated "on-the-fly". The amount of storage associated with the separate proton and neutron basis and the combined proton-neutron vectors is modest. NuShellX makes use of OpenMP to use many cores with nearly 100% efficiency for the Lanczos iterations. The power of this method is illustrated in Fig. 1.9 where the J-scheme and M-scheme dimensions for 8 protons and 8 neutrons in the pf model space. The eigenvectors associated with the very large J-scheme combined dimensions are generated from the pre-calculated information on the separate proton and neutron basis states that each have relative small dimensions. The proton and neutron basis states are obtained by J-projection on M-states with the code NuShell. NuShell is a modern Fortran replacement for the original JT projection code Oxbash.

The layout of NuShellX@MSU is shown in Fig. 1.10. NuShellX is surrounded by a wrapper codes that bring in information from a library of previously derived Hamiltonians (bottom left). Or the Hamiltonians may be obtained from ab-initio

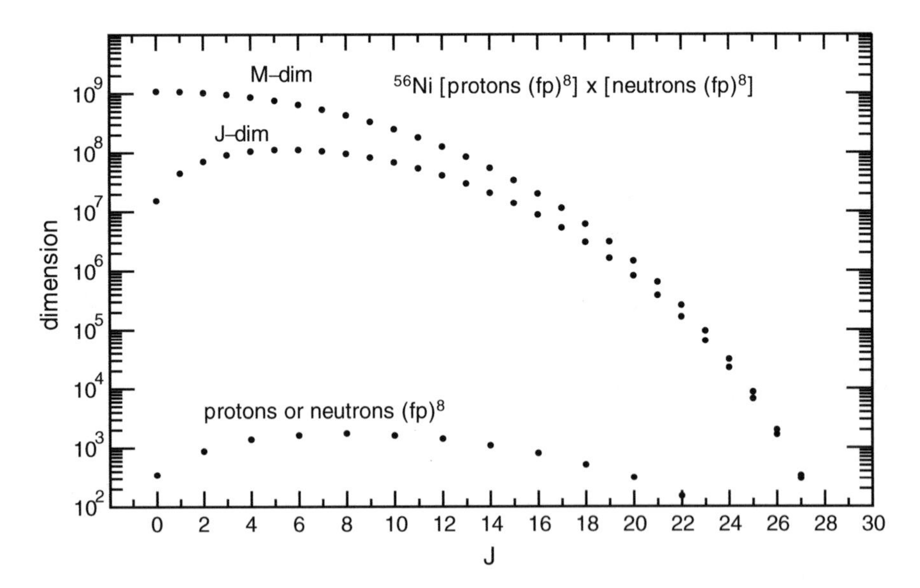

Fig. 1.9 Basis dimensions for the 8 protons or 8 neutron in the pf model space (bottom). The combined proton-neutron J-scheme and M-scheme dimensions are shown at the top

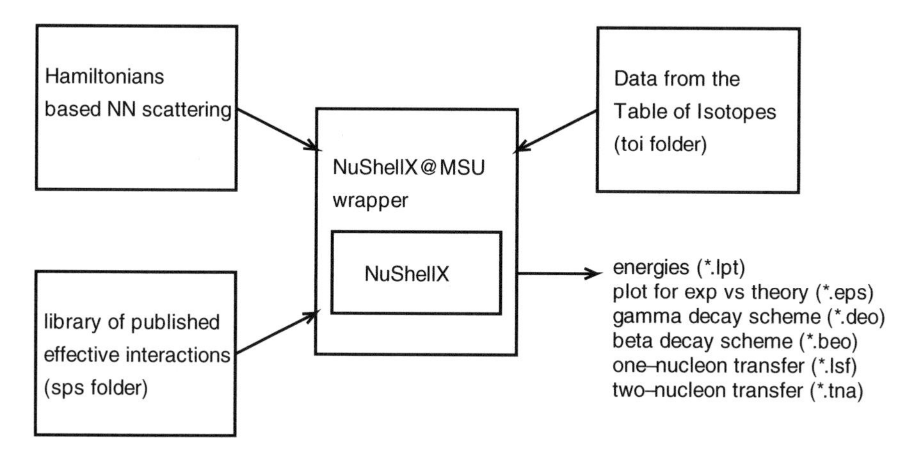

Fig. 1.10 Schematic layout of the NuShellX@MSU codes

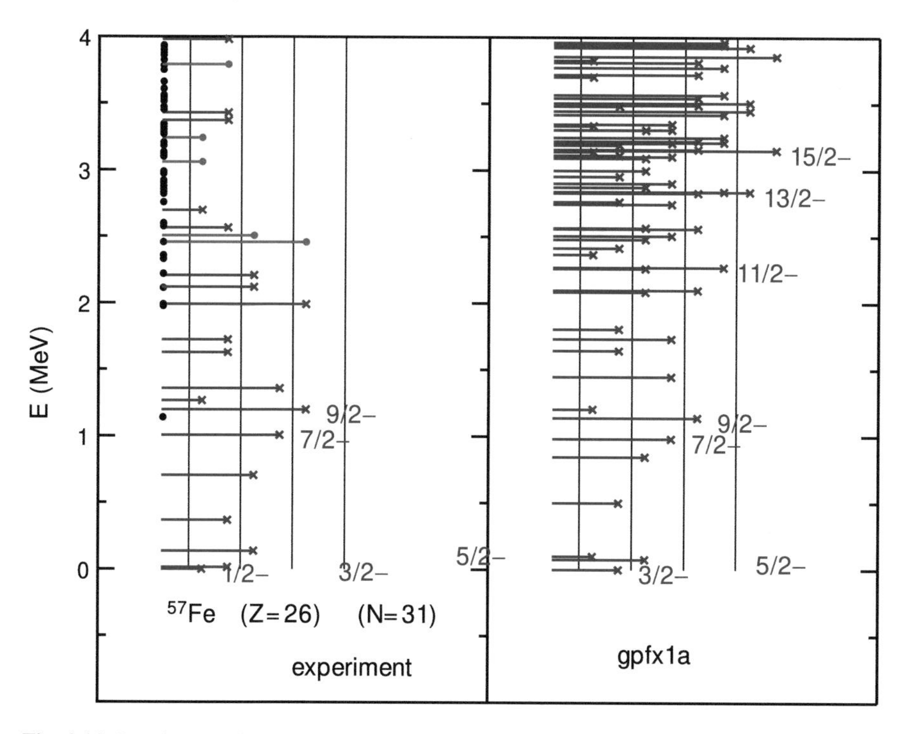

Fig. 1.11 Levels up to 3 MeV in excitation obtained with the GPFX1A Hamiltonian in the full *pf* model space are shown in the right-hand side. The length of each line indicates the *J* value. These are compared with the experimental energies from NNDC on the left-hand side. The blue lines are states with negative parity, and the red line are states with positive parity. If the spin-parity is uncertain it is shown as a black circle on the y axis

approaches such as those available from [73] (top left). At the end the wrapper combines the theoretical energies with the information from the ENSDF files to produce a figure for the comparison between experiment and theory. An example is shown in Fig. 1.11 for the levels of ^{57}Fe obtained in the pf model space with the GPFX1A Hamiltonian [52].

As a second step NuShellX takes the overlaps for the operators a^+, a^+a^+ and a^+a. The wrapper converts these into spectroscopic factors, two-nucleon transfer amplitudes and one-body transition densities, respectively. The one-body transition densities are then used together with the program DENS to calculate the matrix elements for $M\lambda$, $E\lambda$ and Gamow-Teller (GT) operators. The $M1$ and $E2$ results are used to obtain a gamma-ray decay scheme together with the magnetic and quadrupole moments for all states in the calculation. The GT matrix elements are used to obtain an allowed beta-decay scheme. The program DENS allows one to use radial wavefunctions from harmonic-oscillator, Woods-Saxon or Skyrme energy-density functions methods for the matrix elements.

References

1. B.A. Brown, Physics **3**, 104 (2010)
2. M.G. Mayer, Phys. Rev. **75**, 1969 (1949)
3. O. Haxel, J.H.D. Jensen, H.E. Suess, Phys. Rev. **75**, 1766 (1949)
4. K.E. Johnson, Maria Goeppert Mayer: atoms, molecules and nuclear shells. Phys. Today **39**, 44 (1986)
5. L.H. Thomas, Nature (London) **117**, 514 (1926)
6. W.D.M. Rae, A. Etchegoyen, B.A. Brown, *OXBASH, the Oxford-Buenos Aries-MSU Shell-Model Code*, Michigan State University Laboratory Report no. 524
7. R.R. Whitehead, A. Watt, B.J. Cole, I. Morrison, *Advances in Nuclear Nuclear Physics*, vol. 9, p. 123 (1977); L.M. Mackenzie, A.M. Macleod, D.J. Berry, R.R. Whitehead, Comput. Phys. Comm. **48**, 229 (1988)
8. R.D. Lawson, *Theory of the Nuclear Shell Model* (Clarendon Press, 1980)
9. T. Sebe, M. Harvey, Enumeration of many-body states of the nuclear shell model with definite angular momentum and isobaric spin with mixed single particle orbits. ARCL. Chalk River **3007**, (1968)
10. D.K. Sunko, Phys. Rev. C **33**, 1811 (1986)
11. J.P. Draayer, H.T. Valdes, Comput. Phys. Comm. **36**, 313 (1985)
12. P. Navratil, J.P. Vary, B.R. Barrett, Phys. Rev. Lett. **84**, 5728 (2000) (and references therein)
13. W.C. Haxton, C.L. Song, Phys. Rev. Lett. **84**, 5484 (2000)
14. S.K. Bogner, T.T.S. Kuo, A. Schwenk, Phys. Rept. **386**, 1 (2003)
15. S.K. Bogner, R.J. Furnstahl, A. Schwenk, Prog. Part. Nucl. Phys. **65**, 94 (2010)
16. R.L. Lawson, F.J.D. Serduke, H.T. Fortune, Phys. Rev. C **14**, 1245 (1976)
17. C. De Coster, B. Decroix, K. Heyde, Phys. Rev. C **61**, 067306 (2000)
18. E.K. Warburton, B.A. Brown, Phys. Rev. C **46**, 923 (1992)
19. E.K. Warburton, J.A. Becker, B.A. Brown, Phys. Rev. C **41**, 1147 (1990)
20. F. Nowacki, A. Poves, E. Caurier, B. Bounthong, Phys. Rev. Lett. **117**, 272501 (2016)
21. I. Hamamoto, Phys. Rev. C **76**, 054319 (2007)
22. H. Sagawa, B.A. Brown, H. Esbensen, Phys. Lett. B **309**, 1 (1993)
23. A. Lepailleur et al., Phys. Rev. C **92**, 054309 (2015)
24. T. Otsuka et al., Phys. Rev. Lett. **95**, 232502 (2005)

25. T. Otsuka et al., Phys. Rev. Lett. **104**, 012501 (2010)
26. R.R. Whitehead, A. Watt, B.J. Cole, I. Morrison, Adv. Nucl. Phys. **9**, 123 (1977)
27. E. Caurier, F. Nowacki, Acta Phys. Pol. B **30**, 705 (1999)
28. J.P. Vary, *The Many-Fermion Dynamics Shell-Model Code* (Iowa State University, 1992) (unpublished); J.P. Vary, D.C. Zheng, ibid., (1994) (unpublished); P. Sternberg, E. Ng, C. Yang, P. Maris, J.P. Vary, M. Sosonkina, H. Viet Le, Accelerating configuration interaction calculations for nuclear strucure, in *IEEE Conference on Supercomputing the Proceedings of the 2008 ACM* (IEEE Press, Piscataway, NJ). https://doi.org/10.1145/1413370.1413386
29. C.W. Johnsona, W.E. Ormandb, P.G. Krastev. https://arxiv.org/pdf/1303.0905.pdf
30. B.A. Brown, A. Etchegoyen, W.D.M. Rae, N.S. Godwin, W.A. Richter, C.H. Zimmerman, W.E. Ormand, J.S. Winfield, MSU-NSCL Report No. 524, 1985
31. B.A. Brown, W.D.M. Rae, Nucl. Data Sheets **120**, 115 (2014)
32. https://en.wikipedia.org/wiki/Lanczos_algorithm
33. B.H. Wildenthal, Prog. Part. Nucl. Phys. **11**, 5 (1984)
34. W.E. Ormand, B.A. Brown, Nucl. Phys. A **491**, 1 (1989)
35. W.E. Ormand, B.A. Brown, Nucl. Phys. A **440**, 274 (1985)
36. W.E. Ormand, B.A. Brown, Phys. Lett. B **174**, 128 (1986)
37. A. de Shalit, I. Talmi, *Nuclear Shell Theory* (Academic Press, 1963)
38. W. Kutschera, B.A. Brown, K. Ogawa, *La Rivista del Nuovo Cimento* vol. 11, p. 1 (1978)
39. A. Arima, S. Cohen, R.D. Lawson, M.H. MacFarlane, Nucl. Phys. A **108**, 94 (1968)
40. W. Chung, Ph.D. thesis, Michigan State University, 1976
41. B.A. Brown, W.A. Richter, Phys. Rev. C **74**, 034315 (2006)
42. M. Honma, T. Otsuka, B.A. Brown, T. Mizusaki, Phys. Rev. C **65**, 061301(R) (2002)
43. A. Poves, A.P. Zuker, Phys. Rep. **70**, 235 (1981)
44. A. Poves, J. Sanchez-Solano, E. Caurier, F. Nowacki, Nucl. Phys. A **694**, 157 (2001)
45. B.A. Brown, W.A. Richter, R.E. Julies, B.H. Wildenthal, Ann. Phys. **182**, 191 (1988)
46. A.G.M. van Hees, P.W.M. Glaudemans, Z. Phys. A **315**, 223 (1984)
47. I. Talmi, Helv. Phys. Acta **25**, 185 (1952)
48. S. Cohen, D. Kurath, Nucl. Phys. **73**, 1 (1965); Nucl. Phys. **A101**, 1 (1967)
49. R.E. Julies, W.A. Richter, B.A. Brown, South Afr. J. Phys. **15**, 35 (1992)
50. J.B. McGrory, Phys. Rev. C **8**, 693 (1973)
51. W.A. Richter, M.G. Van der Merwe, R.E. Julies, B.A. Brown, Nucl. Phys. A **523**, 325 (1991)
52. M. Honma, T. Otsuka, B.A. Brown, T. Mizusaki, Euro. Phys. J. A **25**(Suppl. 1), 499 (2005)
53. E.K. Warburton, J.A. Becker, D.J. Millener, B.A. Brown, BNL report **40890** (1987)
54. J. Retamosa, E. Caurier, F. Nowacki, A. Poves, Phys. Rev. C **55**, 1266 (1997)
55. Y. Utsuno, T. Otsuka, T. Mizusaki, M. Honma, Phys. Rev. C **60**, 054315 (1999)
56. D.J. Dean, M.T. Ressell, M. Hjorth-Jensen, S.E. Koonin, A.P. Zuker, Phys. Rev. C **59**, 2474 (1999)
57. F. Nowacki, A. Poves, Phys. Rev. C **79**, 014310 (2009)
58. M. Honma, T. Otsuka, T. Mizusaki, M. Hjorth-Jensen, Phys. Rev. C **80**, 064323 (2009)
59. A.F. Lisetskiy, B.A. Brown, M. Horoi, H. Grawe, Phys. Rev. C **70**, 044314 (2004)
60. D. Verney et al., Phys. Rev. C **76**, 054312 (2007)
61. K.T. Flanagan et al., Phys. Rev. C **82**, 041302(R) (2010)
62. A. Gade et al., Phys. Rev. C **81**, 064326 (2010)
63. B. Cheal, Phys. Rev. Lett. **104**, 252502 (2010)
64. P. Vingerhoets et al., Phys. Rev. C **82**, 064311 (2010)
65. A.D. Becerril et al., Phys. Rev. C **84**, 041303(R) (2011)
66. S.J.Q. Robinson, L. Zamick, Y.Y. Sharon, Phys. Rev. C **83**, 027302 (2011)
67. D. Verney et al., Phys. Rev. C **87**, 054307 (2013)
68. B.A. Brown, D.L. Fang, M. Horoi, Phys. Rev. C **92**, 041301(R) (2015)
69. A.C. Dombos et al., Phys. Rev. C **93**, 064317 (2016)
70. F. Recchia et al., Phys. Rev. C **94**, 054324 (2016)
71. S. Mukhopadhyay, B.P. Crider, B.A. Brown, S.F. Ashley, A. Chakraborty, A. Kumar, E.E. Peters, M.T. McEllistrem, F.M. Prados-Estevez, S.W. Yates, Phys. Rev. C **95**, 014327 (2017)
72. E. Caurier et al., Shell-model code NATHAN (unpublished)
73. T. Engeland, M. Hjorth-Jensen, G.R. Jansen, unpublished; M. Hjorth-Jensen, T.T.S. Kuo, E. Osnes, CENS, computational environment for nuclear structure. Phys. Rep. **261**, 125 (1995)

Chapter 2
Introduction to Nuclear-Reaction Theory

Pierre Capel

Abstract These notes summarise the lectures I gave during the summer school "International Scientific Meeting on Nuclear Physics" at La Rábida in Spain in June 2018. They offer an introduction to nuclear-reaction theory, starting with the basics in quantum scattering theory followed by the main models used to describe breakup reactions: the Continuum Discretised Coupled Channel method (CDCC), the Time-Dependent approach (TD) and the eikonal approximation. These models are illustrated on the study of the exotic structure of halo nuclei.

Introduction

Nuclear reactions are used for a variety of goals. They can be used to study the structure of nuclei; sometimes, they can be the only way to probe nuclear structure, especially far from stability. Nuclear reactions also provide information about the interaction between nuclei, either to study the fundamentals of the nuclear force, or to measure reaction rates, which are major inputs in other fields of physics, like nuclear astrophysics, or in a broad range of nuclear applications, like nuclear power or the production of radioactive isotopes for medical purposes.

To correctly analyse and exploit data of reaction measurements, it is important to know the basics in nuclear-reaction theory. The present notes offer an introduction to this exciting discipline. Sect. 2.1 presents the basics of non-relativistic scattering theory for two colliding particles, which interact through a potential. In this section, the notion of *cross section* is introduced and its calculation from the solution of the stationary Schrödinger equation is explained. In particular, the method based on the *partial-wave* expansion of the wave function is presented in detail. To close this first chapter, I introduce the *optical model*, which enables to account for other reaction channels that can take place during the collision of the particles. In these developments, I closely follow the Chapter VIII of the textbook on quantum mechan-

P. Capel (✉)
Institut für Kernphysik, Johannes Gutenberg-Universität Mainz,
55099 Mainz, Germany

Physique Nucléaire et Physique Quantique (CP 229),
Université libre de Bruxelles (ULB), 1050 Brussels, Belgium
e-mail: pcapel@uni-mainz.de

J.-E. García-Ramos et al. (eds.), *Basic Concepts in Nuclear Physics: Theory, Experiments and Applications*, Springer Proceedings in Physics 225,
https://doi.org/10.1007/978-3-030-22204-8_2

ics by Cohen-Tannoudji, Diu and Laloë [1]. For interested readers, a more detailed presentation of quantum reaction theory can be found in [2].

In Sect. 2.2, I give a brief overview of the main methods used to describe *breakup* reactions. That sections starts with a presentation of *halo nuclei*, which are one of the most exotic quantal structures found far from stability, and which are studied mostly through reactions, like breakup. I then pursue with the three main models of breakup: the Continuum Discretised Coupled Channel method (CDCC), the Time-Dependent approach (TD) and the eikonal approximation. The section closes with a comparison between them that emphasises the advantage and drawbacks of each of these models, and which gives their respective range of validity. More advanced developments on nuclear-reaction theory can be found in [3, 4].

In Sect. 2.3, I review the information about the structure of halo nuclei that can be inferred from the analysis of breakup measurements. We will see in this section what can be expected from experimental data, and, most importantly, what cannot be inferred from experiments. This section is built mostly from recent articles published in the literature. Their selection of course reflects my personal biases on the subject as well as my own research activity.

I do not believe this paper exhausts the vast subject of nuclear-reaction theory, but I hope that it will give an incentive to some of the readers to pursue their journey in the landscape of nuclear physics within this exciting and flourishing field of research. Without further ado, let us start this introduction with the basics in quantum collision theory.

2.1 Quantum Collision Theory

2.1.1 Types of Collisions

Quantum collisions are used in various applications. They are sometimes one of the only way to study the interaction between particles. For example, the potentials used in nuclear-structure calculations to simulate the interaction between the nucleons is deduced mostly from observables measured in nucleon-nucleon collisions [5] (see Sect. 2.1.4). Collisions are also used to infer information about the structure of quantal objects. The famous experiment of Rutherford, Geiger and Marsden performed in 1909 is a good example. This experiment, in which alpha particles were fired at a gold foil, enabled the discovery of the structure of atoms. Nowadays, reactions are measured to study the structure of nuclei throughout the whole nuclear chart. In a more natural way, collisions can also be used to obtain reaction rates of particular interest, e.g., for reactions that take place in stars or that are needed in technological applications, like nuclear reactors or to produce radioactive isotopes of medical use.

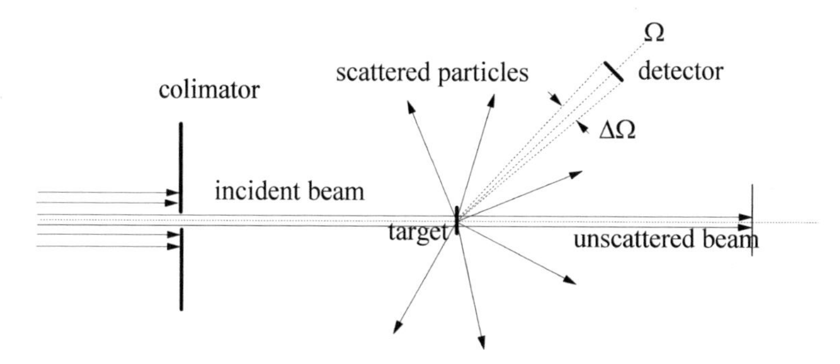

Fig. 2.1 Typical measurement scheme of nuclear reactions

The usual measurement scheme of these reactions in nuclear physics is schematically illustrated in Fig. 2.1. An incident beam made up of projectile particles, coming from the left, is first collimated before impinging on a fixed target. The particles produced during the collision are then scattered in all possible directions $\Omega \equiv (\theta, \varphi)$ and measured in the detectors surrounding the target. These detectors have a finite size and are seen from the target under a solid angle $\Delta\Omega$. Note that a significant amount of the incoming particles will not react and continue straight ahead, undeflected by the target. Due to this unscattered beam, measuring reactions at very small scattering angle can be quite difficult, when not impossible.

When two quantal objects collide, various reactions can take place. Let us consider the collision of two "particles" a and b, where the term particle is taken in a broad sense and can mean molecules, atoms, nuclei, nucleons,...

1. $a + b \to a + b$ (elastic scattering)
2. $\to a + b^*$ (inelastic scattering)
3. $\to c + f + b$ (breakup)
4. $\to d + e$ (rearrangement or transfer)

The first and most evident one is the *elastic scattering*, in which the two particles merely scatter while remaining in their initial states. Second, if the incident energy is high enough, some energy can be transferred from the relative motion of a and b towards one of the—or both—particles, which then leave(s) the collision in an excited state. Such a state is denoted by the asterisk (*) next to the excited particle. The scattering is then said *inelastic*. A third possibility is that this energy transfer is high enough to *break up* one of the particles into its more elementary constituents (e.g. in two ions in the case of a collision of molecules, into an ion and an electron if the collision involves atoms, or into nucleon clusters in a nuclear collision). The fourth case envisaged here is the case of *rearrangement* or *transfer*, in which some sub-particles are transferred from the projectile to the target or vice versa (like atoms in molecular collisions, electrons in atomic collisions, or nucleons or groups of nucleons in nuclear reactions).

As an example in the realm of nuclear physics, let us consider the collision of ^{11}Be on ^{208}Pb, which can be 1. elastically scattered, 2. inelastically scattered (e.g. if ^{11}Be is excited to its $\frac{1}{2}^-$ state during the collision), 3. broken up into a ^{10}Be and a neutron (see Sect. 2.2), or 4. have that valence neutron transferred to the ^{208}Pb target to form ^{209}Pb:

1. $^{11}\mathrm{Be}+^{208}\mathrm{Pb}$	$\rightarrow$ $^{11}\mathrm{Be}+^{208}\mathrm{Pb}$	(elastic scattering)
2.	$\rightarrow$ $^{11}\mathrm{Be}^*(1/2^-)+^{208}\mathrm{Pb}$	(inelastic scattering)
3.	$\rightarrow$ $^{10}\mathrm{Be}+\mathrm{n}+^{208}\mathrm{Pb}$	(breakup)
4.	$\rightarrow$ $^{10}\mathrm{Be}+^{209}\mathrm{Pb}$	(transfer)

Although the reactions taking place during the collision can significantly affect the structure of the colliding particles, some physical values are conserved. One particular case is the total energy of the system, which remains the same before and after the collision. Its conservation means that the sum of the mass energy and total kinetic energy in any outgoing channel must equal that in the incoming channel:

$$m_a\,c^2 + m_b\,c^2 + T_{\rm in} = \sum_i m_i\,c^2 + T_{\rm out} \tag{2.1}$$

where m_a and m_b are the masses of the colliding particles a and b respectively, and $T_{\rm in}$ is the total kinetic energy in the incoming channel. In the right-hand side of equation (2.1) m_i are the masses of all the particles produced during the collision in one particular channel, and $T_{\rm out}$ is their total kinetic energy.

From this expression, one can define the *Q value* of a particular reaction, which corresponds to the energy "produced" by this reaction

$$Q = m_a\,c^2 + m_b\,c^2 - \sum_i m_i\,c^2. \tag{2.2}$$

If $Q > 0$ the reaction is said *exoenergetic* as it *produces energy*. Energetically, the reaction is then always possible. On the contrary, if $Q < 0$, the reaction is said *endoenergetic* and requires a minimal initial kinetic energy to take place. On the sole energy viewpoint, a channel will be said *open* if $T_{\rm in} > -Q$. If this is not the case, i.e. if the reaction is endoenergetic and the incident kinetic energy is lower than $|Q|$, the channel is said *closed*. Note that the elastic-scattering channel is always open, since in that case $Q = 0$.

2.1.2 Notion of Cross Section

To characterise a reaction and measure its probability to take place during the collision of two particles, we use the notion of *cross section*. To introduce this observable, let us go back to the measurement scheme pictured in Fig. 2.1. The number of particles Δn detected in the direction Ω per unit time within the solid angle $\Delta\Omega$ covered by the detector will be naturally proportional to the flux F_i of incoming particles and the number N of particles within the target:

$$\Delta n = F_i \; N \; \Delta\sigma. \tag{2.3}$$

the factor of proportionality $\Delta\sigma$ is the *cross section* for the outgoing channel considered in the measurement. The dimensions of Δn are that of a number of event per unit time, those of F_i that of a number of particles per unit time and unit area, and N is of course just a number of particles. Consequently, the dimensions of $\Delta\sigma$ are that of an area. Its units are usually expressed in *barns* (b) : $1\text{ b} = 10^{-24}\text{ cm}^2 = 100\text{ fm}^2$.

The *differential cross section* corresponds to the limit

$$\frac{\mathrm{d}\sigma}{\mathrm{d}\Omega} = \lim_{\Delta\Omega\to 0} \frac{\Delta\sigma}{\Delta\Omega} \tag{2.4}$$

$$= \lim_{\Delta\Omega\to 0} \frac{\Delta n}{F_i N \Delta\Omega}. \tag{2.5}$$

The direction $\hat{\mathbf{Z}}$ is often chosen along the beam axis. In that case, thanks to the cylindrical symmetry of the problem, the cross section $\mathrm{d}\sigma/\mathrm{d}\Omega$ depends only on the colatitude θ and is independent of the azimuthal angle φ.

This schematically explains how these values can be measured. To see how they can be calculated, let us consider the elastic scattering of two particles, a and b, which interact through a potential V. This potential depends on the a-b relative coordinate $\mathbf{R} = \mathbf{R}_b - \mathbf{R}_a$ (for the sake of clarity, the spin of the particles is neglected in this development). The Hamiltonian of the system hence reads

$$\mathscr{H}(\mathbf{R}_a, \mathbf{R}_b) = T_a + T_b + V(\mathbf{R}), \tag{2.6}$$

where the kinetic energy of particles a and b, respectively, read

$$T_a = \frac{p_a^2}{2m_a} = -\frac{\hbar^2 \Delta_{R_a}}{2m_a} \tag{2.7}$$

$$T_b = \frac{p_b^2}{2m_b} = -\frac{\hbar^2 \Delta_{R_b}}{2m_b}, \tag{2.8}$$

with $\mathbf{p}_a$ and $\mathbf{p}_b$ the momenta of a and b, respectively.

Since V depends only on the relative coordinate $\mathbf{R}$ it is useful to change coordinates and use, instead of $\mathbf{R}_a$ and $\mathbf{R}_b$, their relative coordinate $\mathbf{R}$ and the coordinate of their centre of mass

$$\mathbf{R}_{\text{cm}} = \frac{m_a \mathbf{R}_a + m_b \mathbf{R}_b}{M}, \tag{2.9}$$

were $M = m_a + m_b$ is the total mass in the incoming channel. Within this new set of coordinates, the Hamiltonian of the system reads

$$\mathscr{H}(\mathbf{R}_a, \mathbf{R}_b) = T_{\mathrm{cm}} + T_R + V(\mathbf{R}), \tag{2.10}$$

where

$$T_{\mathrm{cm}} = \frac{P_{\mathrm{cm}}^2}{2M} = -\frac{\hbar^2 \Delta_{R_{\mathrm{cm}}}}{2M} \tag{2.11}$$

is the kinetic energy of the centre of mass of a and b, with $\mathbf{P}_{\mathrm{cm}}$ its momentum, and

$$T_R = \frac{P^2}{2\mu} = -\frac{\hbar^2 \Delta_R}{2\mu}, \tag{2.12}$$

is the kinetic energy of the relative motion between a and b, with $\mathbf{P}$ the corresponding momentum and $\mu = m_a m_b / M$ the *reduced mass* of a and b.

It follows from (2.10) that $\mathscr{H}$ is the sum of two Hamiltonians, which are functions of two independent variables $\mathbf{R}_{\mathrm{cm}}$ and $\mathbf{R}$:

$$\mathscr{H}(\mathbf{R}_a, \mathbf{R}_b) = H_{\mathrm{cm}}(\mathbf{R}_{\mathrm{cm}}) + H(\mathbf{R}). \tag{2.13}$$

Accordingly the wave function that describes this two-particle system can be factorised into

$$\Psi_{\mathrm{tot}}(\mathbf{R}_a, \mathbf{R}_b) = \Psi_{\mathrm{cm}}(\mathbf{R}_{\mathrm{cm}})\, \Psi(\mathbf{R}). \tag{2.14}$$

The wave function Ψ_{cm} describes the motion of the centre of mass of a and b. It is solution of the Schrödinger equation

$$H_{\mathrm{cm}}\, \Psi_{\mathrm{cm}}(\mathbf{R}_{\mathrm{cm}}) = E_{\mathrm{cm}}\, \Psi_{\mathrm{cm}}(\mathbf{R}_{\mathrm{cm}}), \tag{2.15}$$

where $H_{\mathrm{cm}} = T_{\mathrm{cm}}$ [see (2.10)], which describes the motion of a free particle of mass M. For a particle of initial momentum $\mathbf{P}_{\mathrm{cm}} = \hbar \mathbf{K}_{\mathrm{cm}}$, Ψ_{cm} corresponds simply to a plane wave

$$\Psi_{\mathbf{K}_{\mathrm{cm}}}(\mathbf{R}_{\mathrm{cm}}) = (2\pi)^{-3/2} e^{i\mathbf{K}_{\mathrm{cm}} \cdot \mathbf{R}_{\mathrm{cm}}}, \tag{2.16}$$

with $\hbar^2 K_{\mathrm{cm}}^2 / 2M = E_{\mathrm{cm}}$, the kinetic energy of the a-b centre of mass in the reference frame in which $\mathbf{R}_a$ and $\mathbf{R}_b$ are defined. The normalisation factor $(2\pi)^{-3/2}$ is chosen such that

$$\langle \Psi_{\mathbf{K}'_{\mathrm{cm}}} | \Psi_{\mathbf{K}_{\mathrm{cm}}} \rangle = \delta(\mathbf{K}_{\mathrm{cm}} - \mathbf{K}'_{\mathrm{cm}}) \tag{2.17}$$

This wave function (2.16) hence describes a centre of mass in uniform translation, as we would have expected from Galilean invariance. Accordingly, this motion does not add anything and can thus be ignored. The physics of the problem is thus entirely captured within the Hamiltonian

$$H(\mathbf{R}) = T_R + V(\mathbf{R}), \tag{2.18}$$

which describes the relative motion of particles a and b. In scattering theory, the meaningful eigenstates of H are the *stationary scattering states*.

2.1.3 Stationary Scattering States

A stationary scattering state $\Psi_{K\hat{\mathbf{Z}}}$ is a solution of

$$H\,\Psi_{K\hat{\mathbf{Z}}}(\mathbf{R}) = E\,\Psi_{K\hat{\mathbf{Z}}}, \tag{2.19}$$

which exhibits the following asymptotic behaviour

$$\Psi_{K\hat{\mathbf{Z}}}(\mathbf{R}) \underset{R\to\infty}{\longrightarrow} (2\pi)^{-3/2}\left[e^{iKZ} + f_K(\theta)\,\frac{e^{iKR}}{R}\right]. \tag{2.20}$$

In (2.19) and (2.20), $\hat{\mathbf{Z}}$ has been chosen as the beam axis, for which choice the expression does not depend on the azimuthal angle φ as explained above.

The solutions of equation (2.19) we are looking for behave asymptotically as the sum of a plane wave e^{iKZ} and an outgoing spherical wave $f_K(\theta)e^{iKR}/R$, whose amplitude is modulated as a function of the scattering angle θ by the function f_K, which is called the *scattering amplitude*. Note that, for both terms in equation (2.20), the momentum $\hbar K$ is related to the energy $E = \hbar K^2/2\mu$. To interpret the physical meaning of this asymptotic behaviour, let us recall the operator of the current of probability

$$\mathbf{J}(\mathbf{R}) = \frac{1}{\mu}\Re[\Psi^*(\mathbf{R})\,\mathbf{P}\,\Psi(\mathbf{R})], \tag{2.21}$$

and let us compute this operator on each term of equarion (2.20). For the plane wave, we obtain

$$\mathbf{J}_i(\mathbf{R}) = (2\pi)^{-3}\frac{\hbar K}{\mu}\hat{\mathbf{Z}} = (2\pi)^{-3}\,v\,\hat{\mathbf{Z}} \tag{2.22}$$

where $v = \hbar K/\mu$ is the incoming a-b relative velocity. We can thus interpret this term as the incoming current of probability, describing the projectile impinging on the target at a velocity v along the beam axis.

For the spherical wave, we get

$$\mathbf{J}_s(\mathbf{R}) = (2\pi)^{-3}\,v\,|f_K(\theta)|^2\frac{1}{R^2}\hat{\mathbf{R}} + \mathcal{O}\left(\frac{1}{R^3}\right), \tag{2.23}$$

which, at large distance, is purely radial and directed outwards. This current is still proportional to v, but its magnitude varies with θ according to the square modulus of the scattering amplitude $|f_K(\theta)|^2$. It can thus be seen as the scattered current that describes the relative motion of the two particles after they have interacted.

To obtain a formal expression of the cross section, let us go back to its definition (2.5). Following the physical interpretation of the asymptotic behaviour of the scattering state $\Psi_{K\hat{\mathbf{K}}}$, we can assume that in a quantal description of the process, the incoming flux F_i [see (2.3)] will be proportional to the incoming current J_i (2.22)

$$F_i = C\ J_i. \tag{2.24}$$

In the same line of thought, the flux of particle scattered in direction Ω is related to the scattered current J_s (2.23) by

$$F_s = C\ J_s. \tag{2.25}$$

For a single scattering centre, i.e. $N = 1$, the number of particles Δn observed in a given direction per unit time in a detector of section ΔS will thus be

$$\Delta_n = F_s\ \Delta S \tag{2.26}$$

$$= C\ J_s\ R^2\ \Delta\Omega, \tag{2.27}$$

where $\Delta\Omega$ is the solid angle under which the detector is seen from the target when it is placed at a distance R. Following the definition (2.4), the differential cross section for the scattering of a by b hence reads

$$\frac{d\sigma}{d\Omega} = \lim_{\Delta\Omega\to 0} \frac{\Delta n}{F_i\ \Delta\Omega} \tag{2.28}$$

$$= \frac{R^2 J_s}{J_i} \tag{2.29}$$

$$= |f_K(\theta)|^2. \tag{2.30}$$

The cross section is therefore just the square modulus of the scattering amplitude. This means that all the effect of the interaction between a and b—viz. of the potential V—on the scattering process is included in f_K, as expected from the physical interpretation we gave of J_s (2.23). In order to obtain that scattering amplitude, it is necessary to solve the Schrödinger equation (2.19) under the asymptotic condition (2.20). There exist various techniques to do that. In the next section, we will see one that works for central potentials and that is often used for low-energy scattering: the partial-wave expansion.

2.1.4 *Partial-Wave Expansion and Phasesift*

2.1.4.1 Partial-Wave Expansion

When the potential is central, i.e. when it depends only on the distance R between a and b, the Hamiltonian H commutes with the orbital angular momentum operators $\mathscr{L}^2$ and $\mathscr{L}_Z$. The wave function $\Psi_{K\hat{\mathbf{Z}}}$, solution of equation (2.19), which describes the a-b relative motion can then be expanded upon the eigenfunctions of $\mathscr{L}^2$ and $\mathscr{L}_Z$, which are the *spherical harmonics* Y_L^M

$$\Psi_{K\hat{\mathbf{Z}}}(\mathbf{R}) = (2\pi)^{-3/2}\frac{1}{KR}\sum_{L=0}^{\infty} c_L\, u_{KL}(R)\, Y_L^0(\theta), \tag{2.31}$$

taking into account the aforementioned cylindrical symmetry. Including this expansion in equation (2.19), we obtain the following equation for the reduced radial wave functions u_{KL}

$$\left(\frac{\mathrm{d}^2}{\mathrm{d}R^2} - \frac{L(L+1)}{R^2} - \frac{2\mu}{\hbar^2}V(R) + K^2\right) u_{KL}(R) = 0. \tag{2.32}$$

This equation can be solved using numerical techniques. One advantage of this decomposition is to reduce the three-dimensional problem (2.19) to one dimension. Of course, (2.32) will have to be solved for all the values of L. However, as we will see later, especially at low energy E, the sum over L in the expansion (2.31) can be truncated to a limited number of terms.

2.1.4.2 Phaseshift

For now, let us assume that (2.32) can be solved for all the values of L that are needed. Our goal being to calculate the cross section (2.30), we need to evaluate the scattering amplitude f_K, which appears in the asymptotic expression of the stationary scattering states (2.20). To see how to obtain f_K from the solution of equation (2.32), let us study the behaviour of u_{KL} at large R. If we assume the interaction potential V to be short-ranged, viz. $R^2\, V(R) \underset{R\to\infty}{\longrightarrow} 0$, the asymptotic solution of equation (2.32) u_{KL}^{as} is solution of

$$\left(\frac{\mathrm{d}^2}{\mathrm{d}R^2} - \frac{L(L+1)}{R^2} + K^2\right) u_{KL}^{\mathrm{as}}(R) = 0, \tag{2.33}$$

whose solutions are known analytically and exhibit a well-known asymptotic behaviour

$$u_{KL}^{\mathrm{as}}(R) = A\, KR\, j_L(KR) + B\, KR\, n_L(KR) \tag{2.34}$$

$$\underset{R\to\infty}{\longrightarrow} A\, \sin(KR - L\pi/2) + B\, \cos(KR - L\pi/2) \tag{2.35}$$

where j_L and n_L are the spherical Bessel functions of the first and second kinds, respectively. We can re-define the constants A and B as $A = C \cos\delta_L$ and $B = C \sin\delta_L$, respectively. This gives us

$$u^{\text{as}}_{KL}(R) \underset{R\to\infty}{\longrightarrow} C \sin(KR - L\pi/2 + \delta_L), \tag{2.36}$$

where C is just an overall normalisation constant, whereas δ_L, which is called the *phaseshift*, contains all the information about the scattering potential. To better grasp the physical meaning of this phaseshift, let us re-write (2.36) as the sum of an incoming and an outgoing spherical waves

$$u_{KL}(R) \underset{R\to\infty}{\longrightarrow} i\,C\frac{e^{-i\delta_L}}{2}\left[e^{-i(KR-L\pi/2)} - S_L\, e^{i(KR-L\pi/2)}\right] \tag{2.37}$$

where

$$S_L = e^{2i\delta_L} \tag{2.38}$$

is the *scattering matrix*.

Equation (2.37) shows that, when we interpret the asymptotic behaviour of these radial scattering wave functions u_{KL} as the sum of an incoming and an outgoing spherical waves, we observe that the outgoing wave is shifted in phase by $2\delta_L$ from the incoming wave. Intuitively, the former can be seen as describing the particles in the incoming channel, while the latter corresponds to the particles leaving one another after having interacted. This phaseshift therefore contains all the information available to us on the interaction potential V. Accordingly, following what was said earlier, it can be used to compute the elastic-scattering amplitude, and hence the cross section.

To obtain the scattering amplitude from the phaseshifts, we have to compute the coefficients c_L of the partial-wave expansion (2.31). This can be achieved formally by comparing the asymptotic behaviour of that expression to that of the stationary scattering wave function (2.20) taking (2.37) into account. Using the partial-wave expansion of plane waves

$$e^{iKZ} \underset{R\to\infty}{\longrightarrow} \sum_{L=0}^{\infty}(2L+1)i^L P_L(\cos\theta)\frac{i}{2KR}\left[e^{-i(KR-L\pi/2)} - e^{i(KR-L\pi/2)}\right] \tag{2.39}$$

we obtain

$$c_L = \sqrt{4\pi}\sqrt{2L+1}i^L e^{i\delta_L} \tag{2.40}$$

and finally

$$f_K(\theta) = \frac{1}{2iK}\sum_{L=0}^{\infty}(2L+1)(S_L - 1)P_L(\cos\theta). \tag{2.41}$$

Following (2.30), the differential elastic-scattering cross section therefore reads

$$\frac{d\sigma}{d\Omega} = \left|\frac{1}{2K}\sum_{L=0}^{\infty}(2L+1)(e^{2i\delta_L} - 1)P_L(\cos\theta)\right|^2. \tag{2.42}$$

After integration over Ω the total scattering cross section reads

$$\sigma = \frac{4\pi}{K^2}\sum_{L=0}^{\infty}(2L+1)\sin^2\delta_L. \tag{2.43}$$

These expressions show that all the partial waves contribute to the total cross section, however with variable importance depending on the value of the phaseshift δ_L. To understand this point, let us have a look at the effective potential V_L^{eff}, which is the sum of the actual potential V with the centrifugal term $\frac{\hbar^2}{2\mu}\frac{L(L+1)}{R^2}$.

$$V_L^{\text{eff}}(R) = V(R) + \frac{\hbar^2}{2\mu}\frac{L(L+1)}{R^2}. \tag{2.44}$$

This effective potential is displayed in Fig. 2.2 for a typical nucleus-nucleus interaction considering different values of the orbital angular momentum for the projectile-target relative motion L. The purely repulsive centrifugal term combines with the (mostly) attractive potential V to form a *centrifugal barrier* at intermediate distance, where the potential V becomes negligible (here at $R \sim 4$ fm). In a nucleus-nucleus collision as pictured here, this would correspond roughly to the distance at which the surfaces of both nuclei touch one another. As L increases, the centrifugal barrier

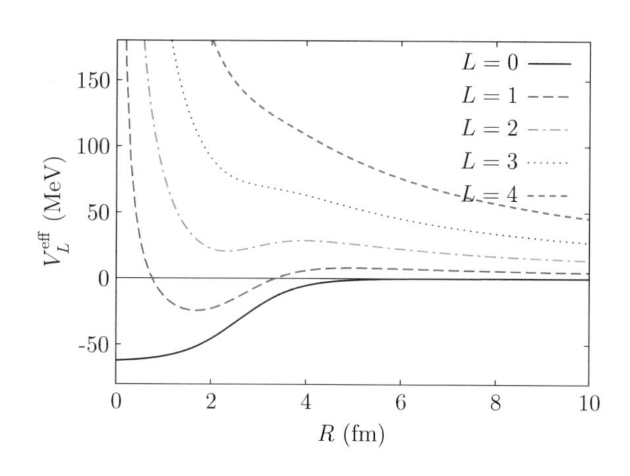

Fig. 2.2 Effective potential V^{eff} (2.44) plotted for $L = 0$ (i.e. V), 1, 2, 3 and 4. We observe that the influence of the interaction potential V on the projectile-target motion decreases as L increases

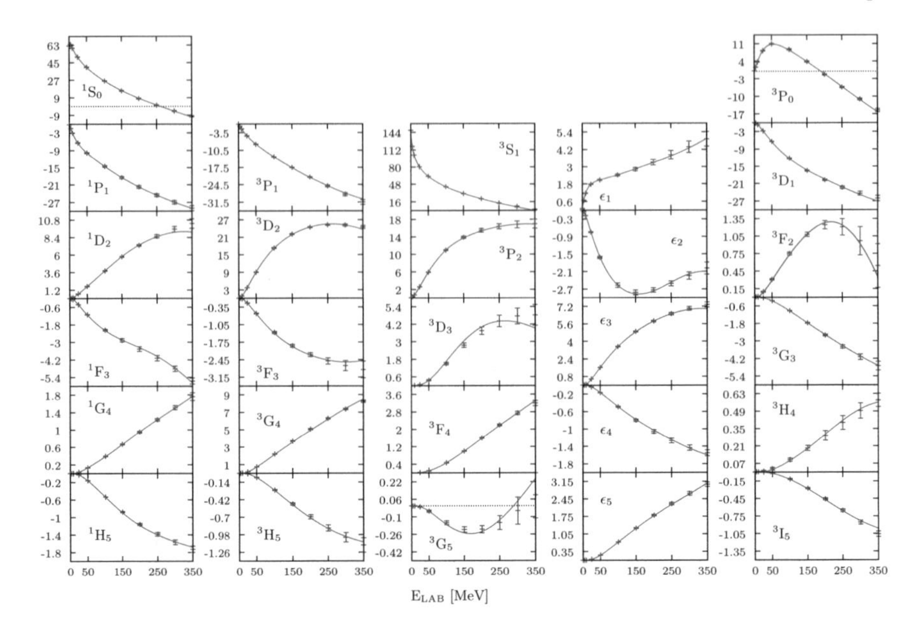

Fig. 2.3 Proton-neutron phase shift in various partial waves. The experimental data (red crosses) are compared with the theoretical prediction of Navarro-Pérez, Amaro and Ruiz-Arriola [5]. Reprinted figure with permission from [5] Copyright (2013) by Elsevier

increases in height and eventually overcomes the pulling effect of the actual interaction. Therefore, above a certain value of L, the centrifugal term will prevent the colliding particles to come close to one another and therefore to interact with each other. In that case, the effect of V on the radial wave function u_{KL} will diminish significantly, the phaseshift δ_L will become very small and hence the contribution of these partial waves to the cross sections (2.42) and (2.43) can be neglected. This is the mechanism that limits the *a priori* infinite sum in the expansion (2.31).

Figure 2.2 also illustrates that the repulsive effect of the centrifugal term will be larger when the a-b relative kinetic energy E is lower. When E gets smaller, the centrifugal barrier gets wider and its top gets higher relative to that incoming energy. The maximum number of partial waves to include in the expansion hence gets smaller at lower energy. Eventually, at very low energy, e.g. for reactions of astrophysical interest or for fissions induced by thermal neutrons in conventional nuclear reactors, only one partial wave will matter, the one for which there is no centrifugal barrier, i.e. $L = 0$.

To illustrate this effect, I reproduce in Fig. 2.3 the proton-neutron phase shift expressed in degrees in different partial waves; this figure is extracted from [5]. Each partial wave is identified by the notation $^{2S+1}L_J$, where $S = 0$ or 1 is the total spin of the two nucleons, L is their relative orbital angular momentum denoted by a letter ($S \Leftrightarrow L = 0$, $P \Leftrightarrow L = 1$, $D \Leftrightarrow L = 2$, $F \Leftrightarrow L = 3$, $G \Leftrightarrow L = 4,\ldots$) and J is the total angular momentum obtained from the coupling of S and L.

The actual value of these phaseshifts are not important to us. What matters in this example is to note that when L increases, the range of variation in the phaseshift decreases, going from 180° in the $^3S_1$ partial wave down to less that 2° in the H and I ones. This illustrates that the influence of the different partial waves generally decreases as L increases and that this is particularly true at low energy. At very low energy, the G, H and I phaseshifts are indeed negligibly small and can therefore be ignored in the computation of the cross sections.

Before concluding on this partial-wave expansion, let us comment on the analysis performed in [5]. As mentioned in Sect. 2.1.1, the study of the collision between two particles can provide information on the interaction between them. This is exactly what Navarro-Pérez, Amaro and Ruiz-Arriola have done here: use the experimental information on the phaseshift in different partial waves in nucleon-nucleon elastic scattering to build a potential that will reliably describe the nucleon-nucleon interaction. All accurate nucleon-nucleon potentials have been constrained in this way, from the phenomenological ones, like Argonne V18 [6] or CD-Bonn [7] to the more recent nucleon-nucleon interactions derived in chiral effect field theory [8].

In conclusion, we should therefore see the partial-wave expansion as a low-energy method. When the incoming energy increases, the number of partial waves to include in the expansion becomes large, and it becomes sensible to consider approximations, like the Born Approximation or the eikonal model of reactions, which have been developed for high-energy reactions (see [4] and Sect. 2.2.2.5).

2.1.4.3 Resonances

In the previous section, we have seen that due to the growing influence of the centrifugal term with L, the contributions of the partial waves tend to decrease with the orbital angular momentum. However, there are cases in which one partial wave can have an unexpected dominant contribution over the other ones within a short energy range. This can happen if the phaseshift in that partial wave δ_L goes quickly from somewhere close to zero to $\pi/2$ and then up to π. In that case, following (2.43), we see that the corresponding contribution will go from a small contribution (when δ_L is small) to its maximum contribution (when $\delta_L = \pi/2$) and then back to something small again (when $\delta_L \sim \pi$). This behaviour, which corresponds to a significant variation of a cross section within a short energy range, is called a *resonance*.

Because they correspond to a well defined orbital angular momentum L and other quantum numbers (total angular momentum J, parity etc.), they can be assimilated to structures in the continuum similar to bound states. In a sense, they can be seen as a structure composed of the two colliding nuclei sticking to, or orbiting around, one another for a while before separating.

To illustrate this point, Fig. 2.4 (left) reproduces the $^3S_1$ phaseshift for the elastic scattering of a proton off ^{13}C as a function of the proton energy E_p. We observe a sharp increase in that phaseshift at an energy of about $E_p = 550$ keV. The presence of that resonant state eases the capture of the proton by ^{13}C to form ^{14}N (a photon γ is then emitted to conserve the total energy and momentum of the system). This is

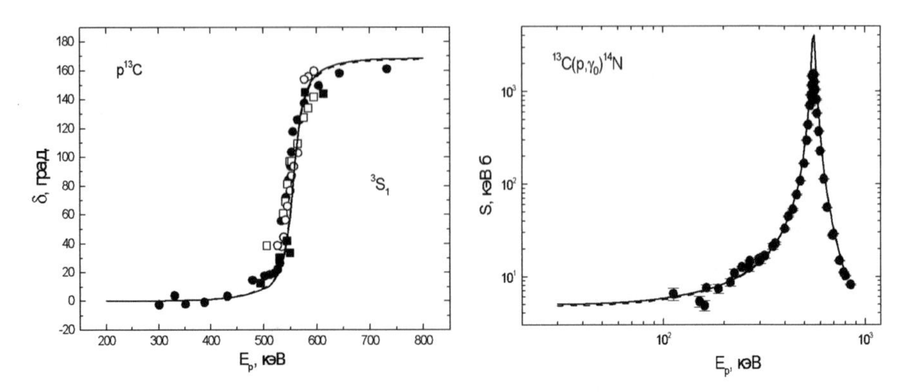

Fig. 2.4 Example of a resonance. Left, the $^3S_1$ phaseshift in the p-^{13}C elastic scattering (in degrees) as a function of the proton energy E_p (in keV). It varies quickly from about 0° to nearly 180° around $E = 550$ keV. Right, this behaviour explains the sudden increase in the cross section for the radiative capture ^{13}C(p,γ)^{14}N [9]. Reprinted figure with permission from [9] Copyright (2012) by Springer Nature

shown in the right panel of Fig. 2.4, where the cross section for that reaction exhibits a sudden and very significant increase at that energy. This reaction p+^{13}C→ ^{14}N+γ, also noted ^{13}C(p,γ)^{14}N in short, is called a *radiative capture*.

2.1.5 Optical Model

So far, we have focussed only on the elastic-scattering reaction, i.e. when the colliding particles come out of the collision in their initial state, merely scattering each other without change neither in their internal structure nor in their internal energy. However, as seen in Sect. 2.1.1, other channels can be open. A cross section can be equally defined for these other channels. For example, a differential cross section for the transfer $a + b \to d + e$ can be defined as

$$\frac{d\sigma}{d\Omega}(a + b \to d + e) = \lim_{R\to\infty} \frac{R^2 J_{d+e}}{J_i}, \tag{2.45}$$

where J_{d+e} corresponds to the probability current for the outgoing channel $d + e$, while J_i is the probability flux for the incoming channel (2.22).

We can also define a cross section for various channels. In particular, the *reaction cross section* (σ_r) corresponds to the sum of the cross sections for all the channels, but the elastic one:

$$\sigma_r = \sum_{\text{channel}\backslash a+b} \sigma(a + b \to \text{channel}). \tag{2.46}$$

The *interaction cross section* (σ_I) corresponds to all channels, but the elastic and inelastic scatterings:

$$\sigma_I = \sum_{\text{channel}\setminus(a+b)\cup(a+b^*)\cup(a^*+b)\cup(a^*+b^*)} \sigma(a + b \rightarrow \text{channel}). \tag{2.47}$$

This cross section corresponds therefore to all the channels in which the projectile or the target—or both—change their internal structure. This includes fusion, transfer, breakup, fragmentation, spallation,...

The existence of other open channels can affect the elastic-scattering process. In particular, since probability flux appears in these non-elastic channels, some probability flux has to be removed from the elastic-scattering one. The *optical model* enables us to account for these other channels phenomenologically.

When the interaction between the two particles a and b is described by a real potential, the Hamiltonian H (2.18) is unitary and hence the total probability flux J is conserved. Since that Hamiltonian accounts only for the elastic channel, that probability flux stays in that very channel, even if it shifts from the incoming plane wave to an outgoing spherical wave during the collision [see (2.20)]. This conservation is expressed as the continuity equation, which, for a stationary state, reads

$$\nabla \mathbf{J}(\mathbf{R}) = 0 \tag{2.48}$$

To simulate absorption of the probability flux towards other channels, it has been suggested to use a complex—or *optical*—potential

$$U_{\text{opt}}(R) = V(R) + i\, W(R). \tag{2.49}$$

Usually, but not always, optical potentials are referred to by the symbol U, instead of V, which is often kept for real potentials or, as in (2.49), for the real part of complex potentials. The imaginary part of these potentials is usually denoted by W.

When the interaction between the colliding particles is complex, the divergence of the probability flux is no longer nil. Accounting for the Schrödinger equation (2.19) written with the optical potential U_{opt} instead of the purely real one used so far, we obtain

$$\begin{aligned}
\nabla \mathbf{J}(\mathbf{R}) &= \nabla \frac{1}{\mu} \Re\{\Psi^*(\mathbf{R})\, \mathbf{P}\, \Psi(\mathbf{R})\} \\
&= -i \frac{\hbar}{2\mu} \nabla[\Psi^*(\mathbf{R})\, \nabla\Psi(\mathbf{R}) - \Psi(\mathbf{R})\, \nabla\Psi^*(\mathbf{R})] \\
&= i \frac{\hbar}{2\mu} [\Psi^*(\mathbf{R})\, \frac{2\mu}{\hbar^2} U_{\text{opt}}(R)\Psi(\mathbf{R}) - \Psi(\mathbf{R})\, \frac{2\mu}{\hbar^2} U^*_{\text{opt}}(R)\Psi^*(\mathbf{R})] \\
&= \frac{2}{\hbar} W(R)\, |\Psi(\mathbf{R})|^2 .
\end{aligned} \tag{2.50}$$

For that divergence to model *absorption* from the elastic channel, it has to be negative. Therefore, the imaginary part of optical potentials has to be negative (or nil): $W(R) \leq 0\ \forall \mathbf{R}$.

To understand how this affects the wave function, let us have a look at the partial-wave expansion explained in the previous section. From (2.32), we can see that using a complex optical potential in the calculation will lead to a complex phaseshift δ_L. It can be shown that since the imaginary part of the potential is negative, the imaginary part of the phaseshift is positive $[\Im(\delta_L) \geq 0]$. The scattering matrix (2.38) then reads

$$S_L = \eta_L\, e^{2i\Re(\delta_L)}, \tag{2.51}$$

where $\eta_L = e^{-2\Im(\delta_L)} \leq 1$. Accordingly, the asymptotic behaviour of the radial wave function u_{KL} (2.37) exhibits an outgoing spherical wave with a reduced amplitude compared to the incoming one

$$u_{KL} \underset{R\to\infty}{\longrightarrow} \propto \left[e^{-i(KR-L\pi/2)} - \eta_L\, e^{2i\Re(\delta_L)}\, e^{i(KR-L\pi/2)}\right]. \tag{2.52}$$

This clearly corresponds to a loss of flux from the entrance channel, simulating the presence of other open channels that can be populated during the collision. The name *optical model* comes from optics, where a complex index of refraction can be used to simulate the absorption of light by the medium.

Within this model, one can compute an *absorption cross section* σ_a that corresponds to all non-elastic channels. It can be computed as the elastic-scattering cross section in equation (2.29) or the transfer cross section in equation (2.45). Alternatively, we can also integrate the loss of flux (2.50) over the whole space:

$$\sigma_a = \frac{-\int \nabla \mathbf{J}\, d\mathbf{R}}{J_i} \tag{2.53}$$

$$= \frac{-\lim_{R\to\infty} \oint \mathbf{J}\cdot\hat{\mathbf{R}}\, R^2 d\Omega}{J_i}$$

$$= \frac{\pi}{K^2}\sum_{L=0}^{\infty}(2L+1)\left(1-\eta_L^2\right). \tag{2.54}$$

By analogy to equation (2.52), let us note that since η_L^2 measures the flux of probability that remains in the elastic channel, $1-\eta_L^2$ corresponds to what has been absorbed from that channel. At the limit of a real potential, $\eta_L = 1\ \forall L$ and the absorption cross section is nil. From what has been introduced at the beginning of this section, we see that σ_a can be compared to the reaction cross section σ_r, which can be measured experimentally.

Note that all the developments performed in this section have been made for a short-range potential, i.e. for which $R^2 V(R) \underset{R\to\infty}{\longrightarrow} 0$. This is not valid for a Coulomb potential, which is non-negligible in most of the nuclear-physics problems, since nuclei are charged. Fortunately, although it leads to more complicated calculations, the Coulomb part of the interaction can be treated exactly and results similar to what has been shown here can be performed seeing the nuclear interaction as a short-ranged perturbation to the Coulomb interaction. I refer the interested readers to more detailed references to study the corresponding developments [2–4].

2.2 Breakup Models

2.2.1 An Introduction to Halo Nuclei

Stable nuclei exhibit a compact structure with a density (for both matter and charge) roughly constant from the centre of the nucleus until its surface. Beyond that point, the density drops quickly with a decay similar throughout the whole nuclear chart. This structure leads to a *liquid-drop* model of the nucleus [10], in which the nucleons are seen tightly packed, interacting mostly to their closest neighbours and forming a nuclear droplet of constant density. Their volume being directly proportional to the number A of nucleons, the radius of these drops hence vary linearly with $A^{1/3}$.

In the mid-80s, when the first Radioactive Ion Beams became available, Tanihata and his collaborators have undertaken to test if this property remains true away from stability. Because radioactive nuclei are, by definition, unstable, it is not possible to measure their radii with usual techniques, like electron scattering. Therefore, to estimate the size of these nuclei, they have measured their interaction cross section σ_I (see Sect. 2.1.5) on various targets at high energy [11, 12]. In a very geometric model, where the projectile P and the target T are seen as hard spheres, they will *interact* in the sense of σ_I when they touch each other. The cross section for this process is thus expressed as

$$\sigma_I(P,T) = \pi[R_I(P) + R_I(T)]^2, \qquad (2.55)$$

where R_I is the *interaction radius*. In a first approximation, this radius can be used to estimate the size of the nucleus.

Figure 2.5 shows the interaction radii obtained by Tanihata et al. for various isotopes of He (full circles), Li (open circles), Be (full squares) and B (open squares) [13]. In addition to the data, the $A^{1/3}$ usual behaviour is plotted as a dotted line. We can see that most of the nuclei studied here exhibit more or less the radius predicted by this empirical law. However, a few of them, like ^{11}Li, ^{11}Be or ^{14}Be stick out of this trend and seem larger than expected. One possibility to explain this unusual feature could be a significant collective deformation. However, an exotic structure is actually the reason for these variations from a well established property of stable nuclei.

Subsequent studies have shown that the valence neutrons play a significant role in the sudden increase of the interaction cross section for these nuclei. For example, it has been observed that the difference between the interaction cross section of one of

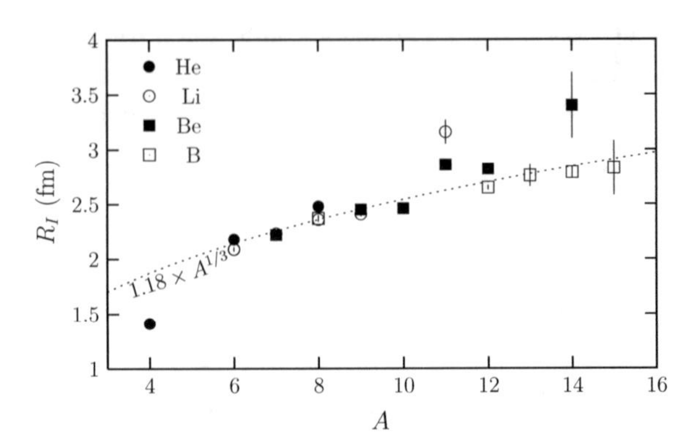

Fig. 2.5 Interaction radius for isotopes of He, Li, Be and B. Data are from [13]

these exotic nuclei and that of the isotope one or two neutron down the isotopic line roughly equals the one- or two-neutron removal cross section for that very nucleus [13]. This suggests that most of the increase in the interaction cross section is due to the additional valence neutrons.

Other groups have then measured one-neutron knockout (KO) on these nuclei. In that reaction, one neutron is removed from the projectile during a high-energy collision on a light target, like C or Be (see, e.g., [14]). In these measurements, only the $A - 1$ nucleus produced by the knockout is measured, not the kicked out neutron. For the nuclei that exhibit an unusually large interaction radius, experimentalists have observed narrow parallel-momentum distributions of the remaining $A - 1$ nucleus. Since the reaction takes place at high energy, one can assume that the measured distribution reveals the momentum distribution the remnant cluster had within the initial nucleus. That distribution being narrow, it means, from the Heisenberg uncertainty principle, that the spatial neutron-core distribution must be extended.

These different results—large interaction cross section, special role of the valence neutrons, narrow momentum distribution of the $A - 1$ nucleus in KO,...—have led to the notion of *halo nuclei* [15]. Halo nuclei are light, neutron-rich nuclei that are unusually large compared to their isobars, viz. their matter radius does not follow the empirical $A^{1/3}$ rule. This large size is qualitatively understood as resulting from their small one- or two-neutron separation energies ($S_{\rm n}$ or $S_{\rm 2n}$). Thanks to this loose binding, the valence nucleons can tunnel far into the classically forbidden region, i.e. far away from the range of the nuclear interaction. They hence form a sort of diffuse *halo* around the core of the nucleus, which exhibits a usual nuclear structure, being tightly bound and compact. Quantum-mechanically, this translates by a long-range tail in the wave function that describes their relative motion to the other nucleons.

Two types of halo nuclei have been observed so far: the one-neutron halo nuclei, and those with two neutrons in their halo. ^{11}Be and ^{15}C exhibit a one neutron halo; they are thus seen as a ^{10}Be, respectively ^{14}C, core to which one neutron is loosely bound. ^{6}He and ^{11}Li are typical examples of two-neutron halo nuclei; they exhibit a clear three-body structure with an α, respectively ^{9}Li, core surrounded by two valence neutrons.

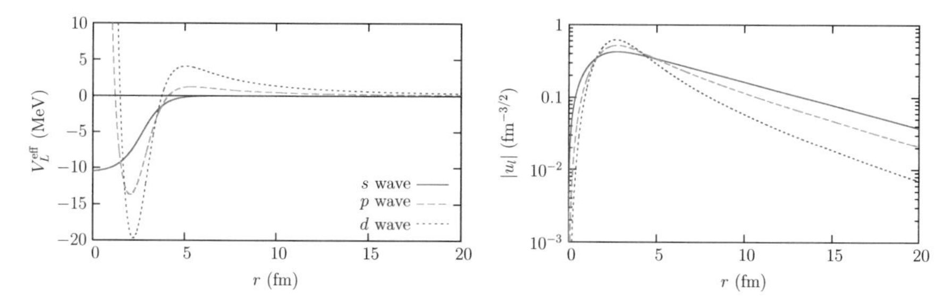

Fig. 2.6 Left: Effective potential $V_L^{\rm eff}$ (2.44) in S, P and D waves fitted to host a state bound by 0.5 MeV. Right: Reduced radial wave functions of the corresponding bound states

In order for a halo to develop in a nucleus, not only does $S_{\rm n}$ or $S_{\rm 2n}$ need to be small, but also nothing may hinders the extension of the wave function to large distances. Therefore halos are observed mostly when the valence neutrons sit in an orbital corresponding to a low orbital angular momentum L, i.e. usually in an S or P wave. This is illustrated in Fig. 2.6, where effective potentials $V_L^{\rm eff}$ (2.44) in the S, P and D waves are plotted as a function of the radial distance R (left). Each of them is fitted to host a state bound by 0.5 MeV, which is the one-neutron separation energy of ^{11}Be, the archetypical one-neutron halo nucleus. The corresponding radial wave functions are depicted in Fig. 2.6 (right). We observe that a higher orbital angular momentum will lead to a higher centrifugal barrier, which will force the wave function inside the nucleus, hence reducing the probability that the valence neutron can be found far from the other nucleons.

For this reason, although not impossible, the development of a halo in proton-rich nuclei is less probable. In that case, in addition to the centrifugal barrier, the valence nucleon always feels a Coulomb barrier, that also reduces the density of probability at large distance. The ground state of ^{8}B and the first excited state of ^{17}F are often seen as candidates for one-proton halo nuclei.

Note that in addition to exhibiting a halo, two-neutron halo nuclei, like ^{6}He or ^{11}Li, are also *Borromean nuclei*. This means that although the three-body system composed of the core plus two neutrons is bound, none of the two-body subsystems is: although ^{6}He exists, neither the dineutron ^{2}n nor ^{5}He exist [16]. This name was coined by Zhukov et al. after the Borromean rings, which are three rings entangled in such a way that when one of the rings is broken, the other two get loose. These rings appear on the coat of arms of the Borromeans, a noble family from Northern Italy, hence their name.

As mentioned above, halo nuclei are usually located close to the neutron dripline, i.e. on the edge of the valley of stability. These nuclei are therefore (very) short-lived and cannot be studied through usual spectroscopic methods, like electron or proton elastic scattering. To probe their internal structure, we must then rely on indirect techniques, like reactions. To infer reliable nuclear-structure information about the projectile from the experimental data, a precise model of the reaction coupled to a realistic description of the projectile must be used. In the next section, I describe different models that exist for *breakup reactions*.

2.2.2 Breakup Reaction

2.2.2.1 Introduction

In breakup reactions the projectile dissociates into its more elementary constituents during its collision with a target (see also Sect. 2.1.1). Such process takes place because the constituents of the projectile interact differently with the target, leading to a tidal force that can be sufficient to break the nucleus apart. This reaction hence reveals the internal structure of the projectile. It is particularly well suited to study the cluster structure of loosely bound nuclei, like halo nuclei. When performed on heavy targets, i.e. when they are dominated by the Coulomb interaction, breakup reactions can also be used to infer reaction rates of astrophysical interest [17, 18].

In what follows, we will focus on what is called the *elastic* or *diffractive* breakup, i.e. the reaction in which all the clusters are measured in coincidence. In that case we talk of *exclusive* measurements, in opposition to the *inclusive* measurements, in which only some of the outgoing particles are detected, like in knockout.

2.2.2.2 Theoretical Framework

Since breakup leads to the dissociation of the projectile into two or more parts, the most basic descriptions of that reaction must include these parts as degrees of freedom. The simple model presented in Sect. 2.1, in which the internal structure of the colliding nuclei is neglected, is therefore not sufficient to describe the dissociation of the projectile. In this introduction to breakup modelling, we will assume the simplest case of a two-cluster projectile: a core c to which a fragment f, e.g. a valence nucleon, is loosely bound. This corresponds to reactions involving one-nucleon halo nuclei, like ^{11}Be or ^{8}B.

Such a projectile is described phenomenologically by the one-body Hamiltonian

$$H_0 = T_r + V_{cf}(\mathbf{r}), \tag{2.56}$$

where $\mathbf{r}$ is the coordinate of the fragment relative to the core and $T_r = -\hbar^2\Delta_r/2\mu_{cf}$ is the operator corresponding to the c-f kinetic energy, with $\mu_{cf} = m_c m_f/m_P$ the c-f reduced mass, where m_c and m_f are the masses of the core and the fragment, respectively, and $m_P = m_c + m_f$ is the mass of the projectile. In (2.56), V_{cf} is a phenomenological potential that simulates the interaction that binds the fragment to the core.

In this simple picture, the c-f relative motion is described by the eigenstates of the Hamiltonian H_0 (2.56). Since the potential V_{cf} is usually chosen as central, the c-f

orbital angular momentum l and its projection m are good quantum numbers (the spin is ignored here for clarity; although a bit tedious, the extension of the developments of this section to the case of particles with spin is not difficult). In partial waves lm these states are thus solution of

$$H_0\, \phi_{lm}(\mathbf{r}) = E\, \phi_{lm}(\mathbf{r}), \tag{2.57}$$

where E is the c-f relative energy. The threshold $E = 0$ corresponds to the fragment separation from the core. Negative energies ($E < 0$) correspond to states in which the fragment is bound to the core, while positive-energy states ($E > 0$) describe the projectile broken up into its core and fragment. The former states are discrete; we add the number of nodes in the radial wave function n to l and m to enumerate them: ϕ_{nlm}. Since the latter states describe the fragment separated from the core, they correspond to a *continuum* of energies, possibly including resonances (see Sect. 2.1.4.3). We add the wave number k ($E = \hbar^2 k^2/2\mu_{cf}$) to l and m to distinguish them and remind that they belong to the continuum: ϕ_{klm}.

The c-f potential V_{cf} usually exhibits a Woods-Saxon form sometimes with a spin-orbit coupling term. Its parameters (mostly its depth) are adjusted to reproduce the know low-energy spectrum of the nucleus, i.e. the binding energy of the fragment to the core, the spin and parity of the ground state and maybe some of the excited states as well, including sometimes resonances.

As an example, let us mention the case of ^{11}Be, which I will use later to illustrate different models of reactions. Being the archetypical one-neutron halo nucleus, ^{11}Be is usually described as a ^{10}Be core to which a neutron is bound by a mere $S_{\rm n}(^{11}\text{Be}) = 501.64 \pm 0.25$ keV [19]. Its ground state has spin and parity $\frac{1}{2}^+$. In addition, it also has an excited $\frac{1}{2}^-$ bound state with $S_{\rm n}(^{11}\text{Be}^*) = 181.60 \pm 0.35$ keV [19, 20]. Above the one-neutron separation threshold, at a c-f energy $E = 1.281$ MeV ± 4 keV [20], ^{11}Be exhibits a $\frac{5}{2}^+$ resonance. Assuming that the ^{10}Be core is in its 0^+ ground state, the $\frac{1}{2}^+$ state of ^{11}Be is described as a neutron bound to the core in the $1s_{1/2}$ orbit, where we have added to the number of radial nodes $n = 1$ and the orbital angular momentum $l = 0$, the total angular momentum j obtained from the coupling of l to the spin $s = \frac{1}{2}$ of the valence neutron, hence $j = \frac{1}{2}$ here. In that model, the $\frac{1}{2}^-$ excited state is seen as a $0p_{1/2}$ neutron bound to ^{10}Be(0^+) and the $\frac{5}{2}^+$ resonance is usually reproduced in the $d_{5/2}$ orbital. In this way the energy relative to the one-neutron separation threshold, the total angular momenta and the parity of these states agree with the known low-energy spectrum of ^{11}Be. Note that this enables us to constrain the potential V_{cf} in partial waves $s_{1/2}$, $p_{1/2}$, and $d_{5/2}$, but not in the $p_{3/2}$ or higher. This has significant implications in the analysis of actual breakup data (see Sect. 2.3).

The third body in the model of the reaction is the target T. It is usually described as a structureless particle, whose interaction with the projectile constituents are simulated by optical potentials U_{cT} and U_{fT} (see Sect. 2.1.5). Once the three-body centre of mass motion has been removed, we are left with the Jacobi set of coordinates illustrated in Fig. 2.7. It is composed of the coordinate of the projectile centre of mass relative to the target $\mathbf{R}$, in addition to the c-f coordinate $\mathbf{r}$ used in (2.56).

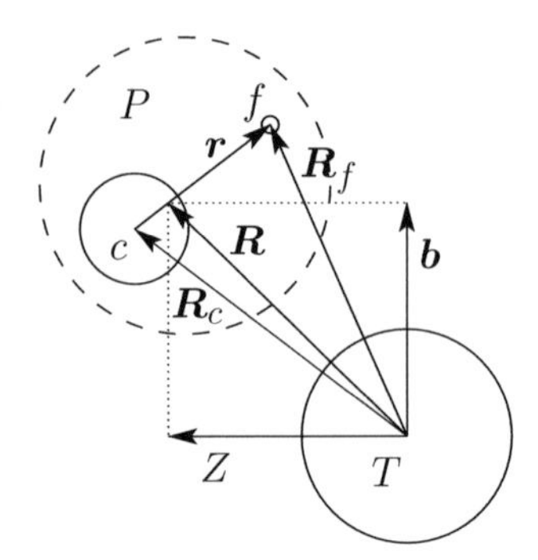

Fig. 2.7 Jacobi set of coordinates ($\mathbf{r}, \mathbf{R}$) used to describe the breakup on a target T of a projectile P composed of two internal clusters, a core c and a fragment f. The beam axis is usually chosen as the Z axis. The transverse coordinate of $\mathbf{R}$ is denoted by $\mathbf{b}$. The c-T ($\mathbf{R}_c$) and f-T ($\mathbf{R}_f$) relative coordinates are shown for completeness

Within this three-body model, describing the P-T collision reduces to solving the following Schrödinger equation

$$\left[T_R + H_0 + U_{cT}(\mathbf{R}_c) + U_{fT}(\mathbf{R}_f)\right] \Psi(\mathbf{r}, \mathbf{R}) = E_T \Psi(\mathbf{r}, \mathbf{R}), \qquad (2.58)$$

where $T_R = -\hbar^2 \Delta_R / 2\mu$ is the P-T kinetic-energy operator, with μ the P-T reduced mass, and $\mathbf{R}_c$ and $\mathbf{R}_f$ are the c-T and f-T relative coordinates, respectively (see Fig. 2.7). In the center-of-mass rest frame, the total energy is related to the initial P-T relative momentum $\hbar K$ and the binding energy $E_{n_0 l_0}$ of the projectile ground state $\phi_{n_0 l_0 m_0}$: $E_T = \hbar^2 K^2 / 2\mu + E_{n_0 l_0}$.

The (2.58) has to be solved with the condition that the projectile, initially in its ground state $\phi_{n_0 l_0 m_0}$, is impinging on the target:

$$\Psi^{(m_0)}(\mathbf{r}, \mathbf{R}) \underset{Z \to -\infty}{\longrightarrow} e^{iKZ} \, \phi_{n_0 l_0 m_0}(\mathbf{r}). \qquad (2.59)$$

It is not possible to solve this three-body problem exactly. Therefore, various methods have been developed over the years to solve it numerically with more or less sophistication and using different levels of approximation. We will see three of them in the next sections: the Continuum Discretised Coupled Channel method (CDCC), the time-dependent approach (TD) and the eikonal approximation. A recent and more complete review of these methods can be found in [21].

2.2.2.3 Continuum Discretised Coupled Channel Model (CDCC)

Since the eigenstates $|\phi_i\rangle$ of the Hamiltonian H_0 (2.56) form a basis in the vector space of the projectile internal coordinate $\mathbf{r}$, the idea of the *Continuum Discretised Coupled Channel* method (CDCC) is to expand the three-body wave function Ψ

upon that basis

$$\Psi(\mathbf{r},\mathbf{R}) = \sum_i \chi_i(\mathbf{R})\langle \mathbf{r}|\phi_i\rangle. \quad (2.60)$$

Introducing this expansion in the Schrödinger equation (2.58) leads to

$$\sum_i T_R\chi_i(\mathbf{R})|\phi_i\rangle + \chi_i(\mathbf{R})H_0|\phi_i\rangle + \left[U_{cT}(\mathbf{R}_c) + U_{fT}(\mathbf{R}_f)\right]\chi_i(\mathbf{R})|\phi_i\rangle$$
$$= \sum_i E_T\chi_i(\mathbf{R})|\phi_i\rangle. \quad (2.61)$$

Accounting for the fact that $H_0|\phi_i\rangle = E_i|\phi_i\rangle$ and projecting each member of the equality on $\langle\phi_j|$ one gets the equations the functions χ_j of $\mathbf{R}$ must satisfy

$$T_R\,\chi_j(\mathbf{R}) + \sum_i \langle\phi_j|U_{cT} + U_{fT}|\phi_i\rangle\,\chi_i(\mathbf{R}) = (E_T - E_j)\,\chi_j(\mathbf{R}). \quad (2.62)$$

This is a set of *coupled equations* in which the coupling terms are the matrix elements of the optical potentials $U_{cT} + U_{fT}$ within the basis of the eigenstates of H_0 $|\phi_i\rangle$. These terms simulate the interaction between the projectile constituents and the target. Without them nothing would happen and the projectile would pass by the target unscathed, and the P-T system would stay in the initial channel $e^{iKZ}\phi_{n_0l_0m_0}$. As mentioned above, because of these interactions, the system can shift from this initial (elastic) channel to other channels, where the projectile is in other eigenstates of H_0. A stated in Sect. 2.2.2.2, these states can correspond to excited bound states of the projectiles or, in the case of breakup, to the c-f continuum, i.e. one of the states ϕ_{klm}. Because they describe a continuum part of the spectrum, these states are not numerically tractable. The symbolic expansion (2.60) should actually include an integral over the value of the c-f relative momentum $\hbar k$. To circumvent this issue, Rawitscher has suggested to *discretise* this continuum [22], viz. to approximate the continuum functions ϕ_{klm} by a set of discrete ϕ_{ilm}. This hence leads to the Continuum Discretised Coupled Channel method or CDCC [23, 24], see [25] for a recent review.

Various methods have been suggested to discretise the continuum. The simplest idea is to divide the continuum into small energy intervals, also known as *bins*: $[E_i - \Delta E_i/2, E_i + \Delta E_i/2]$ and choose to describe each interval by the continuum wave function at the midpoint:

$$\phi_{ilm}(\mathbf{r}) = \phi_{k_ilm}(\mathbf{r}), \quad (2.63)$$

with $E_i = \hbar^2k_i^2/2\mu_{cf}$.

Unfortunately, these mid-point functions are not square integrable (as seen in Sect. 2.1.4.2 they oscillate indefinitely when $r \to \infty$). It has then been suggested to choose as bin functions the average

$$\phi_{ilm}(\mathbf{r}) = \frac{1}{W_{il}} \int_{E_i-\frac{\Delta E_i}{2}}^{E_i+\frac{\Delta E_i}{2}} f_l(E)\, \phi_{klm}(\mathbf{r})\, \mathrm{d}E, \tag{2.64}$$

where f_l is a weight function that can differ from one partial wave to another and the normalisation factor $W_{il} = \int_{E_i-\frac{\Delta E_i}{2}}^{E_i+\frac{\Delta E_i}{2}} f_l(E) dE$. This method produces square-integrable wave functions, which ease the calculation of the coupling terms in (2.62) [22–24]. However, these bin wave functions may extend over a long distance before becoming negligible, especially if the bins are narrow.

A third way to obtain a discretised continuum is to use *pseudo-states*. These states can be generated in different ways. A first one is by diagonalising the projectile Hamiltonian H_0 within a finite set of square-integrable basis states, e.g. using an R-matrix formalism [26]. The resulting eigenstates of H_0 are thus also discrete and square integrable, even for positive energies E. Another way to obtain pseudo-states is to use a Transformed Harmonic Oscillator (THO) basis [27]. The idea of this method is to build a mathematical transformation between the harmonic-oscillator states and the eigenstates of H_0 (2.56). Since the eigenstates of the harmonic oscillator are naturally discrete and square integrable, the transformation produces the desired discretised continuum.

The pseudo-states produced by these different methods usually vanish faster at large r than the radial wave function obtained by averaging (2.64), which is useful in the calculation of the coupling matrix elements of equation (2.62). However, unlike in the binning technique, the eigenenergies E_i cannot be chosen at will because they are the direct outcome of the diagonalisation of the Hamiltonian H_0, and hence depend on the choice of the basis states considered, or the way the THO is built.

Except for this discretisation of the continuum, the CDCC method does not make any other approximation and solves the three-body Schrödinger equation (2.58) "exactly", viz. numerically. It is therefore fully quantal and makes no approximation on the P-T relative motion. It can thus be used at all energies. However, it may be quite computationally expensive, especially at high energies. This is the reason why other models based on different approximations of the P-T relative motion have been suggested (see Sects. 2.2.2.4 and 2.2.2.5).

Various codes have been written to solve the CDCC coupled equations. The code FRESCO written by Thomson is the best known [28]; it can be freely downloaded from www.fresco.org.uk.

To illustrate the CDCC method, I display in Fig. 2.8 the elastic-scattering cross section of ^{11}Be on Zn at 24.5 MeV [29, 30]. This measurement is part of a larger effort to measure the elastic scattering of beryllium isotopes ^{9,10,11}Be on Zn around the P-T Coulomb barrier to see if the presence of the one-neutron halo in ^{11}Be has any effect on the elastic process. The stable ^{9}Be and near-to-stability ^{10}Be exhibit usual elastic-scattering cross sections, which can be easily described with usual optical potentials (see Sect. 2.1.5). The halo nucleus ^{11}Be, however, exhibits a significant

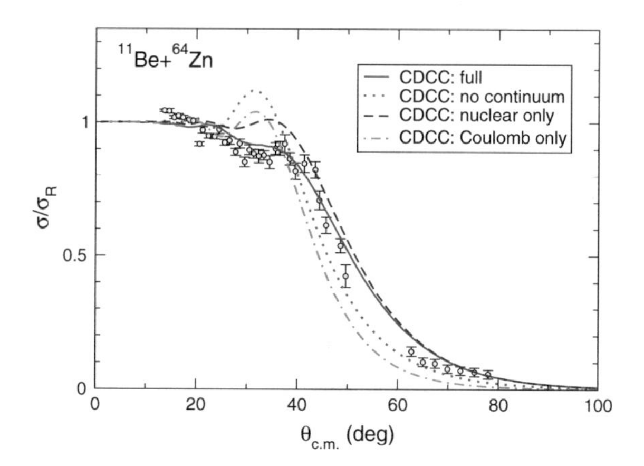

Fig. 2.8 Elastic scattering of ^{11}Be on Zn at 24.5 MeV measured at ISOLDE (CERN) [29, 30]. The coupling to the ^{10}Be-n breakup channel included in the "full" CDCC calculation must be accounted for in order to properly reproduce the measured cross section. Reprinted figure with permission from [30] Copyright (2012) by the American Physical Society

suppression of its elastic-scattering cross section around $\theta \sim 30°$–$40°$ (see Fig. 2.8), which can only be reproduced if a long-range term is added to the imaginary part of the optical potential [29]. This suggests that the halo indeed influences the elastic scattering. To investigate in more details this effect, precise CDCC calculations of this collision have been performed [30].

To properly reproduce the data, a full calculation, i.e., including the coupling to and within the ^{10}Be-n continuum needs to be included (solid blue line in Fig. 2.8). Solely accounting for the extended wave function of the projectile without the couplings to the continuum (red dotted line in Fig. 2.8) does not exhibit the strong suppression of the elastic-scattering cross section at 30°. This shows the importance to account for the breakup channel in collision involving loosely bound systems, like halo nuclei, even if the breakup process is not the primarily study of the experiment.

2.2.2.4 Time-Dependent Approach

To reduce the computational complexity of the CDCC framework, various approximations have been suggested. The first one I present in this short review is the *Time-Dependent* approach (TD). The main idea is to approximate the P-T relative motion by a classical trajectory $\mathbf{R}(t)$, while maintaining a quantum description of the projectile. This approach was first developed to describe Coulomb excitation in heavy-ion collisions, in which a nucleus is excited during its collision with a heavy target [31]. The high Z of the target ensures that the process is dominated by the Coulomb interaction. The reaction can thus be described as resulting from the exchange of virtual photons between the colliding nuclei. It has naturally be extended to describe the Coulomb breakup of halo nuclei, where the excitation takes place between the initial ground state of the projectile (2.59) and its continuum [32–36].

In this approximation, the projectile follows a classical trajectory $\mathbf{R}(t)$ along which it feels a time-dependent potential simulating its interaction with the target. This leads to the resolution of the time-dependent Schrödinger equation

$$i\hbar\frac{\partial}{\partial t}\Psi(\mathbf{r},b,t)=\left\{H_0+U_{cT}\left[\mathbf{R}_c(t)\right]+U_{fT}\left[\mathbf{R}_f(t)\right]-V_{\text{traj}}\left[\mathbf{R}(t)\right]\right\}\Psi(\mathbf{r},b,t), \tag{2.65}$$

where V_{traj} is the potential used to generate the trajectories and b is the impact parameter, which characterises each trajectory. The time dependence of the optical potentials U_{cT} and U_{fT} arises from the time dependence of $\mathbf{R}$, upon which $\mathbf{R}_c$ and $\mathbf{R}_f$ depend (see Fig. 2.7).

Equation (2.65) has to be solved for each value of the impact parameter b with the condition that the projectile is initially in its ground state:

$$\Psi^{(m_0)}(\mathbf{r},b,t\to-\infty)=\phi_{n_0l_0m_0}(\mathbf{r}) \qquad \forall b\in\mathbb{R}^+. \tag{2.66}$$

The usual way to solve the time-dependent equation (2.65) is to numerically compute the wave function Ψ by small time steps Δt using an approximation of the time-evolution operator U:

$$\Psi^{(m_0)}(\mathbf{r},b,t+\Delta t)=U(t+\Delta t,t)\,\Psi^{(m_0)}(\mathbf{r},b,t) \tag{2.67}$$

with

$$U(t',t)=\exp\left[\frac{-i}{\hbar}\int_t^{t'}\left[H_0+U_{cT}(\tau)+U_{fT}(\tau)-V_{\text{traj}}(\tau)d\tau\right]\right], \tag{2.68}$$

starting from $\phi_{n_0l_0m_0}$ at a large negative time $t\to-\infty$.

Since each trajectory is treated separately, the computational cost of the TD approach is strongly reduced compared to CDCC's. However, as we will see in Sect. 2.2.3, because of that some quantal effects are lost, in particular the model lacks the interferences between neighbouring trajectories.

Various codes—mostly unpublished—have been written to solve (2.65). In [32–34], the wave function Ψ is expanded in partial waves and the P-T optical potentials are decomposed into multipoles. This technique has the advantage to lead to a diagonal representation of H_0, which is simpler to use in the expression of the time-evolution operator U. Reference [35] makes use of a three-dimensional cubic mesh in $\mathbf{r}$ upon which Ψ is expanded. This enables to treat the kinetic-energy term of H_0 using a fast Fourier transform. Finally, in [36], the wave function is expanded on a three-dimensional spherical mesh, which leads to a diagonal representation of the P-T optical potentials.

Figure 2.9 illustrates this method; it displays the cross section obtained by Esbensen within his TD code for the breakup of ^{15}C into ^{14}C and a neutron while impinging on Pb at 68AMeV [38, 39], which correspond to the conditions of the RIKEN experiment performed by Nakamura et al. [37]. Once folded with the experimental resolution (solid red lines in Fig. 2.9), the TD calculations are in excellent

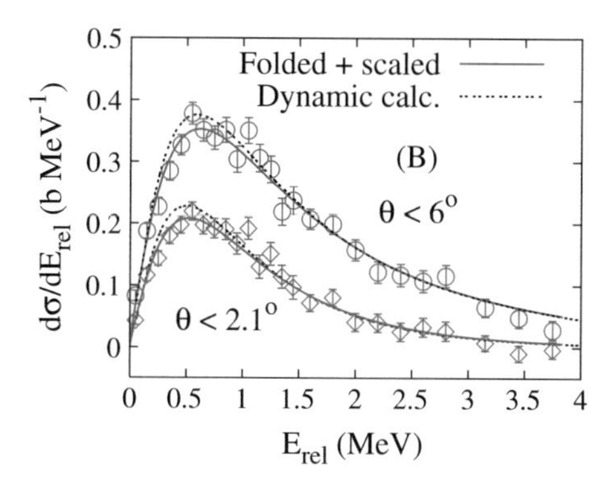

Fig. 2.9 Coulomb breakup cross section for ^{15}C impinging on Pb at 68A MeV measured at RIKEN [37] with two angular cutoffs. The TD calculation of Esbensen is shown with the dotted lines [38, 39], the solid lines correspond to these calculations folded with the experimental energy resolution. Reprinted figure with permission from [38] Copyright (2009) by the American Physical Society

agreement with the data, which confirms, besides the validity of the TD approach for this kind of observable, the clear two-cluster structure of ^{15}C.

2.2.2.5 Eikonal Approximation

The second approximation, which is often used to model reactions involving halo nuclei, is the *eikonal* approximation. This approximation was first derived by Glauber [40] and is suited to describe high-energy reactions. The cornerstone of this approximation is to realise that at sufficiently high energy, the P-T relative motion is not very different from the incoming plane wave e^{iKZ} [see (2.59)]. At high energy, the fragments of the projectile are emitted at a velocity close to the projectile one and are scattered at very forward angles. The idea of the approximation is thus to factorise the plane wave out of the three-body wave function by posing

$$\boldsymbol{\Psi} = e^{iKZ}\widehat{\boldsymbol{\Psi}}. \tag{2.69}$$

Inserting this in the Schrödinger equation (2.58), the only term that requires some attention is the kinetic operator T_R

$$T_R\boldsymbol{\Psi} = e^{iKZ}\left(T_R + vP_Z + \frac{\mu_{PT}}{2}v^2\right)\widehat{\boldsymbol{\Psi}}, \tag{2.70}$$

where $v = \hbar K/\mu_{PT}$ is the asymptotic P-T relative velocity. Since the major dependence on $\mathbf{R}$ has been removed from the wave function through the factorisation (2.69), $\widehat{\Psi}$ is smoothly varying with $\mathbf{R}$ and hence its second-order derivatives $T_R\widehat{\Psi}$ can be neglected compared to its first-order derivative $vP_Z\widehat{\Psi}$. Accounting for the energy conservation $E_T = \mu_{PT}v^2/2 + E_{n_0l_0}$, (2.58) then reads

$$i\hbar v\frac{\partial}{\partial Z}\widehat{\Psi}(\mathbf{r},\mathbf{b},Z)=\left[H_0-E_{n_0l_0}+U_{cT}(\mathbf{R}_c)+U_{fT}(\mathbf{R}_f)\right]\widehat{\Psi}(\mathbf{r},\mathbf{b},Z),\quad(2.71)$$

where $\mathbf{b}$ is now the transverse component of $\mathbf{R}$ (see Fig. 2.7), not to be mixed up with the classical impact parameter of the previous section. Note that $\mathbf{b}$ is still a quantal variable here; unlike in the TD approach, no semiclassical hypothesis has been made in the derivation of (2.71).

Following the initial condition (2.59), (2.71) has to be solved knowing that

$$\widehat{\Psi}^{(m_0)}(\mathbf{r},\mathbf{b},Z)\underset{Z\to-\infty}{\longrightarrow}\phi_{n_0l_0m_0}(\mathbf{r})\qquad\forall\mathbf{b}\in\mathbb{R}^2.\quad(2.72)$$

At this level of approximation, we obtain the *Dynamical Eikonal Approximation* (DEA) [41, 42]. For completeness, note that a similar approximation called the *Eikonal-CDCC* (E-CDCC) has been derived by the Kyushu group [43]. It is based on the CDCC expansion but exploits the eikonal approximation to simplify the coupled equations (2.62).

Note that (2.71) and its initial condition (2.72) are mathematically equivalent to equations (2.65) and (2.66), respectively, which means that it can be solved using the same codes as that time-dependent Schrödinger equation. The DEA has thus a computational cost similar to TD calculations, hence significantly lower than CDCC. The major difference with the TD approach is that $\mathbf{b}$ being a quantal variable, interference effects between neighbouring trajectories can be taken into account. The DEA hence extends the TD technique by including part of the quantal interferences that are neglected within the semiclassical approximation. We will come back to that in the next section.

What people usually call the eikonal approximation performs a subsequent *adiabatic*—or *sudden*—approximation to simplify (2.71). Since the factorisation (2.69) is applied at high energy, the idea of this additional approximation is to say that the collision time will be short and that during this time, the projectile structure will not evolve much. One can thus see the internal coordinates of the projectile, i.e. $\mathbf{r}$, as being frozen during the collision. This amounts to neglect the influence of the projectile Hamiltonian and set

$$H_0\sim E_{n_0l_0}.\quad(2.73)$$

With that adiabatic approximation, the solutions of equation (2.71) that corresponds to the incoming condition (2.72) can be easily computed and read

$$\widehat{\Psi}^{(m_0)}_{\rm eik}(\mathbf{r},\mathbf{b},Z)=\exp\left\{-\frac{i}{\hbar v}\int_{-\infty}^{Z}\left[U_{cT}(\mathbf{r},\mathbf{b},Z')+U_{fT}(\mathbf{r},\mathbf{b},Z')\right]dZ'\right\}\phi_{n_0l_0m_0}(\mathbf{r}).\quad(2.74)$$

This expression can be easily interpreted: the projectile is seen as following a straight-line trajectory ($\mathbf{R}=\mathbf{b}+Z\widehat{\mathbf{Z}}$) along which it accumulates a phase through its interaction with the projectile. As can be seen from expression (2.74), this model of the reaction is quite simple to implement and has a significantly shorter computational cost compared to CDCC, and even to TD methods. The fact that the eikonal wave function (2.74) includes the initial ground state of the projectile $\phi_{n_0 l_0 m_0}$ translates the fact that the internal coordinate $\mathbf{r}$ of the projectile is seen as frozen during the collision.

This approximation will provide accurate results only if the adiabatic approximation is well justified. If the dynamics of the projectile plays a significant role, then the DEA or the E-CDCC have to be considered. This is the case for Coulomb-dominated reactions. The long—actually infinite—range of the Coulomb interaction makes the sudden approximation invalid. In that case, the collision time cannot be considered as brief, since the projectile will never end interacting with the Coulomb field of the target and hence couplings within the continuum, like post-acceleration of the core, must be accounted for in the description of the reaction. This is illustrated in Fig. 2.10, where the breakup cross section for ^{11}Be impinging on Pb at 69A MeV is plotted as a function of the scattering angle θ of the centre of mass of the ^{10}Be and n constituents after dissociation with a relative energy $0 \leq E \leq 1$ MeV [42]. The DEA calculation (solid line) is in excellent agreement with the data of RIKEN [44]. The contributions of the main partial waves to the cross section (s wave with short-dashed line, p wave with dash-dotted line and d waves with dotted line) show that at forward angle the process is dominated by a transition towards the ^{10}Be-n continuum in the p wave, i.e. a direct E1 transition from the s ground state of ^{11}Be. The usual eikonal approximation (long-dashed line), however, diverges at forward angles, confirming that the adiabatic approximation does not hold in this Coulomb case. This result emphasises the importance to know the range of validity of the reaction model one uses when analysing experimental data.

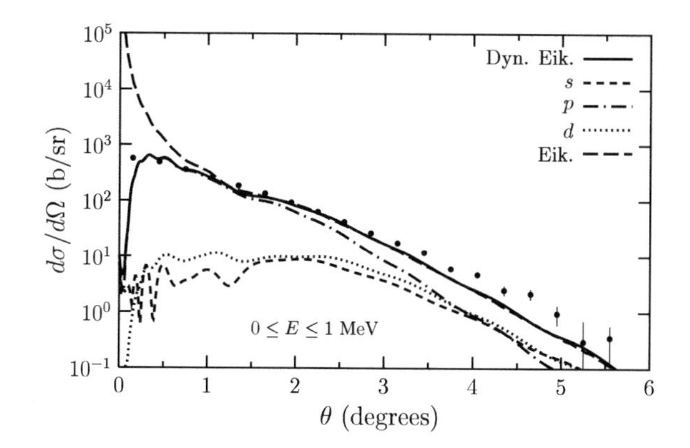

Fig. 2.10 Angular distribution for the Coulomb breakup of ^{11}Be on Pb at 69A MeV measured at RIKEN [44]. The DEA calculation is in excellent agreement with the data. The usual eikonal approximation diverges at forward angle due to its additional adiabatic approximation, which is invalid for the infinitely ranged Coulomb force [42]. Reprinted figure with permission from [42] Copyright (2006) by the American Physical Society

2.2.3 Benchmark of Breakup Models

Various comparisons between reaction models have been performed recently [45, 46]. In this section, I present the results of [45], where the three models presented in the previous section, CDCC, TD and DEA, have been compared to one another on the test case of the breakup of ^{15}C on Pb at 68A MeV, which has been measured at RIKEN [37].

Using the same ^{14}C-n potential V_{cf} and the same optical potentials U_{cT} and U_{fT}, the breakup cross section has been computed within each of the reaction models. The results are summarised in Fig. 2.11. The energy distribution displayed in the left panel shows that all three models provide nearly identical cross sections when the same two-body potentials are considered in input. This shows that, at this beam energy and for this observable, all three models are equivalent. Incidentally, their predictions are in excellent agreement with the RIKEN data [37], which confirms the two-body structure of ^{15}C.

The differences appear in the right panel of Fig. 2.11, which presents the breakup cross section as a function of the scattering angle θ of the ^{14}C-n centre of mass after dissociation. For that observable, we observe oscillations in both the CDCC (solid line) and DEA (dashed line) cross sections, whereas the TD calculation (dotted line) provides a smooth observable. This is the sign of the missing quantal effects within the semi-classical approximation mentioned in Sect. 2.2.2.4. Both CDCC and the DEA includes these interferences. They also show excellent agreement with one another at this energy, although the DEA is much less time consuming than CDCC. Note however, that the TD model provides a cross section that follows the general trend of the quantal results. This explains that the agreement of the TD energy distribution with the other two observed in Fig. 2.11 (left) is not accidental. Since that observable is obtained after integration over the scattering angle θ, the oscillations around the general trend cancel out, which leads to identical energy distributions.

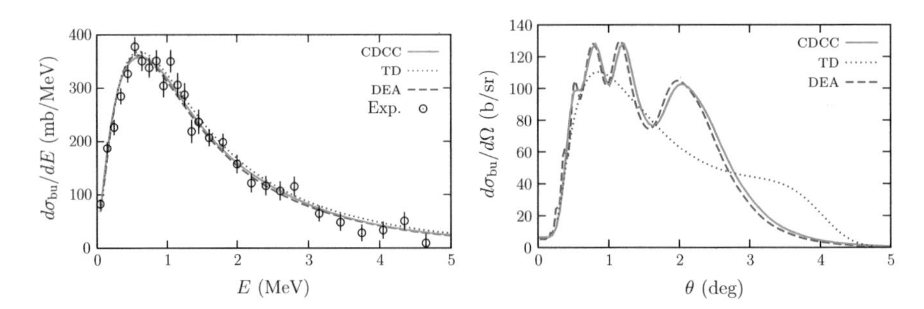

Fig. 2.11 Comparison of breakup models (CDCC, TD and DEA) on the Coulomb breakup of ^{15}C on Pb at 68A MeV measured at RIKEN [37]. Left: energy distribution. Right: angular distribution [45]. Reprinted figures with permission from [45] Copyright (2012) by the American Physical Society

In [45], the comparison has also been extended to lower beam energy, viz. 20*A* MeV, where the eikonal approximation is no longer supposed to provide a reliable description of the reaction. As expected, it is observed that the DEA is no longer in agreement with CDCC: the DEA breakup cross section is too large and focused at too forward a scattering angle compared to CDCC's. This comes from the hypothesis that the P-T relative motion does not differ much from the incoming plane wave (see Sect. 2.2.2.5). The semi-classical interpretation of this eikonal approximation, in which the projectile follows a straight-line trajectory, suggests that the projectile is then forced to pass through the high-field zone of the target, in which it would otherwise not enter at low beam energy due to the significant Coulomb repulsion by the target. The TD approximation, which naturally includes the P-T repulsion through the semi-classical (non-linear) trajectory, does not exhibit such a flaw and hence provides an agreement with CDCC similar to what is seen in Fig. 2.11: a nearly identical energy distribution and an angular distribution which lacks the quantal oscillatory pattern but otherwise follows the trend of the CDCC prediction [45].

To account for the P-T Coulomb repulsion within the DEA (or E-CDCC), it has been suggested to use a semi-classical correction, in which the norm of the transverse part $\mathbf{b}$ of the P-T relative coordinate $\mathbf{R}$ is replaced by the distance of closest approach in the corresponding Coulomb trajectory [4]. With that correction, the DEA and E-CDCC calculations fall in perfect agreement with the CDCC prediction, including the effect of the quantal interferences [47]. In this way, at least for Coulomb-dominated reactions, the eikonal approximation in its most general expression, i.e. without the adiabatic approximation, can be safely extended to low energies. Since this model of reactions is significantly less time consuming than CDCC, this is not a vain gain.

2.3 Application of Breakup Reactions to Nuclear-Structure Study

2.3.1 Binding Energy

The first observable to which breakup reactions are sensitive is the binding energy of the projectile. This is illustrated in Fig. 2.12, where the Coulomb-breakup cross section of ^{19}C—another candidate one-neutron halo nucleus—is displayed as a function of the ^{18}C-n energy E after dissociation. The beam energy is 67*A* MeV and the target is Pb. The experiment was performed at RIKEN [48] and the theoretical analysis illustrated in Fig. 2.12 is due to Typel and Shyam using a TD approach (see Sect. 2.2.2.4) [49]. In this analysis, ^{19}C is described as an inert ^{18}C core to which a $1s_{1/2}$ neutron is loosely bound. At the time, the binding energy of ^{19}C was poorly known, which is why two series of calculations were made, corresponding to two realistic choices of the one-neutron separation energy for ^{19}C: $S_n(^{19}\text{C}) = 530$ keV (top panel) and 650 keV (bottom panel).

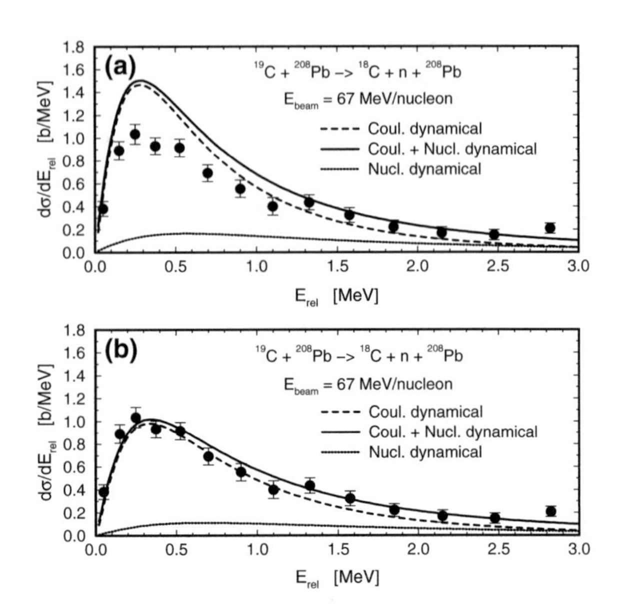

Fig. 2.12 Coulomb breakup of ^{19}C on Pb at 67A MeV measured at RIKEN [48]. The theoretical analysis of Typel and Shyam illustrates the influence of the binding energy of the system on the breakup energy distribution [49]. $S_n(^{19}\text{C}) = 530$ keV (top) and 650 keV (bottom). Reprinted figures with permission from [49] Copyright (2001) by the American Physical Society

Typel and Shyam's analysis clearly shows that the breakup cross section is significantly sensitive to the binding energy of the system. This can be easily understood at a qualitative level: the less a system is bound the easier it will be to break. We indeed see that calculations for a ^{19}C bound by 530 keV (top panel of Fig. 2.12), the breakup cross section is about 50% higher than when it is bound by 650 keV. In addition to the change in magnitude, we also witness a variation in the dependence of the cross section on the continuum energy E. The RIKEN data are better reproduced in both magnitude and shape when $S_n(^{19}\text{C})$ is set to 650 keV. Since this analysis, the mass of ^{19}C has been accurately measured and its binding energy has been precisely determined to be 580 ± 9 keV [19]. This value is clearly higher than 530 keV, but it is lower than the suggested value of Typel and Shyam of 650 keV.

In addition to its sensitivity to the binding energy of ^{19}C, the analysis of Typel and Shyam also shows how the breakup cross section is affected by the Coulomb and nuclear parts of the interaction between the projectile and the target. The thick solid (higher) lines in Fig. 2.12 correspond to the full calculation, i.e. containing both the Coulomb and nuclear parts of the P-T interaction. Purely Coulomb—yet fully dynamical—calculations are shown by the dashed lines. The cross sections obtained with the sole nuclear interaction are displayed by the thin solid (lower) lines. As expected, this reaction is strongly dominated by the Coulomb interaction. Yet, the nuclear part of the interaction has non-negligible effects, especially at high ^{18}C-n energy.

A systematic analysis of the sensitivity of the Coulomb-breakup cross section to the projectile structure (binding energy, orbital angular momentum in the ground state $l_0,\ldots$) has been made by Typel and Baur within a perturbative solution of the TD approach assuming a purely Coulomb P-T interaction [50, 51]. This seminal work provides an analytical expression for the cross section. Although the influence of the nuclear interaction is entirely neglected, and the effects of orders beyond the first one, i.e. beyond a mere one-step transition between the ground state and the continuum, this enables them to explain in simple terms how the shape of the energy distribution depends on the structure of the projectile. This helps identifying the key degrees of freedom to properly describe Coulomb breakup.

2.3.2 Spectroscopy

All the models of breakup presented in Sect. 2.2 are based on a single-particle description of the projectile [see (2.56)], where the internal structure of the core is neglected and the bound states of the projectile are described within a single configuration with a wave function of norm 1. In reality the structure of any nucleus ${}^{A}P$ in its state of spin and parity J^{π} is an admixture of various configurations in which the core ${}^{A-1}c$ can be in different states ci of spin and parity $J_{ci}^{\pi_{ci}}$ (see A. Brown lectures to this summer school)

$$ {}^{A}P(J^{\pi}) = \sum_{ci} \left[{}^{A-1}c(J_{ci}^{\pi_{ci}}) \otimes \psi_{l}^{ci}\right]^{JM} \tag{2.75} $$

For each configuration, the *overlap wave function* ψ_{lm}^{ci} describes the relative motion of the halo neutron to the core in a given state ci. The square of the norm of this overlap wave function measures the probability to find the system in the configuration ci. It is called the *spectroscopic factor* (SF)

$$ \mathscr{S}_{l}^{ci} = \|\psi_{lm}^{ci}\|^{2}. \tag{2.76} $$

The structure of halo nuclei is usually dominated by one configuration $c0$, which provides the largest contribution to the breakup cross section. It has therefore been suggested to extract spectroscopic factors from the comparison of reaction calculations to actual breakup data. This idea is based on the approximation that the overlap wave function $\psi_{l_0 m_0}^{c0}$ can be well approximated by the single-particle wave function $\phi_{n_0 l_0 m_0}$ (2.57)

$$ \psi_{l_0 m_0}^{c0}(\mathbf{r}) = \sqrt{\mathscr{S}_{l_0}^{c0}}\, \phi_{n_0 l_0 m_0}(\mathbf{r}). \tag{2.77} $$

This single-particle approximation leads to the idea that the spectroscopic factor can be obtained through the simple ratio

$$ \mathscr{S}_{l_0}^{c0} = \frac{\sigma_{\mathrm{bu}}^{\mathrm{exp}}}{\sigma_{\mathrm{bu}}^{\mathrm{th}}}, \tag{2.78} $$

where the experimental cross section $\sigma_{\text{bu}}^{\text{exp}}$ is compared to the theoretical prediction $\sigma_{\text{bu}}^{\text{th}}$ starting from the initial single-particle ground-state wave function $\phi_{n_0 l_0 m_0}$.

The question I would like to raise here is whether this approach is valid for breakup. Because of the large extension of the halo, this reaction is expected to be rather *peripheral*, in the sense that it should probe mostly the halo structure, i.e. the tail of the projectile wave function. If this is the case, we would expect the breakup cross section to scale with the normalisation of the tail of the c-f wave function [see Fig. 2.6 (right)] rather than with the norm of the whole overlap wave function, i.e. the spectroscopic factor $\mathscr{S}_{l_0}^{c0}$. At large distance, i.e. beyond the centrifugal barrier, the effect of the short-range nuclear c-f potential is negligible and the behaviour of the radial wave function is exactly known, but for its normalisation. For the effective single-particle wave function $\phi_{n_0 l_0 m_0}$ it reads (the radial part of the actual overlap wave function $\psi_{l_0 m_0}^{c0}$ has an identical behaviour)

$$u_{\kappa_{n_0 l_0} l_0}(r) \underset{R\to\infty}{\longrightarrow} \mathscr{C}_{n_0 l_0}\, e^{-\kappa_{n_0 l_0} r}, \tag{2.79}$$

where the wave number $\kappa_{n_0 l_0}$ is related to the projectile binding energy $E_{n_0 l_0} = -\hbar^2 \kappa_{n_0 l_0}^2 / 2\mu_{cf}$, and $\mathscr{C}_{n_0 l_0}$ is the Asymptotic Normalisation Coefficient (ANC), which measures the probability strength in the halo. The value of this ANC depends on the short-range physics, i.e. the particulars of the effective V_{cf} potential in the single-particle viewpoint.

To test which part of the wave function is actually probed, breakup calculations have been performed using two choices for V_{cf} that lead to radial single-particle wave functions with identical asymptotics, i.e. identical ANCs, but with significant different interiors [52]. This was achieved using supersymmetric transformations of deep V_{cf} potentials [53, 54]. The deep potentials are chosen to host, in addition to the physical loosely bound state, a deeply spurious bound state in the ground-state partial wave of orbital angular momentum l_0. The presence of that state adds a node in the radial wave function of the physical loosely bound state. Through supersymmetric transformations the spurious deeply bound state can be removed without affecting the long-range physics of the projectile Hamiltonian H_0, viz. the ANC of the other bound states and the phaseshift in the continuum.

This provides us with two descriptions of the projectile with identical ANCs but strongly different interiors: one with a node and one without a node. This is illustrated in Fig. 2.13 (left) in the case of ^{8}B, the best known one-proton halo candidate, which is thus described as a ^{7}Be core to which a valence proton is loosely bound. The ^{7}Be-p radial wave function obtained with the deep potential exhibits a node at $r \simeq 2$ fm (solid red line), which disappears once the spurious deeply bound state is removed by the sypersymmetric transformations (blue dashed line). Note that by construction, both wave functions exhibit exactly the same tail, i.e. the same ANC.

If the reaction process is sensitive to the internal part of the projectile wave function, significant differences should appear in the breakup cross section. However, the calculations performed with such pair of descriptions show no sensitivity to the choice of the potential. This is illustrated in Fig. 2.13 (right) in the particular case of

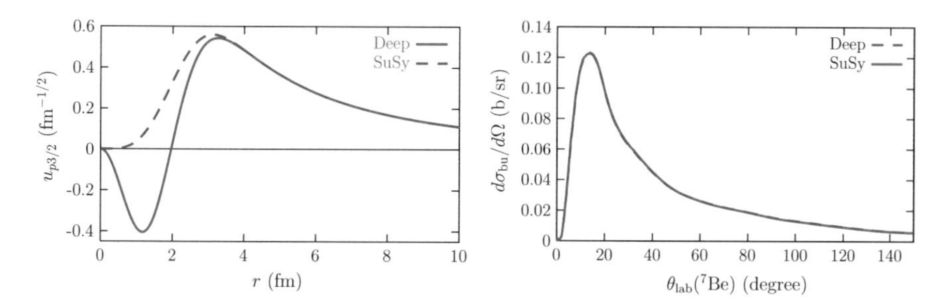

Fig. 2.13 Influence of the internal part of the projectile wave function upon breakup calculations [52]. Using two descriptions of ^{8}B with identical asymptotics but strongly different interiors (left), the breakup calculations provide identical cross sections, shown here on the right as an angular distribution obtained on a nickel target at 26 MeV. Reprinted figures with permission from [52] Copyright (2007) by the American Physical Society

the angular distribution for the breakup of ^{8}B on Ni at 26 MeV, which correspond to the conditions of the Notre-Dame experiment [55]. The calculations were performed within the CDCC framework as in [56]. Although both descriptions of ^{8}B exhibit significant differences [see Fig. 2.13 (left)], the corresponding breakup cross sections are superimposed on one another. This result is very general as it is observed for both Coulomb- and nuclear-dominated reactions, at low and high beam energies, for one-proton and one-neutron halo nuclei, and is valid for various kinds of breakup observables (energy and angular distributions) [52]. This clearly shows that the reaction process is purely peripheral and hence probes only the ANC of the initial bound state. Since the reaction process is not sensitive to the internal part of the wave function, it cannot be sensitive to the norm of the whole wave function. The spectroscopic factors extracted from such measurements are thus highly questionable.

2.3.3 Resonances

In addition to its two bound states, the ^{11}Be spectrum hosts a resonance at low energy in its continuum. It is a $\frac{5}{2}^+$ resonance, which is interpreted as a single-particle resonance in the $d_{5/2}$ partial wave [57] (see Sect. 2.1.4.3). Measuring the breakup of ^{11}Be on carbon at 67A MeV, Fukuda et al. have observed a large peak in the ^{10}Be-n continuum at the energy of this resonance [44], see Fig. 2.14.

Within a TD technique, it has been confirmed that the increase of the breakup strength in that region could be explained when the ^{10}Be-n potential is fitted to host such a resonance in the $d_{5/2}$ partial wave [57], see Fig. 2.14. The sharp peak observed in the theoretical cross section (dashed line) is entirely due to the contribution of the $d_{5/2}$ partial wave that hosts the resonance (dotted line). Interestingly, the width of that peak matches that of the resonance, confirming that nuclear-dominated breakup can provide significant information about structures within the continuum. Once folded

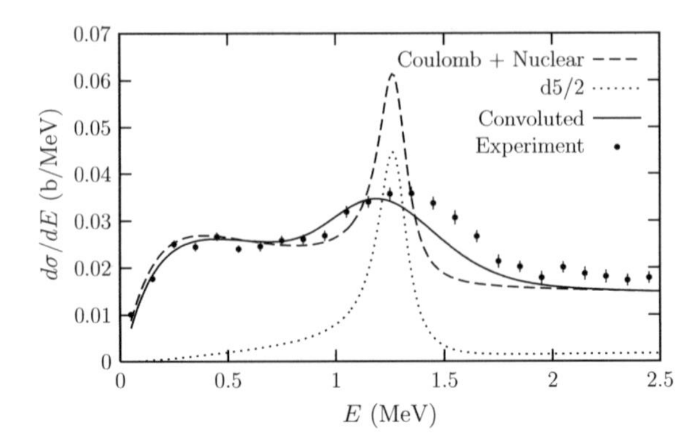

Fig. 2.14 Influence of a resonance within the low-energy ^{10}Be-n continuum upon the breakup of ^{11}Be on C at 67A MeV [44, 57]. The contribution of the $d_{5/2}$ partial wave, which hosts the resonance, is responsible for the sharp peak in the theoretical cross section. Once folded with the experimental energy resolution, the results of this TD calculation fit nicely with the data of [44]

with the experimental energy resolution, these TD calculations come quite close to the data of [44].

2.3.4 Role of the Non-resonant Continuum

In the previous section, we have seen that resonant structures within the continuum could affect nuclear-dominated breakup reactions. In this section, let us explore the sensitivity of breakup calculations to the non-resonant part of the continuum. One way to do so is to fit various c-f potentials, e.g. with different geometries, to the same nuclear-structure inputs. This was done, for example in [58], where different ^{10}Be-n potentials of Woods-Saxon shape were fitted to reproduce the $\frac{1}{2}^+$ ground state of ^{11}Be within the $1s_{1/2}$ orbit, its first excited state $\frac{1}{2}^-$ in the $0p_{1/2}$, and the $\frac{5}{2}^+$ resonance in the $d_{5/2}$ partial wave.

The breakup cross section on Pb at 69A MeV computed with these potentials within a TD approach are shown in Fig. 2.15 [58]. To remove the significant dependence on the ANC mentioned in Sect. 2.3.2, they have been divided by the square of the ANC of the single-particle wave function of the ground state (denoted in Fig. 2.15 by $b_{1s1/2}$) We observe that although the main dependence on the ANC has been removed, there remains a large variation in the breakup cross sections obtained with the different potentials. A detailed analysis shows that most of that sensitivity comes from the $p_{3/2}$ partial wave, which is not constrained by any physical observable. The variation between the $p_{3/2}$ contributions to the breakup cross sections obtained with the^{10}Be-n potentials built in [58] can be traced back to the changes induced in the $p_{3/2}$ phaseshifts.

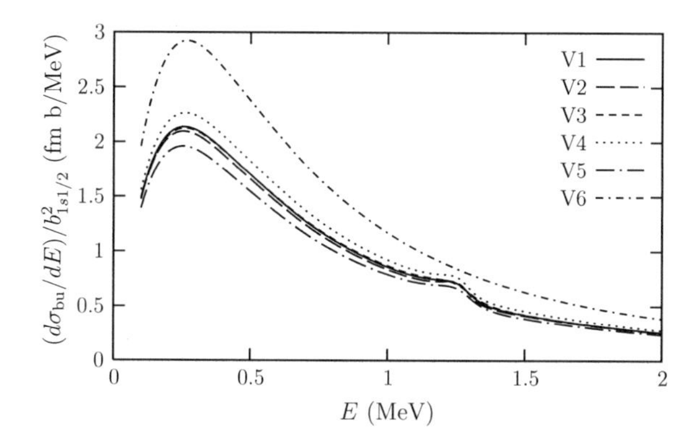

Fig. 2.15 Influence of the non-resonant continuum upon the breakup cross section. Example on the Coulomb breakup of ^{11}Be on Pb at 69A MeV [58]. The sensitivity to the ground state ANC has been removed by dividing the cross section by the square of the ANC. The remaining difference comes mostly from the differences in the $p_{3/2}$ partial wave, which has not been constrained in the construction of the different ^{10}Be-n potentials. Reprinted figure with permission from [58] Copyright (2006) by the American Physical Society

Such an influence of the non-resonant continuum upon the breakup cross section is very general. Similar results have also been observed for the breakup of a ^{8}B projectile at low energy [58]. This effect should be taken into account in the analysis of experiments, as it can significantly affect theoretical predictions. In particular, this could spoil the extraction of an ANC for the initial ground state from the direct—and naive—comparison of calculations to data. A similar conclusion has been drawn in a recent work using a Halo-EFT description of the projectile [59].

2.3.5 Effect of Core Excitation

So far, we have seen models in which the core was seen as a structureless body, of which the spin and parity are neglected. Although this is usually a good approximation, especially when the ground state of the core is a 0^+ state and the energy of its first excited state is large, other configurations are possible, as mentioned in Sect. 2.3.2. Including these other configurations in reaction models is a tricky business, not only because it increases the complexity of the structure of the projectile, but also because the reaction mechanism must then include the possible dynamical excitation of the core, i.e. reaction channels in which the core is excited by the target during the collision.

The first attempt to include the core excitation within the CDCC framework was performed by Summers et al. [60–62]. It led to the development of the XCDCC model, which stands for eXtended CDCC. However, the effect of this core excitation in the case of ^{11}Be seemed then rather small [63, 64].

More recently, Moro and Lay have extended a DWBA code to include the core excitation [65]. This enabled them to study the resonant breakup of ^{11}Be on C at 67*A* MeV, already mentioned in Sect. 2.3.3 . Their results are illustrated in Fig. 2.16, which displays the breakup cross sections as angular distributions at the energy of the $\frac{5}{2}^+$ (top) and $\frac{3}{2}^+$ (bottom) resonances within the ^{11}Be continuum. Their calculations (black solid lines) are in very good agreement with the data. In particular, the angular dependence of the cross sections perfectly matches that of the RIKEN data [44]. Such a good result is obtained only if both the excitation of the valence neutron to the continuum (red dash-dotted lines) and the core excitation (blue dashed lines) are included together in the reaction model. For the $\frac{5}{2}^+$ resonance (Fig. 2.16 top), we see that albeit small, the effect of the core excitation is required to obtain an angular dependence that fits the data. For the $\frac{3}{2}^+$ state, however, the core excitation is the dominant process in the reaction, suggesting that the structure of this resonance is dominated by a configuration in which the ^{10}Be core is in its 2^+ excited state.

These results indicate that for some observables, the core excitation can be a significant part of the breakup process, especially at higher energy in the continuum.

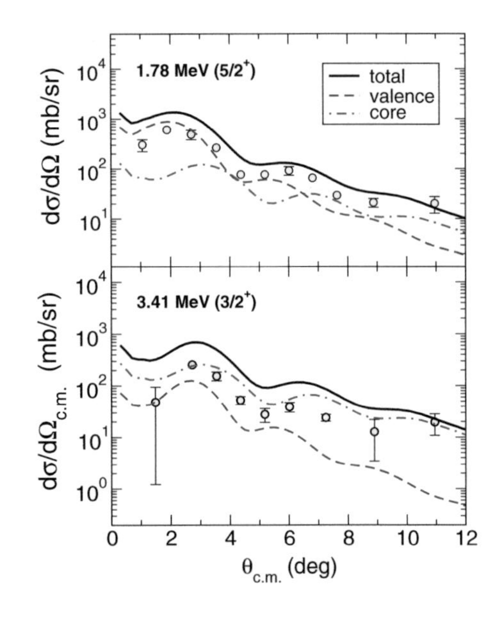

Fig. 2.16 Influence of the core excitation on the breakup of ^{11}Be on C at 67*A* MeV [65]. The shape of the data for these angular distributions can be reproduced only if the core excitation is included in these DWBA calculations. Data are from [44]. Reprinted figure with permission from [65] Copyright (2012) by the American Physical Society

More efforts should therefore be made to include it within models to improve the description of the collision and increase our understanding of reactions involving exotic nuclei.

Conclusion

Reactions are used in many fields of quantum physics. Whether in the field of molecular, atomic, nuclear, or particle physics, significant amounts of information can be gathered through the study of collisions. The present notes offer an introduction to quantum scattering theory explained in the realm of nuclear physics. The extension of that theory to reactions in which the projectile can break up into two more elementary constituents and its particular application to the study of halo nuclei has also been presented. The three major models of breakup reaction have been detailed, the CDCC method, the TD approach and the eikonal approximation. A benchmark of these models on the special case of ^{15}C shows the limitation of each model, which enables us to estimate the range of validity of each of these approaches.

In the last section, an analysis of the nuclear-structure information about halo nuclei one can get from breakup measurements has been provided. Through examples chosen from the literature, the complexity of the reaction mechanism and its sensitivity to the structure of the projectile has been presented. By showing which part of the projectile wave function is probed during breakup, these results emphasise the importance to study in detail to which structure observable the reaction is actually sensitive. The blind application of an accepted recipe can lead to misinterpretation of experimental results.

Two key points should be taken away from this brief summary: the first is to pay attention to the domain of validity of the reaction model used to analyse data. Does the model fit the experimental conditions under which they have been gathered? For example, it does not make sense to use an eikonal model of reaction to analyse an experiment performed at the Coulomb barrier. The second is to know to what the reaction is sensitive. For example, it does not make sense to extract a spectroscopic factor from a peripheral experiment.

With this short review, I hope to have delivered a basic introduction to quantum-reaction modelling within the realm of nuclear physics. Hopefully, it will have triggered the interest of the reader in reaction theory and provide him/her with a list of useful references to deepen his/her knowledge in this exciting field of physics.

Acknowledgements This project has received funding from the European Unions Horizon 2020 research and innovation program under grant agreement No 654002, the Deutsche Forschungsgemeinschaft within the Collaborative Research Centers 1245 and 1044, and the PRISMA (Precision Physics, Fundamental Interactions and Structure of Matter) Cluster of Excellence. I also acknowledges the support of the State of Rhineland-Palatinate.

References

1. C. Cohen-Tannoudji, B. Diu, F. Laloë, *Quantum Mechanics* (Wiley, Paris, 1977)
2. J.R. Taylor, *Scattering Theory: The Quantum Theory of Nonrelativistic Collisions* (Dover, New York, 1972)
3. F.M. Nunes, I.J. Thompson, *Nuclear Reactions for Astrophysics: Principles, Calculation and Applications of Low-Energy Reactions* (Cambridge University Press, Cambridge, 2009)
4. C.A. Bertulani, P. Danielewicz, *Introduction to Nuclear Reactions* (Institute of Physics Publishing, Bristol, 2004)
5. R. Navarro-Pérez, J. Amaro, E. Ruiz-Arriola, Phys. Lett. B **724**(1), 138 (2013). https://doi.org/10.1016/j.physletb.2013.05.066. http://www.sciencedirect.com/science/article/pii/S0370269313004486
6. R.B. Wiringa, V.G.J. Stoks, R. Schiavilla, Phys. Rev. C **51**, 38 (1995). https://doi.org/10.1103/PhysRevC.51.38
7. R. Machleidt, Phys. Rev. C **63**, 024001 (2001). https://doi.org/10.1103/PhysRevC.63.024001
8. E. Epelbaum, H.-W. Hammer, U.G. Meißner, Rev. Mod. Phys. **81**, 1773 (2009). https://doi.org/10.1103/RevModPhys.81.1773
9. S.B. Dubovichenko, Phys. At. Nucl. **75**, 173 (2012). https://doi.org/10.1134/S1063778812020044. https://link.springer.com/article/10.1134%2FS1063778812020044
10. K.S. Krane, *Introductory Nuclear Physics* (Wiley, New York, 1987)
11. I. Tanihata, H. Hamagaki, O. Hashimoto, S. Nagamiya, Y. Shida, N. Yoshikawa, O. Yamakawa, K. Sugimoto, T. Kobayashi, D. Greiner, N. Takahashi, Y. Nojiri, Phys. Lett. B **160**(6), 380 (1985). https://doi.org/10.1016/0370-2693(85)90005-X. http://www.sciencedirect.com/science/article/pii/037026938590005X
12. I. Tanihata, H. Hamagaki, O. Hashimoto, Y. Shida, N. Yoshikawa, K. Sugimoto, O. Yamakawa, T. Kobayashi, N. Takahashi, Phys. Rev. Lett. **55**, 2676 (1985). https://doi.org/10.1103/PhysRevLett.55.2676
13. I. Tanihata, J. Phys. G **22**(2), 157 (1996). http://stacks.iop.org/0954-3899/22/i=2/a=004
14. E. Sauvan, F. Carstoiu, N. Orr, J. Anglique, W. Catford, N. Clarke, M.M. Cormick, N. Curtis, M. Freer, S. Grvy, C. LeBrun, M. Lewitowicz, E. Ligard, F. Marqus, P. Roussel-Chomaz, M. SaintLaurent, M. Shawcross, J. Winfield, Phys. Lett. B **491**(1), 1 (2000). https://doi.org/10.1016/S0370-2693(00)01003-0. http://www.sciencedirect.com/science/article/pii/S0370269300010030
15. P.G. Hansen, B. Jonson, Europhys. Lett. **4**(4), 409 (1987). http://stacks.iop.org/0295-5075/4/i=4/a=005
16. M. Zhukov, B. Danilin, D. Fedorov, J. Bang, I. Thompson, J. Vaagen, Phys. Rep. **231**(4), 151 (1993). https://doi.org/10.1016/0370-1573(93)90141-Y. http://www.sciencedirect.com/science/article/pii/037015739390141Y
17. G. Baur, C. Bertulani, H. Rebel, Nucl. Phys. A **458**(1), 188 (1986). https://doi.org/10.1016/0375-9474(86)90290-3. http://www.sciencedirect.com/science/article/pii/0375947486902903
18. G. Baur, H. Rebel, Ann. Rev. Nucl. Part. Sci. **46**(1), 321 (1996). https://doi.org/10.1146/annurev.nucl.46.1.321
19. National Nuclear Data Centre (2018). http://www.nndc.bnl.gov/
20. J.H. Kelley, E. Kwan, J.E. Purcell, C.G. Sheu, H.R. Weller, Nucl. Phys. A **880**, 88 (2012). https://doi.org/10.1016/j.nuclphysa.2012.01.010. http://www.sciencedirect.com/science/article/pii/S0375947412000413
21. D. Baye, P. Capel, in *Clusters in Nuclei*, ed. by C. Beck, Vol. 2 (Springer, Heidelberg, 2012), pp. 121–163
22. G.H. Rawitscher, Phys. Rev. C **9**, 2210 (1974). https://doi.org/10.1103/PhysRevC.9.2210
23. M. Kamimura, M. Yahiro, Y. Iseri, Y. Sakuragi, H. Kameyama, M. Kawai, Prog. Theor. Phys. Suppl. **89**, 1 (1986). https://doi.org/10.1143/PTPS.89.1

24. N. Austern, Y. Iseri, M. Kamimura, M. Kawai, G. Rawitscher, M. Yahiro, Phys. Rep. **154**(3), 125 (1987). https://doi.org/10.1016/0370-1573(87)90094-9. http://www.sciencedirect.com/science/article/pii/0370157387900949
25. M. Yahiro, K. Ogata, T. Matsumoto, K. Minomo, Prog. Theor. Exp. Phys. **2012**(1), 01A206 (2012). https://doi.org/10.1093/ptep/pts008
26. T. Druet, D. Baye, P. Descouvemont, J.-M. Sparenberg, Nucl. Phys. A **845**(1), 88 (2010). https://doi.org/10.1016/j.nuclphysa.2010.05.060. http://www.sciencedirect.com/science/article/pii/S0375947410005282
27. A.M. Moro, F. Pérez-Bernal, J.M. Arias, J. Gómez-Camacho, Phys. Rev. C **73**, 044612 (2006). https://doi.org/10.1103/PhysRevC.73.044612
28. I.J. Thompson, Comput. Phys. Rep. **7**(4), 167 (1988). https://doi.org/10.1016/0167-7977(88)90005-6. http://www.sciencedirect.com/science/article/pii/0167797788900056
29. A. Di Pietro, G. Randisi, V. Scuderi, L. Acosta, F. Amorini, M.J.G. Borge, P. Figuera, M. Fisichella, L.M. Fraile, J. Gomez-Camacho, H. Jeppesen, M. Lattuada, I. Martel, M. Milin, A. Musumarra, M. Papa, M.G. Pellegriti, F. Perez-Bernal, R. Raabe, F. Rizzo, D. Santonocito, G. Scalia, O. Tengblad, D. Torresi, A.M. Vidal, D. Voulot, F. Wenander, M. Zadro, Phys. Rev. Lett. **105**, 022701 (2010). https://doi.org/10.1103/PhysRevLett.105.022701
30. A. Di Pietro, V. Scuderi, A.M. Moro, L. Acosta, F. Amorini, M.J.G. Borge, P. Figuera, M. Fisichella, L.M. Fraile, J. Gomez-Camacho, H. Jeppesen, M. Lattuada, I. Martel, M. Milin, A. Musumarra, M. Papa, M.G. Pellegriti, F. Perez-Bernal, R. Raabe, G. Randisi, F. Rizzo, G. Scalia, O. Tengblad, D. Torresi, A.M. Vidal, D. Voulot, F. Wenander, M. Zadro, Phys. Rev. C **85**, 054607 (2012). https://doi.org/10.1103/PhysRevC.85.054607. https://link.aps.org/doi/10.1103/PhysRevC.85.054607
31. K. Alder, A. Winther, *Electromagnetic Excitation* (North-Holland, Amsterdam, 1975)
32. T. Kido, K. Yabana, Y. Suzuki, Phys. Rev. C **50**, R1276 (1994). https://doi.org/10.1103/PhysRevC.50.R1276
33. H. Esbensen, G. Bertsch, C.A. Bertulani, Nucl. Phys. A **581**(1), 107 (1995). https://doi.org/10.1016/0375-9474(94)00423-K. http://www.sciencedirect.com/science/article/pii/037594749400423K
34. S. Typel, H.H. Wolter, Z. Naturforsch **54a**, 63 (1999)
35. M. Fallot, J.A. Scarpaci, D. Lacroix, P. Chomaz, J. Margueron, Nucl. Phys. A **700**(1), 70 (2002). https://doi.org/10.1016/S0375-9474(01)01303-3. http://www.sciencedirect.com/science/article/pii/S0375947401013033
36. P. Capel, D. Baye, V.S. Melezhik, Phys. Rev. C **68**, 014612 (2003). https://doi.org/10.1103/PhysRevC.68.014612
37. T. Nakamura, N. Fukuda, N. Aoi, N. Imai, M. Ishihara, H. Iwasaki, T. Kobayashi, T. Kubo, A. Mengoni, T. Motobayashi, M. Notani, H. Otsu, H. Sakurai, S. Shimoura, T. Teranishi, Y.X. Watanabe, K. Yoneda, Phys. Rev. C **79**, 035805 (2009). https://doi.org/10.1103/PhysRevC.79.035805
38. H. Esbensen, Phys. Rev. C **80**, 024608 (2009). https://doi.org/10.1103/PhysRevC.80.024608
39. H. Esbensen, R. Reifarth, Phys. Rev. C **80**, 059904 (2009). https://doi.org/10.1103/PhysRevC.80.059904
40. R.J. Glauber, in *Lecture in Theoretical Physics*, vol. 1, ed. by W.E. Brittin, L.G. Dunham (Interscience, New York, 1959), p. 315
41. D. Baye, P. Capel, G. Goldstein, Phys. Rev. Lett. **95**, 082502 (2005). https://doi.org/10.1103/PhysRevLett.95.082502
42. G. Goldstein, D. Baye, P. Capel, Phys. Rev. C **73**, 024602 (2006). https://doi.org/10.1103/PhysRevC.73.024602
43. K. Ogata, M. Yahiro, Y. Iseri, T. Matsumoto, M. Kamimura, Phys. Rev. C **68**, 064609 (2003). https://doi.org/10.1103/PhysRevC.68.064609
44. N. Fukuda, T. Nakamura, N. Aoi, N. Imai, M. Ishihara, T. Kobayashi, H. Iwasaki, T. Kubo, A. Mengoni, M. Notani, H. Otsu, H. Sakurai, S. Shimoura, T. Teranishi, Y.X. Watanabe, K. Yoneda, Phys. Rev. C **70**, 054606 (2004). https://doi.org/10.1103/PhysRevC.70.054606

45. P. Capel, H. Esbensen, F.M. Nunes, Phys. Rev. C **85**, 044604 (2012). https://doi.org/10.1103/PhysRevC.85.044604
46. N.J. Upadhyay, A. Deltuva, F.M. Nunes, Phys. Rev. C **85**, 054621 (2012). https://doi.org/10.1103/PhysRevC.85.054621
47. T. Fukui, K. Ogata, P. Capel, Phys. Rev. C **90**, 034617 (2014). https://doi.org/10.1103/PhysRevC.90.034617
48. T. Nakamura, N. Fukuda, T. Kobayashi, N. Aoi, H. Iwasaki, T. Kubo, A. Mengoni, M. Notani, H. Otsu, H. Sakurai, S. Shimoura, T. Teranishi, Y.X. Watanabe, K. Yoneda, M. Ishihara, Phys. Rev. Lett. **83**, 1112 (1999). https://doi.org/10.1103/PhysRevLett.83.1112
49. S. Typel, R. Shyam, Phys. Rev. C **64**, 024605 (2001). https://doi.org/10.1103/PhysRevC.64.024605
50. S. Typel, G. Baur, Phys. Rev. Lett. **93**, 142502 (2004). https://doi.org/10.1103/PhysRevLett.93.142502
51. S. Typel, G. Baur, Nucl. Phys. A **759**(3), 247 (2005). https://doi.org/10.1016/j.nuclphysa.2005.05.145. http://www.sciencedirect.com/science/article/pii/S0375947405008493
52. P. Capel, F.M. Nunes, Phys. Rev. C **75**, 054609 (2007). https://doi.org/10.1103/PhysRevC.75.054609
53. D. Baye, Phys. Rev. Lett. **58**, 2738 (1987). https://doi.org/10.1103/PhysRevLett.58.2738
54. D. Baye, J. Phys. A **20**(16), 5529 (1987). http://stacks.iop.org/0305-4470/20/i=16/a=027
55. J.J. Kolata, V. Guimarães, D. Peterson, P. Santi, R.H. White-Stevens, S.M. Vincent, F.D. Becchetti, M.Y. Lee, T.W. O'Donnell, D.A. Roberts, J.A. Zimmerman, Phys. Rev. C **63**, 024616 (2001). https://doi.org/10.1103/PhysRevC.63.024616
56. J.A. Tostevin, F.M. Nunes, I.J. Thompson, Phys. Rev. C **63**, 024617 (2001). https://doi.org/10.1103/PhysRevC.63.024617
57. P. Capel, G. Goldstein, D. Baye, Phys. Rev. C **70**, 064605 (2004). https://doi.org/10.1103/PhysRevC.70.064605
58. P. Capel, F.M. Nunes, Phys. Rev. C **73**, 014615 (2006). https://doi.org/10.1103/PhysRevC.73.014615
59. P. Capel, D.R. Phillips, H.W. Hammer, Phys. Rev. C **98**, 034610 (2018). https://doi.org/10.1103/PhysRevC.98.034610
60. N.C. Summers, F.M. Nunes, I.J. Thompson, Phys. Rev. C **73**, 031603 (2006). https://doi.org/10.1103/PhysRevC.73.031603
61. N.C. Summers, F.M. Nunes, I.J. Thompson, Phys. Rev. C **74**, 014606 (2006). https://doi.org/10.1103/PhysRevC.74.014606
62. N.C. Summers, F.M. Nunes, I.J. Thompson, Phys. Rev. C **89**, 069901 (2014). https://doi.org/10.1103/PhysRevC.89.069901
63. N.C. Summers, F.M. Nunes, Phys. Rev. C **76**, 014611 (2007). https://doi.org/10.1103/PhysRevC.76.014611
64. N.C. Summers, F.M. Nunes, Phys. Rev. C **77**, 049901 (2008). https://doi.org/10.1103/PhysRevC.77.049901
65. A.M. Moro, J.A. Lay, Phys. Rev. Lett. **109**, 232502 (2012). https://doi.org/10.1103/PhysRevLett.109.232502

Chapter 3
Nuclear Reaction Experiments

Alexandre Obertelli

Abstract Nuclear reactions play an essential role to address the many-body problem of nuclear structure. Reactions are used to produce exotic nuclei, to populate specific nuclear states and the so-called direct reactions have been a unique tool to built up our representation of the nuclear shell structure. In this lecture, the basics of radioactive beam production are described and direct reactions are introduced. In particular, spectroscopic factors (SF) are defined and discussed. Transfer and knockout reactions are introduced in the light of today's experiments and detection setups.

3.1 Overview of Nuclear Reactions and Their Relevance to Nuclear Studies

Nuclear reactions have been studied since the discovery of the nucleus: the atomic nucleus was actually evidenced by use of elastic scattering, the simplest nuclear reaction: in 1911, Rutherford and Geiger, in England, investigated the nature of gold atoms by bombarding them with alpha particles emitted from a collimated radioactive source. A fraction of these alpha particles, known at the time to be penetrating particles, were scattered at large angles and even backward. This observation of backward-angle elastic scattering was the first use of nuclear reactions to evidence the existence of a nucleus inside the atom [1].

Since then, with the development of particle accelerators, nuclear structure has grown as a major domain of physics with unique specificities. Nuclear Physics is a non perturbative quantum domain *par excellence* where the strong interaction drives structure and where new phenomena emerge by the addition or removal of few nucleons. The richness of nuclear structure originates in many phenomena steaming from the complexity of the nuclear many-body problem, spanning over a wide range of energy scales, as illustrated on Fig. 3.1. Nuclei have the particularity to be bound systems composed of two kinds of fermions, the protons and neutrons. As finite-size

A. Obertelli (✉)
TU Darmstadt, Darmstadt, Germany
e-mail: aobertelli@ikp.tu-darmstadt.de

J.-E. García-Ramos et al. (eds.), *Basic Concepts in Nuclear Physics: Theory, Experiments and Applications*, Springer Proceedings in Physics 225,
https://doi.org/10.1007/978-3-030-22204-8_3

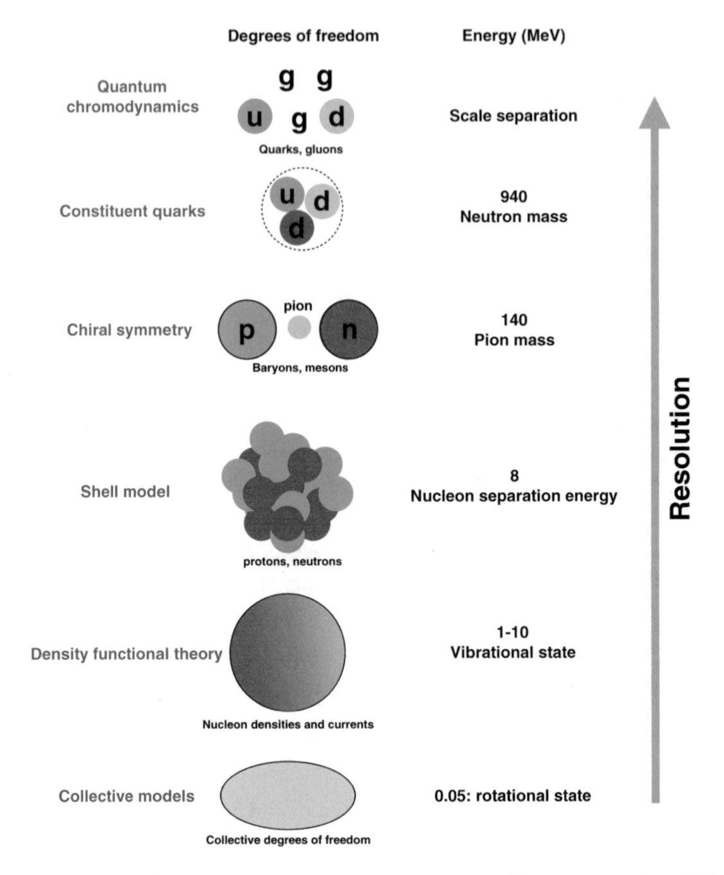

Fig. 3.1 In subatomic physics, phenomena correspond to specific energy scales. Although the energy scales of phenomena can partially overlap, the modelling and the understanding of phenomena requires the choice of a suitable energy scale and degrees of freedom. Inspired from [2]

composite quantum systems, they develop a shell structure. A comprehensive picture of nuclear structure and its dependence with the number of nucleons and isospin is an exciting fundamental question which has not been solved yet, despite the joint efforts of theorists and experimentalists.

The understanding of the structure of a nucleus usually requires several probes. Nuclear reactions play a central role in building our representation of nuclear structure, in several ways:

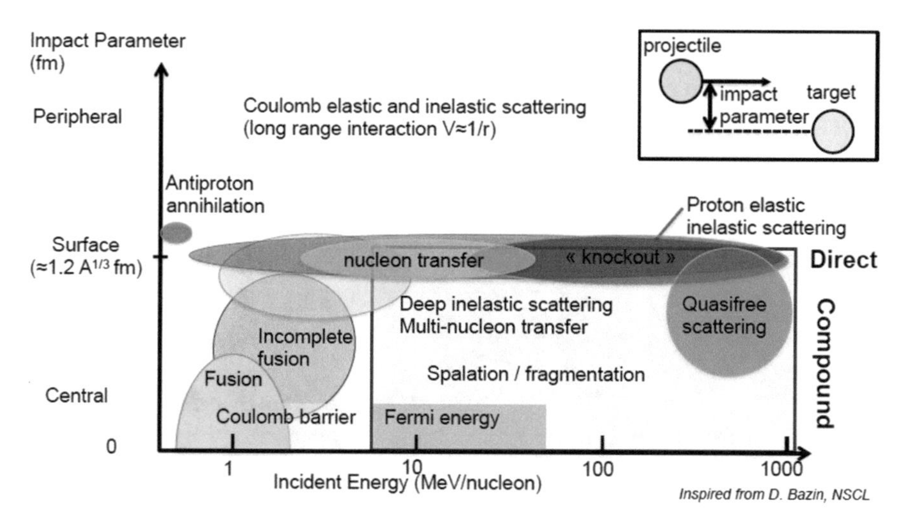

Fig. 3.2 Partial overview of nuclear reactions categorised as a function of the collision energy (in MeV/nucleon) and impact parameter from peripheral to central collisions. Direct reactions take place at the nuclear surface

- as a tool to produce radioactive ions from stable beams,
- as a tool to excite nuclei, a necessary step to go beyond the study of ground-state properties,
- as a probe to investigate nuclear structure,
- as a subject in itself to understand the dynamics of hadronic systems.

A partial overview of nuclear reactions is represented in Fig. 3.2, where reactions are categorised as a function of the collision energy (in MeV/nucleon) and impact parameter from peripheral to central collisions. Direct reactions take place at the nuclear surface.

In the following, we first introduce the concept of cross sections (Sect. 3.2). Then (Sect. 3.3), the production of Radioactive-Ion Beams (RIB) is exposed. Direct reaction cross sections are used to infer information on the overlap of initial and final states involved in the reaction [3] and therefore to learn about the nuclear wave function. We then describe the main features of the most used direct reactions: nucleon transfer and nucleon knockout reactions (Sect. 3.4).

3.2 Cross Sections

A cross section for a given reaction quantifies its probability to happen. It has the unit of a surface. The most common unit used in nuclear physics is the barn:

$$1\,\text{barn} = 1\,\text{b} = 10^{-24}\,\text{cm}^{-2},\ \ 1\,\text{millibarn} = 1\,\text{mb} = 10^{-27}\,\text{cm}^{-2} \qquad (3.1)$$

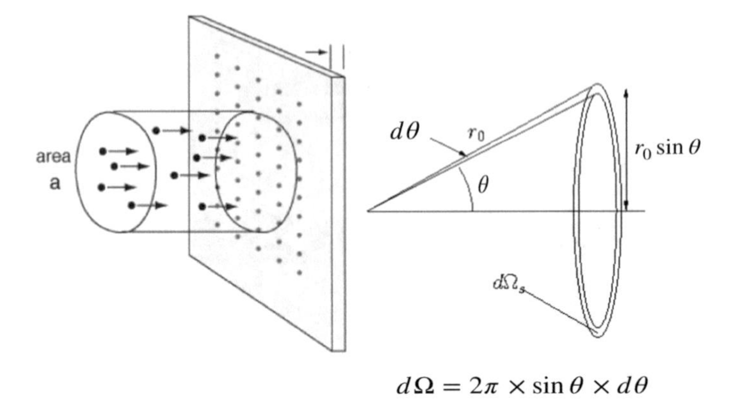

Fig. 3.3 (Left) The cross section of a reaction relates the number of incident particles to the number of reaction events. It quantifies the probability of a given reaction to happen. (Right) Cross sections can be integrated over all angles (called *inclusive*) or expressed as differential cross sections, for example as a function of the scattering angle θ of the reaction products, defining the solid angle $\mathrm{d}\Omega$

The number of events $N_r(\Omega)$ produced by a specific reaction in a solid angle Ω is related to the number of incoming particles N_i and target thickness e (number of scattering centers per surface unit) by the differential cross section $\mathrm{d}\sigma/\mathrm{d}\Omega$:

$$N_r(\Omega) = N_i \times \frac{\mathrm{d}\sigma}{\mathrm{d}\Omega} \times e, \tag{3.2}$$

where $\mathrm{d}\Omega$ is given by the scattering angle θ at which reaction products are detected, as illustrated in Fig. 3.3. The above relation is valid for small target thicknesses, i.e. for which the number of unreacted projectiles at the exit of the target is $\sim N_i$.

We define the following cross section terminology:

- The **total** cross section (σ_T) is the sum of all interaction processes, including the elastic scattering.
- The **elastic** cross section (σ_{el}) gathers interactions where both the projectile and target nuclei remain in their ground state.
- The **reaction** cross section (σ_R) gathers processes where the projectile (or target) nucleus is excited during the process.
- The **interaction** cross section (σ_I) gathers all processes which change the number of protons and/or neutrons of the projectile (or target).

The above cross sections are related to each other:

$$\sigma_T = \sigma_{el} + \sigma_R = \sigma_{el} + \sigma_{\mathrm{inel}} + \sigma_I, \tag{3.3}$$

where σ_{inel} is the inelastic cross section to bound excited states (i.e. N and Z remain unchanged). The interaction cross section σ_I gathers all reaction events for which the final nucleus has a change of nucleons compared to the projectile. It is related to the reaction cross section σ_R as follows

$$\sigma_R = \sigma_I + \sigma_{\text{inel}}. \tag{3.4}$$

If σ_{inel} is small enough, one can assume that $\sigma_I \sim \sigma_R$, depending on the targeted accuracy. As an example, at the relativistic energy of 600 MeV/nucleon, the inelastic cross section for ^{132}Sn+^{12}C is measured to be $\sigma_{\text{inel}} = 100$ mb, amounting to 4% of the reaction cross section σ_R [4].

In the following, we illustrate the use of reactions for nuclear structure with the example of the sensitivity of the inclusive interaction cross section σ_I to the matter radius of nuclei. Indeed, the nuclear interaction is short range and strong: the nuclear potential follows the matter density profile. This feature is used to extract the matter radius of nuclei from interaction cross sections.

One can define an *interaction radius* R_I of a nucleus B with a target A, following a black disk approximation

$$\sigma_I(A, B) = \pi[R_I(A) + R_I(B)]^2. \tag{3.5}$$

The target interaction radius can be extracted from a symmetric reaction,

$$\sigma_I(A, A) = 4\pi[R_I(A)]^2, \tag{3.6}$$

leading to the interaction radius of the projectile

$$R_I(B) = \sqrt{\frac{\sigma_I(A, B)}{\pi}} - \sqrt{\frac{\sigma_I(A, A)}{\pi}}. \tag{3.7}$$

An example of interaction radii extracted from heavy-ion collisions at relativistic energies of several 100 MeV/nucleon is shown in Fig. 3.4.

The above approach can be refined by using a microscopic reaction model in which microscopic density profiles can be used as inputs. Reaction cross sections at relativistic energies can be predicted under the eikonal approximation which assumes that the projectile is not deflected during the reaction process. The reaction cross section can be expressed as a summation over impact parameters b of the probabilities for the projectile to interact

$$\sigma_R = 2\pi \int_0^{+\infty} [1 - T(b)]b\mathrm{d}b, \tag{3.8}$$

where $T(b)$ is the transmission function at impact parameter b. $T(b)$ expresses the probability for projectile impinging at impact parameter b not to react with the target. It can be expressed as

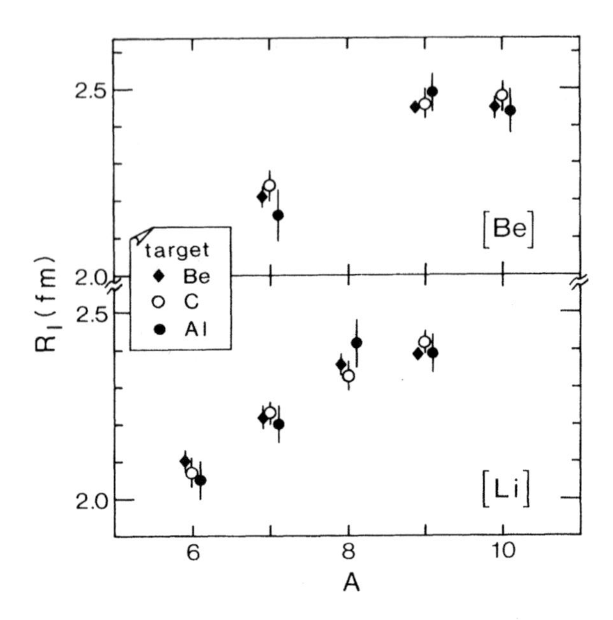

Fig. 3.4 Interaction radii of Be (top panel) and Li (bottom panel) obtained from interaction cross sections with different targets. Reprinted from [5] with permission from the American Physical Society

$$T(b) = \exp\left[-\bar{\sigma}\int_{-\infty}^{+\infty} q(b,z)\mathrm{d}z\right], \tag{3.9}$$

where $\bar{\sigma}$ is the effective NN cross section, taking into account the isospin asymmetry of the target and projectile and $q(b, z)$ is given by

$$q(b,z) = \int_{-\infty}^{+\infty} d\eta 2\pi \int_{0}^{+\infty} \rho^{t}(b,z,b,\eta)\rho^{p}(b,z,r,\eta)r\mathrm{d}r, \tag{3.10}$$

where $\rho^{t,p}$ are the target and projectile matter densities. (3.10) shows the sensitivity of the reaction cross section to the projectile density ρ^p. For a measured reaction cross section, the extracted matter radii $\langle r^2\rangle_m$ would be taken as those of microscopic density profiles that reproduce the experimental reaction cross section within the above formalism (Fig. 3.5).

One known technique to measure interaction cross section is the so called *transmission method* which consists of measuring the unreacted beam particles after transmission through a target of thickness e (cm^{-2}). Assuming that the interaction cross section σ_I does not vary across the target thickness, one gets a number of reactions $\delta N(z)$ produced in a slice of thickness δz at a position z along the beam axis

$$\delta N(z) = -N(z) \times \sigma_I \times \delta z, \tag{3.11}$$

where $N(z)$ is the number of unreacted beam particles at position z. By integrating the above relation over the target length, one gets

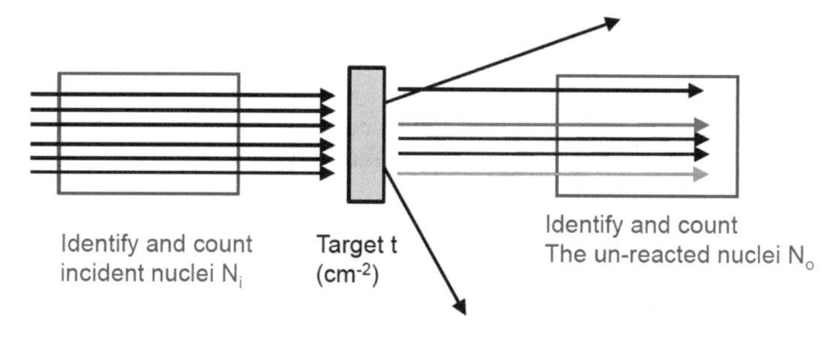

Fig. 3.5 Transmission method for interaction cross section measurements

$$\sigma_I = \frac{1}{t} ln \left(\frac{N_0}{N_i} \right), \tag{3.12}$$

with N_0 the number of unreacted particles and N_i the initial number of beam particle impinging on the target. Note the difference with the thin-target approximation of (3.2). As seen before, the number of unreacted particles is the number of beam particles measured downstream the reaction target and corrected from the inelastic scattering to bound excited states. The inelastic cross section to bound excited states can either be obtained from another measurement, or estimated from theory.

3.3 Radioactive Ion Beam Production

As short-lived nuclei do not exist on earth, they have to be synthesized from stable nuclei, accelerated in an ion state and studied shortly (before they β decay) after production. There are two main techniques to produce accelerated radioactive beams:

- the in-flight method,
- the Isotopic Separation On Line (ISOL) method.

The in-flight method is illustrated on the right-hand side of Fig. 3.6. An accelerated beam impinges onto a thin production target. Radioactive nuclei originate from the fragmentation or induced-fission of projectiles. The production target being thin, the beam and reaction products loose little energy in the target and produced radioactive nuclei have about the same velocity of the beam. This production method has three main advantages:

1. the production method is not sensitive to the chemical properties of the produced nuclei. Therefore the only relevant quantity for the production of a given nucleus is its production cross section,
2. radioactive nuclei are produced in flight, meaning that they can be directed right after production to the experimental zone with no further delay than the time of flight. The in-flight method is particularly suited for the production and study of very short-lived nuclei down to half-lifes of hundreds of nanoseconds,

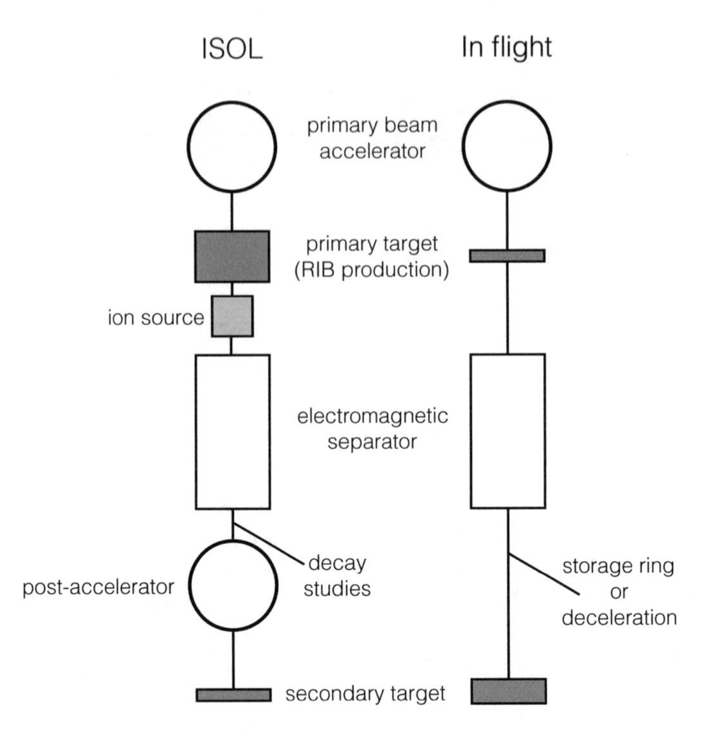

Fig. 3.6 Schematics of an ISOL (left) and in-flight (right) radioactive ion beam production. Inspired from [6]

3. fast beams are naturally produced with the in-flight technique, requiring no further acceleration.

The ISOL method is sketched on the left-hand side of Fig. 3.6. The ISOL production has several advantages compared to the in-flight produced isotopes, and the two methods can be considered as complementary to some extent. The advantages are:

1. beams can be produced from very low energy (keV) to high energy. Of course, high energy ISOL-produced beams would require a very expensive post-accelerator for reaction products. Fast beams (above 20 MeV/nucleon) are today preferentially produced in flight,
2. the beams produced via ISOL and reaccelerated present a high quality emittance, i.e. ensuring a small beam size on target and small energy dispersion,
3. since the reaction products are stopped in the production target and should be extracted towards an ion source to be ionized and transport, and often reaccelerated, the transmission and rates are highly dependent on the isotopes due to chemical selectivity. For each element (isotope) a ionization scheme has to be developed before the beam can be produced. On the other hand, this selectivity can ensure pure isotopic beams.

4. the extraction from the production target and ionization process take time, typically 1 ms. Radioactive nuclei with a half-life lower than 1 ms cannot be efficiently produced via the ISOL method.

3.3.1 The Primary Beam: Ion Sources and Acceleration Systems

Almost[1] all radioactive ion beam production starts with the interaction of an accelerated stable beam with a production target.

The first element of an heavy-ion accelerator is the source. Ions have to be produced at high intensity, in a charged state so that they can then be accelerated by strong electric fields. The ion source consists of two parts: an ion generator and an extraction system.

Ions are produced from the creation of a plasma. There are several techniques to make this plasma and ionize atoms: by electric discharges in a low pressure gas volume, by heating, using lasers or beams of other particles. There are many different types of ion sources. The choice for the design of an ion source are: (i) the ions to be produced, (ii) the ion intensity to be reached, (iii) the stability in time of the beam, (iv) the optical quality of the beam, quantified by the so-called *emittance*.

A supply of atoms must be provided to the plasma to compensate the loss of material. The material can be introduced in a gas phase inside the plasma via a needle valve or solid material can be heated and ionized inside the ion source. As an illustration of how an ion source works, we take here the example of an electron bombardment source, whose principle is sketched in Fig. 3.7. This type of source is at the origin of all ion sources. It was first developed by A. Dempster at the University of Chicago in 1916. It has a very simple design but contains all features of ion sources. In this specific case, the plasma is created by accelerated electrons created from the cathode filament.

Electron bombardment sources are easy to produce but the reached intensities are too low for the requirements of modern accelerators. Modern sources are based on other techniques to generate the plasma and the geometry of the source is made in such a way that the positive ion density close to the extraction is as high as possible. An important type of ion source is called ECR (Electron Cyclotron Resonance) where the plasma is formed by a high power microwave radiofrequency applied to the source. The source is surrounded by a solenoid to generate a magnetic field which confines the plasma. Since the plasma in an ERC source has a relatively long lifetime, it has the advantage to produce efficiently multi-charged ions from multi-collisions of electrons with atoms and ions.

---

[1]One exception is the CARIBU facility at Argonne National Laboratory, USA, which relies on the production of radioactive ions from a high intensity radioactive source of Californium which releases fission fragments. CARIBU does not require any primary beam.

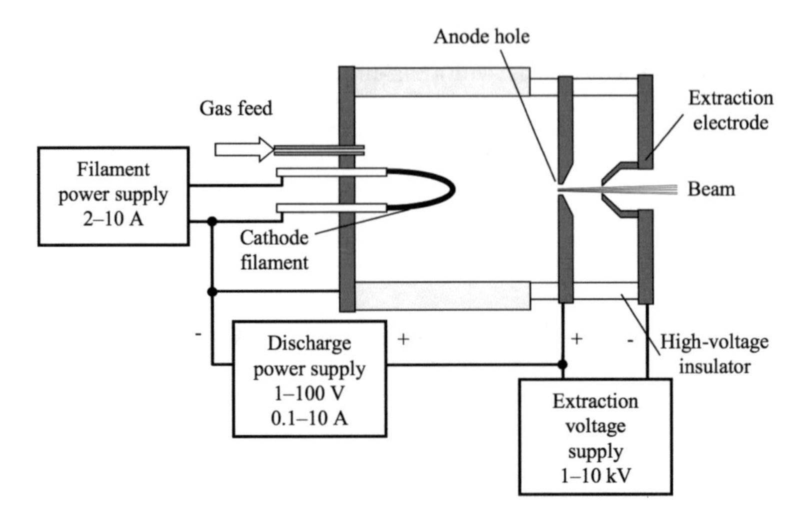

Fig. 3.7 Schematic design of an electron bombardment ion source. Taken from [7]

As examples of intense ion beams produced from ion sources, the SILHI light ion source of CEA reaches stable intensities of 100 mA for protons [8]. The Berkeley ECR ion source produces beams of Bi^{29+} at 0.25 mA. In modern ion sources, The extraction voltage is typically 50 kV, and ranges from few kV to 400 kV.

Ions are accelerated via electric fields. There are different schemes to accelerated ions:

- A linear accelerator composed of a series of cavities excited in a stationary mode with a radio-frequency. Ions are introduced in the linear accelerator and extracted in bunches of few MeV/nucleon.
- A cyclotron in which low energy ions are introduced at the center. A cyclotron works with permanent magnetic field which curves the trajectories of ions. The ions are accelerated in between two regions of a magnetic field by an electric field in phase with the ions. The direction of the electric field changes at each half turn of the ions, leading to an acceleration phase and increase of trajectory radius every half turn. Ions are extracted at a given radius on the outer radius of the cyclotron.
- A synchrotron aims at accelerating ions in closed orbits like a cyclotron, with the difference that it works with a fixed trajectory radius: along the acceleration, the magnetic field of the coils constituting the synchrotron is increased to compensate the increase of velocity of the accelerated ions and to maintain the same rigidity.

Here we give three examples of state-of-the-art Radioactive Ion Beam (RIB) facilities which are based on different acceleration schemes for the primary beam: FRIB in the US accelerated primary beams with a linear accelerator up to 200 MeV/nucleon with a foreseen upgrade to 400 MeV/nucleon, the RIBF in Japan accelerates heavy ions up to 345 MeV/nucleon with a set of coupled cyclotrons (see Fig. 3.8) and GSI accelerate fully stripped heavy ions at relativistic energies (up to 2 GeV/nucleon)

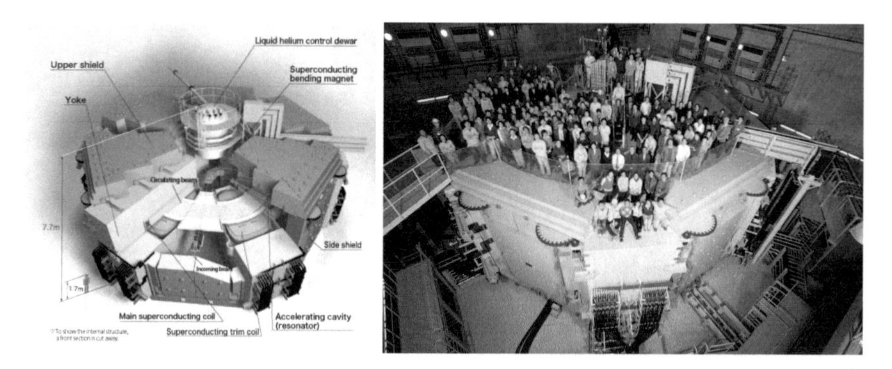

Fig. 3.8 (Left) Cut view of the SRC superconducting cyclotron of RIKEN. (Right) Picture of the SRC team standing on the cyclotron. Pictures taken from the website of the RIKEN Nishina Center (https://www.nishina.riken.jp)

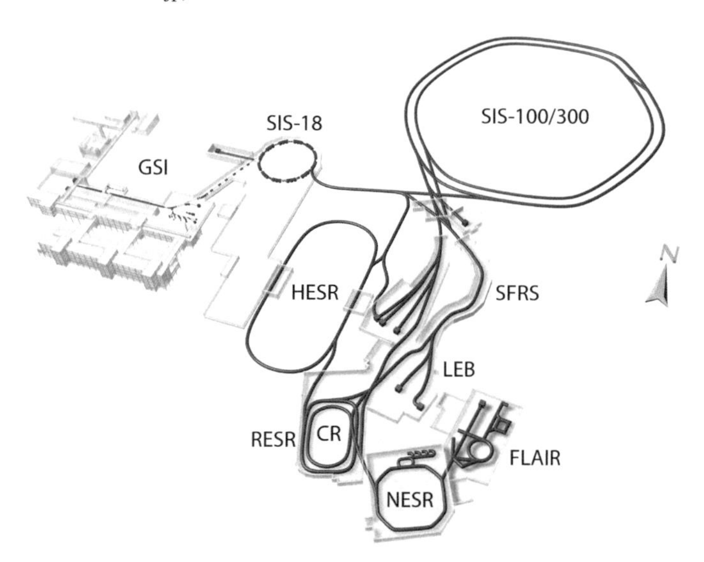

Fig. 3.9 Overview of the FAIR facility. It is composed of a new accelerator stage (red) coupled to the existing GSI facility (blue). At FAIR heavy ions are first accelerated with a LINAC and then fully stripped and sent to the S18. At the future FAIR facility, not fully stripped ions will be accelerated by the SIS18, then fully stripped and eventually accelerated by the SIS100. At the first stage of its development, the last acceleration stage will be limited to 100 Tm. The facility is meant to reach 300 Tm (SIS300) on a longer term. Figure taken from the website of FAIR (https://fair-center.de)

with a synchrotron. In the future, the last stage of the FAIR acceleration scheme will include the SIS100 synchrotron which will enable to accelerate beams up to 100 Tm (see Fig. 3.9). The long-term future of FAIR foresees an additional acceleration stage to 300 Tm with the SIS300 ring.

As an illustration of beam production, we detail here the example of a $^{8}He^{2+}$ ISOL beam at the SPIRAL-GANIL facility, France. Radioactive nuclei produced by the

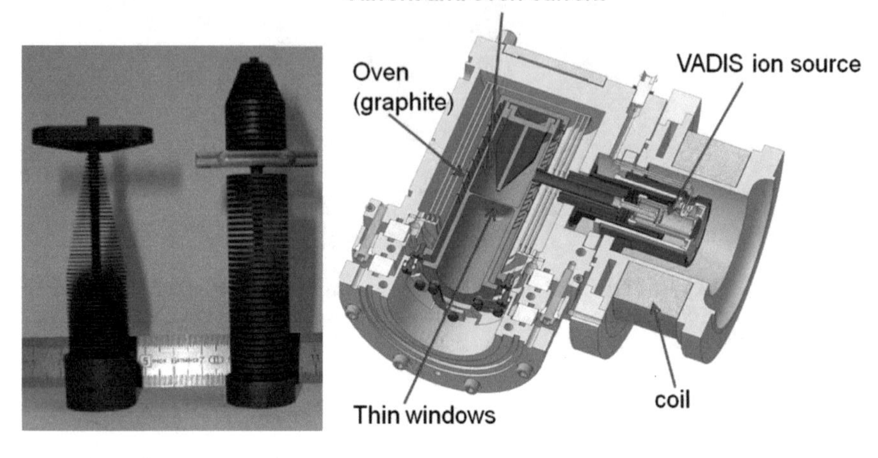

Fig. 3.10 (Left) SPIRAL graphite targets. The target geometry is defined for a shorter diffusion path from the production position to the vacuum of the source chamber. It is optimized on the primary beam and energy to be used, as well as the isotopes to be extracted. The picture shows two targets optimized for the production of neutron-rich He isotopes (left) and Ne isotopes (right). (Right) View of the FEBIAD production ion source where the production target is installed. Reprinted from [9] with permission from Elsevier

ISOL method are stopped in the production target. After extraction from the target, they are ionized and injected after separation into an accelerator to bring them to the desired energy.

A ^{13}C primary beam is produced in an ECR source and accelerated through a set of cyclotrons up to 95 MeV/nucleon after a series of stripping stages. The primary beam is sent onto a carbon target shown in Fig. 3.10. Carbon targets have excellent release properties and a high sublimation temperature. The design of the target guarantees the production of noble gases with reasonable yields and can be used with high power primary beams. The target temperature should be as high and as uniform as possible, in order to minimise the delay time between production and release. The temperature profile within the target is related to the properties of the Bragg peak. A specific geometry divided into two parts was developed for the ^{6,8}He production due to the long range of He in carbon. The ^{13}C primary beam only heats the first part (production target), while the second one stops the fragmentation products, while being heated by an electric current through the axis. The target chamber is to be seen as a first production stage to be followed by an ion source. The downstream part of the target chamber is therefore based on techniques for ion sources. In the case of the SPIRAL target, the radioactive isotopes are transferred via a transfer line to an ECR source for ionization. The transfer tube from the target chamber to the ECR source is heated to minimize the adsorption of atoms on the surface of the tube.

3.4 Direct Reactions

As a first attempt to classify reactions, one can define two kinds of reactions when two nuclei collide: *compound-nucleus* and *direct* reactions. In the first case, the two colliding nuclei form an intermediate system which is most often highly excited and lives a sufficient time so that the excitation energy is shared by all nucleons. This compound nucleus does not keep memory of the initial state and the decay channels of the compound do not depend on the structure of the colliding nuclei. On the other hand, direct reactions are short time reactions (about 10^{-22} s, the typical time for the projectile to pass by the target) which occur at the surface of the target nucleus. In these reactions, the projectile may gain, lose or exchange few nucleons with the target. Due to the short time of the reaction and the few steps involved in the process, one can extend the definition of a direct reaction as a reaction that disturbs few degrees of freedom of the initial wave function, i.e. the populated final state keeps memory of the initial wave function. In other words, direct reaction cross sections to individual states can be used to probe the overlap between the initial and final states.

Direct reactions have been used to study the nucleus and its structure. These reactions can be divided into elastic reactions that leave the nucleus in its initial quantum state after the reaction, inelastic scattering where the initial nucleus is excited to bound and unbound states, and nucleon stripping (pickup) reactions where the final nucleus is obtained from the removal (addition) of one or few nucleons from the initial nucleus.

In most general terms, any scattering problem can be described by a transition amplitude $T_{\beta\alpha}$ from an initial state α to a final state β, as illustrated in Fig. 3.11. In the case of elastic scattering, α and β both represent the same intrinsic state. For the general case, the final state wave function can be described as

$$\Psi_\beta(\mathbf{r}_\beta) \sim e^{i\mathbf{k}_\alpha.\mathbf{r}_\alpha}\delta_{\alpha\beta} + \sum_{\beta'} f_{\beta'\alpha}(\mathbf{k}_{\beta'}, \mathbf{k}_\alpha)\delta_{\beta\beta'}\frac{1}{r_{\beta'}}e^{ik_{\beta'}r_{\beta'}}, \tag{3.13}$$

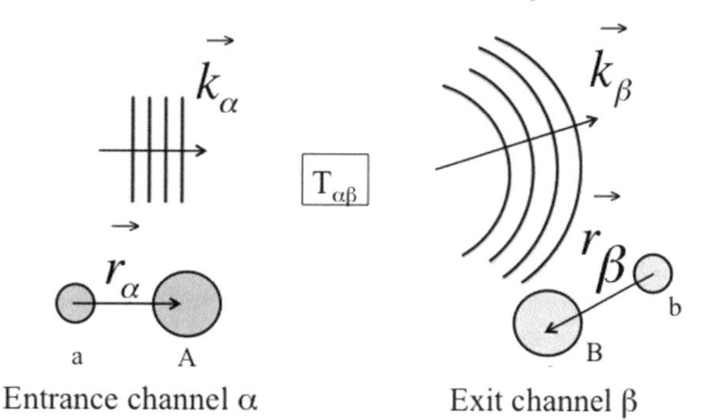

Fig. 3.11 Illustration of a scattering process starting from an entrance channel α and leading to an exit channel β. In a general form, the structure information is contained in the transition amplitude $T_{\beta\alpha}$

where $f_{\beta\alpha}(\mathbf{k}_\beta, \mathbf{k}_\alpha)$ are called the scattering functions from a given initial state α to a given final state β and contain all the nuclear structure information, $\mathbf{r}_{\alpha,\beta}$ ($\mathbf{k}_{\alpha,\beta}$) being the relative coordinates (momenta) in the entrance (α) and exit (β) channels. The form of (3.13) is imposed to respect proper asymptotic conditions. The transition amplitude $T_{\beta\alpha}$ is related to the scattering amplitude by

$$T_{\beta\alpha} = -\frac{2\pi\hbar^2}{\mu_\beta} f_{\beta\alpha}, \tag{3.14}$$

where μ_β is the reduced mass of the system in the exit channel

$$\mu_\beta = \frac{m_b m_B}{m_b + m_B} \tag{3.15}$$

with $m_{b,B}$ the mass of the two nuclei b and B in the exit channel. In the operator formalism, the so called post form of the T matrix elements is given by

$$T_{\beta\alpha} = \langle \Phi_\beta | V | \Psi_\alpha \rangle, \tag{3.16}$$

where Φ_β is the plane wave of the relative movement in the exit channel β, Ψ_α is the full solution of the Schrödinger equation driven by the interaction potential V between the target and projectile nuclei in the entrance channel α. $T_{\beta\alpha}$ contains all the structure and dynamics information for the reaction from channel α to channel β.

The differential cross section for a given reaction channel is given by

$$\frac{d\sigma_{\beta\alpha}}{d\Omega} = \frac{\mu_\alpha \mu_\beta}{(2\pi\hbar^2)^2} \left(\frac{k_\beta}{k_\alpha}\right) |T_{\beta\alpha}(\mathbf{k}_\beta, \mathbf{k}_\alpha)|^2, \tag{3.17}$$

where μ_α and μ_β are the reduced masses in the entrance and exit channel, respectively, as defined in relation (3.15). In one way or another, every direct reaction analysis aims at comparing the experimental cross section to such a calculation. If the reaction mechanism is thought to be sufficiently controlled and benchmarked, one can use such a comparison to extract nuclear structure information about the initial and final wave functions.

Let us first consider the simplest reaction: elastic scattering. Due to the wave nature of nuclear scattering, it has a clear analogy with light diffusion. When light from a point source passes through a small circular aperture, it does not produce a bright dot as an image, but rather a diffuse image, known as Airy's disc for a circular aperture, surrounded by regular, much fainter bright zones. When the light source can be approximated by a plane wave and the detection plane is located far away from the object, the phenomenon is called *Fraunhofer diffraction*, or far-field diffraction. In the case of a square aperture (or an opaque square), the diffraction minima are located at angles θ in both the horizontal and vertical directions such that

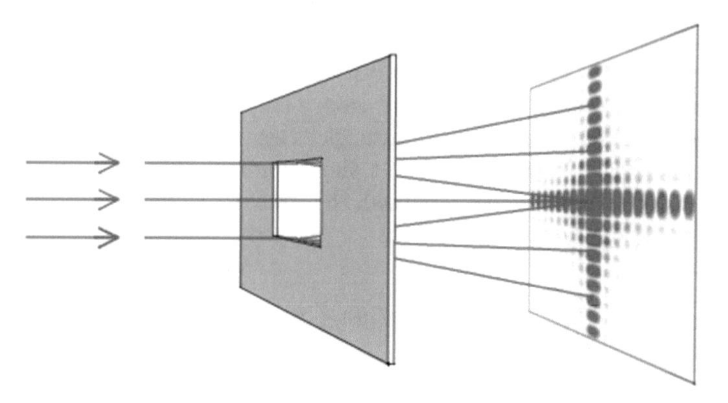

Fig. 3.12 Illustration of a diffraction pattern from a light beam going through a rectangular aperture. The diffraction pattern reflects the size and shape of the aperture

$$\sin(\theta) = \frac{m\lambda}{d}, \tag{3.18}$$

where m is the number of minima starting from 1, λ is the wavelength of the considered light source and d is the dimension of the aperture. The diffraction pattern is sensitive to the size and shape (a square in this example) of the diffractive object. Similarly, the pattern of angular distributions of an elastically scattered particle off a nucleus is sensitive to the size of the nucleus (Fig. 3.12).

3.4.1 *Elastic Scattering*

Electron scattering can be used to learn about charge distribution in nuclei. Electrons indeed constitute optimal probes for the study of atomic nuclei. Their point-like nature, and the fact that the electromagnetic interaction is weak (implying low rescattering rates) and well understood (QED) make the reaction mechanism well under control. Their sensitivity to nuclear structure depends on their incident energy. The electron-probe resolution can be deduced from the De Broglie relation between the electron's momentum p and its wavelength λ:

$$\lambda = \frac{\hbar}{p}. \tag{3.19}$$

An electron of 500 MeV/c will then have a "resolution" of about 1 fm, perfect for nuclear structure, while an electron of 5 GeV/c with wavelength of 0.1 fm will probe the details of the nucleon.

In the case of a point charge target, with no internal constituents nor spatial extension, the electron scattering cross section can be calculated exactly and is known as the Mott cross section σ_{Mott}. In the case of a (unrealistic) spin-less electron in the non relativistic limit, the cross section can be reduced to the so-called Rutherford cross section. The scattering of electrons from a nucleus differs from the Mott cross section due to the spatial extension of the nucleus leading to diffraction and damping. Assuming the elastic scattering acts in a single virtual photon exchange (which is shown to be an excellent approximation), the structure of the nucleus impacts the cross section though a form factor F via

$$\frac{\mathrm{d}\sigma}{\mathrm{d}\Omega} = \sigma_{\text{Mott}} |F(q)|^2, \tag{3.20}$$

where $q = |\mathbf{k}_f - \mathbf{k}_i|$ is the momentum of the exchanged virtual photon, $\mathbf{k}_{i(f)}$ the momentum of the incoming (outgoing) electron. For a given scattering angle θ, q can be calculated from the incident (exit) electron energy E (E'), $q^2 = 4EE' \sin^2(\theta/2)$, neglecting the electron mass relative to the kinetic energy. The form factor is then given by the Fourier transform of the charge density of the target nucleus

$$F(q) = \langle \Phi_{k_f} | V | \Phi_{k_i} \rangle = \int e^{i \frac{\mathbf{q}.\mathbf{r}}{\hbar}} \rho(\mathbf{r}') \mathrm{d}\mathbf{r}', \tag{3.21}$$

showing that the electron elastic scattering, if measured for a large range of momentum transfer q, can provide a full knowledge of the proton distribution. Cross sections at low momentum transfer give information about the outer part of the distribution, high momentum transfer gives details about the full distribution, including the interior of the nucleus. From cross sections at low transferred momenta (of the order of $1\,\text{fm}^{-1}$) the charge radius can be deduced. The state-of-the-art of electron scattering is illustrated in Fig. 3.13 where the elastic scattering of electrons off ^{208}Pb is shown. Charge densities are typically obtained by fitting to the scattering data a calculated cross section based on a parameterized charge form factor, as in (3.20), which is built over an ansatz for the charge distribution such as a two-parameter Fermi distribution.

Although electrons have been shown to be the most precise tool to examine nuclear structure, it is extremely difficult to perform collisions of electrons with unstable nuclei. The proper energy of electrons in the center of mass should be of the order of several hundreds of MeV. The very first facility, called SCRIT (Self-Confining Radioactive Ion Target), dedicated to collisions of unstable ions and electrons has been commissioned very recently in RIKEN, Japan. The concept of SCRIT is new [12, 13]: it is not a collider properly speaking since the unstable ions are trapped and at rest in the electron ring. The first "real" collider for electrons with exotic nuclei is expected to be built at FAIR, in Europe. The facility, called ELISe, aims at luminosities up to $10^{28}\,\text{cm}^{-2}\,\text{s}^{-1}$ (much less than what was achieved with fixed stable-nuclei targets) with a very large detection efficiency for the Lorentz focusing of the heavy ions and an in-beam magnetic spectrometer for high-resolution spectroscopy from electron detection [14].

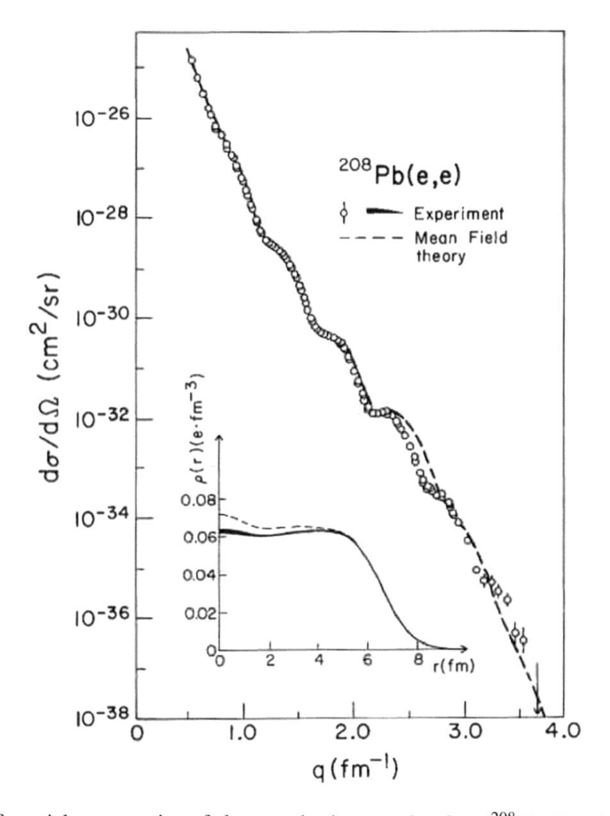

Fig. 3.13 Differential cross section of electron elastic scattering from ^{208}Pb. The electron incident energy was set to 502 MeV. The position of minima and the measured amplitude allow to reconstruct the charge distribution of the target nucleus as illustrated in the inset. The first minimum gives the charge radius, the second minimum leads to the diffusiveness of the charge distribution while higher momenta give further details about the inner part of the nuclear charge density distribution. A precise determination of the charge distribution from the surface down to the interior of the nucleus requires the measurement of the elastic scattering cross section over several orders of magnitude and can therefore only be accessed in stable nuclei. Data from [10], figure reprinted from [11] with permission from Annual Reviews

An alternative solution is to study unstable nuclei via elastic scattering of nuclei in inverse kinematics. In the specific case of proton elastic scattering, in analogy with the identity (3.16), the transition amplitude $T_{\beta\alpha}$ is given in its prior form by

$$T_{\beta\alpha} = \langle \phi_\beta | V | \Psi_\alpha \rangle, \tag{3.22}$$

where the exit channel β is composed of the same nuclei A and a. If one makes the further approximation that the solution of the Schrödinger equation is not distorted by the optical potential, i.e. that the relative motion reduces to the solution of a homogeneous equation, i.e. $\Psi_\alpha = e^{i\mathbf{k}_\alpha \cdot \mathbf{r}_\alpha} \Phi_\alpha$, one gets

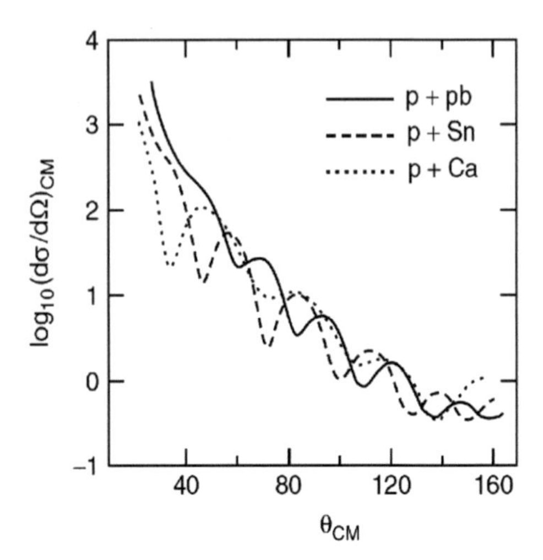

Fig. 3.14 Calculated proton elastic scattering cross sections from Ca, Sn and Pb isotopes. The first minimum position of the angular distribution is correlated to the nuclear radius of the target. Reprinted from [19] with permission from Wiley-VCH Verlag

$$T_{\beta\alpha} = \langle \phi_\beta | V | \phi_\alpha \rangle = \int e^{i\frac{(\mathbf{k}_\beta - \mathbf{k}_\alpha).\mathbf{r}_\alpha}{\hbar}} V(\mathbf{r}_\alpha) \mathrm{d}\mathbf{r}_\alpha. \tag{3.23}$$

The transition amplitudes for the elastic scattering is the exact Fourier transform of the optical potential. Since the nuclear interaction is short-ranged, one expects that the cross section reflects the size and shape of the nuclear matter density, in a similar way that the electron scattering is sensitive to the charge spatial distribution.

Similarly to the diffraction of light in a aperture, the angular distribution obtained from proton elastic scattering is also sensitive to the nuclear size. This can be seen in Fig. 3.14 where such distributions for Ca, Sn and Pb isotopes are plotted. The heavier the nucleus (i.e. the larger radius it has), the smaller is the angle between two diffraction minima. This correlation can be understood in classical physics considering that elastic scattering is a process confined to the surface because at the interior the potential is highly absorptive, as would be the scattering of light by a black disk. In that analogy, the angular distance between two diffractive minima is given by

$$\Delta\theta = \frac{\hbar}{pR}, \tag{3.24}$$

where R is the radius of the target nucleus and p is the momentum of the incoming proton.

Elastic scattering is a very effective tool to extract the matter radius of a nucleus. When the statistics of the measurement is good enough (achieved in experiments of few days with a beam intensity of about 10^4 pps), it is shown that uncertainties down to 0.1 fm can be reached, mostly due to the uncertainties in the interaction potential and in the reaction mechanism itself. To limit the effect of the latter couplings, intermediate energies above 100 MeV/nucleon are believed to be most reliable to extract nuclear radii from elastic scattering. Such elastic scattering data, when combined to measurements of the charge radius of the nucleus, can be used to determine the neutron skin thickness in neutron-rich nuclei, a key information for understanding nuclear structure.

In inverse kinematics, a proper angular resolution can be obtained by measuring the recoil proton elastically scattered from the target. In many cases, polyethylene targets (CH_4) are used. Although complex to develop and operate in experimental conditions, thin pure hydrogen targets have been used [16, 17] and others are under development [18]. Polarized-target experiments, sensitive to the spin orientation, can also be performed for further sensitivity to the spin-dependent terms of the optical potential. In the case of inelastic, transfer or quasi-free scattering experiments, polarized-target or polarized-beam experiments are also used to assign the total momentum of populated final states, while cross sections from non polarized experiments depend only on the transferred angular momentum.

3.4.2 Transfer

The transfer of one nucleon *from* a nucleus (stripping) or *to* a nucleus (pickup) is a powerful tool to determine the nature of the populated states. More quantitatively, the final states after pickup and stripping from the same nucleus and the corresponding population cross sections give access, in principle, to single-particle energies, i.e. energies of orbitals with no correlation effect. Effective single-particle energies are the backbone of many nuclear models and their comparison to experimentally-deduced energies is important. Two-nucleon transfer has been extensively used to study two-nucleon correlations inside the nucleus. Transfer cross sections are often analyzed through the DWBA and coupled-channel formalisms.

A large part of our understanding of nuclear structure is based on transfer reactions. Transfer is a key tool and a work horse for the main low-energy facilities worldwide such as GANIL, ISOLDE in Europe, TRIUMF and ANL in North America. A large part of the physics programs at future facilities is based on transfer reactions: HIE-ISOLDE at CERN, SPIRAL2 (France), SPES (Italy), Re3-12 at FRIB (USA) or the OEDO project at the RIBF, RIKEN (Japan). Transfer reaction experiments with radioactive beams are mostly based on the detection of recoil light charged particles. Several recent or ongoing developments of detectors aim at improving the energy or angular resolution as well as the detection efficiency, compared to the previous generation of detectors. To mention few of them: the ACTAR Time Projection Chamber [20] is a follow up of the MAYA detector [21] first developed at GANIL and the GRIT Si-based telescope array [22] compact enough to fit inside a photon-detection array such as PARIS [23] or AGATA [24] partly issued from the MUST2-array developments [25] (see Fig. 3.15). These instrumentations were primarily developed for transfer and inelastic scattering experiments. These new devices are complemented by the development of the very thin pure hydrogen target CHyMENE [18]. Recently, the project of a storage ring at ISOLDE has been accepted [26]. In the US, a complementary approach based on the use of magnetic field is beginning to be explored. The

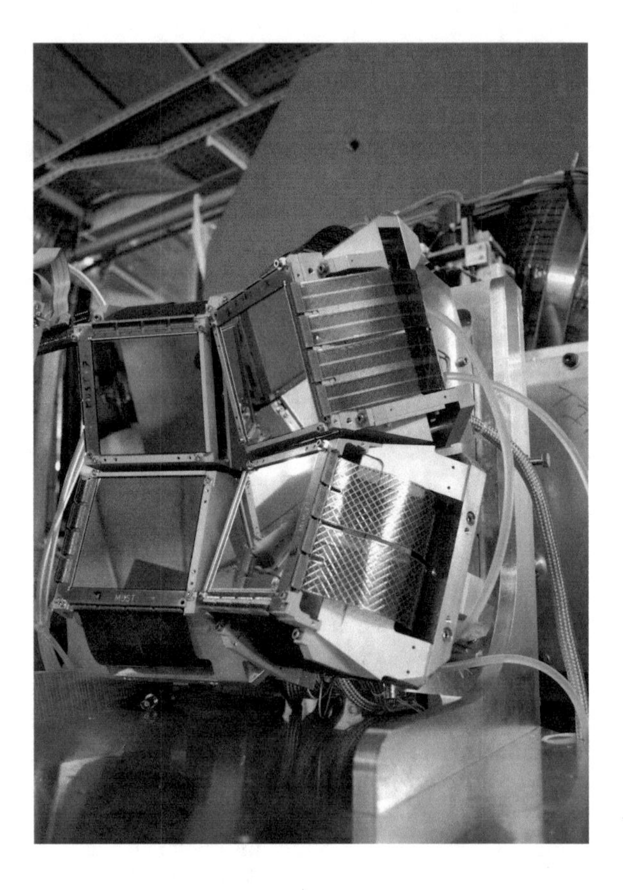

Fig. 3.15 Picture of four MUST2 telescopes arranged in a compact forward angle geometry for an experiment at GANIL. Courtesy F. Flavigny

leading projects focusing on transfer and inelastic scattering studies are AT-TPC [27], a TPC located inside a solenoid, and HELIOS [28], a novel setup using Si detectors inside a solenoid that allows an as so far unreached energy resolution when very thin targets are used.

3.4.3 Single-Particle Energies and Spectroscopic Factors

The independent particle model of nuclear structure supposes nucleons lying on single-particle energy orbitals with no correlation among them. The shell model description of nuclei is based on single-particle configurations on top of which nucleon correlations are built. Mean field models also lead to the definition of single-particle orbitals from which beyond-the-mean-field long-range correlations can be built. These single-particle (uncorrelated) energies are not observables since real nuclei are correlated systems by nature. They nevertheless can be obtained from data following the definition by Baranger [29]:

$$e_p = \frac{\sum_k S_k^{p+}(E_k - E_0) + S_k^{p-}(E_0 - E_k)}{\sum_k S_k^{p+} + S_k^{p-}}, \tag{3.25}$$

where S_k^{p+} (S_k^{p-}) are the spectroscopic factors (square of spectroscopic amplitudes) for the population of a final state k following the annihilation (creation) of a nucleon with quantum numbers $p = \{n\ell\}$, defined as

$$S_k^{p+} = |\langle\Psi_0^A|a_p|\Psi_k^{A+1}\rangle|^2, \; S_k^{p-} = |\langle\Psi_0^A|a_p^\dagger|\Psi_k^{A-1}\rangle|^2, \tag{3.26}$$

where $|\Psi_k^{A(\pm 1)\rangle}$ is the wave function of the nucleus with $A(\pm 1)$ nucleons in the state k. In the case of uncorrelated (unrealistic) nuclei, spectroscopic factors are either 1 or 0. Note that in some notations, the stripping spectroscopic factor can be normalized to $(2j + 1)$, j being the total angular momentum of the orbital from which the nucleon is removed. For example, in ^{18}O, in mean field, the spectroscopic factor for the outermost neutron would be 2, not 1, and the single-particle energies coincide with single neutron excitations (either particle or hole states) from nucleon stripping or pickup reactions. In physical nuclei, one nucleon stripping or pickup from a given state leads to the population of several states in the residual nucleus: the spectroscopic strength of a given orbital is spread over several final states, as illustrated in Fig. 3.16.

If one accesses experimentally all E_k and S_k^p, the single-particle energies would then be accessible from (3.25). Today, direct reactions are used to extract spectroscopic factors from cross sections to individual final states. In case the theoretical description of the reaction mechanism has been sufficiently benchmarked and is under control, one can aim at testing our understanding of the nuclear structure from such measurements. Note that recently, the non observability of single particle energies and their dependence over unitary transformations of the nuclear Hamiltonian have been discussed quite extensively. The basic concept of spectroscopic factors is also not an observable and is model dependent, i.e. should be discussed with care. This notion of "non observability" is fundamental and not a detail one can oversee. We report the reader to [31, 32] for more details.

3.4.4 *A Brief Introduction to Reaction Theory*

In the traditional treatment of transfer reactions, the target or projectile are most often described as two-body systems composed of a core and the nucleon to be transferred. In a nucleon transfer reaction, the optical potential V between the projectile and the target contains two parts: the potential that will distort the waves when the nuclei will approach each other that can be approximated by an optical potential $U(\mathbf{R})$ and the potential $\Delta U(\mathbf{R}, \mathbf{r})$ that will be responsible for the nucleon transfer itself. The potential can be written as:

$$V = U(\mathbf{R}) + \Delta U(\mathbf{R}, \mathbf{r}) \tag{3.27}$$

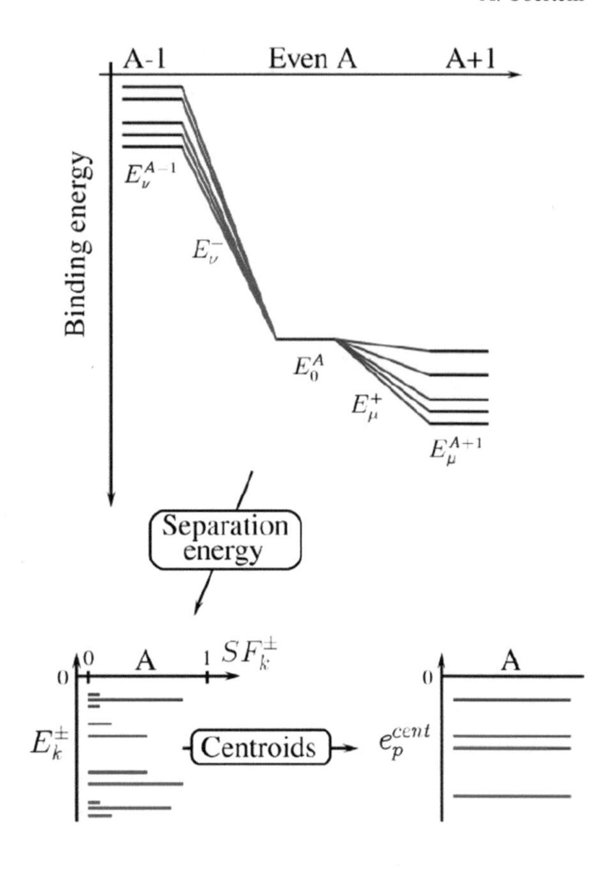

Fig. 3.16 (Top) Total binding energies and associated one-nucleon stripping and pickup energies from the ground state of the initial nucleus. (Bottom) Single-particle energies (right) are obtained from the centroid of the measured spectral function (left) following the relation (3.25). Reprinted from [30] with permission from the American Physical Society

In the DWBA approach, the transfer process is seen as a one step "perturbation" of the distorted trajectory. The homogeneous and inhomogeneous equations are

$$(T + U - E)\chi = 0, (T + U - E)\phi = \Delta U\phi. \tag{3.28}$$

The transition matrix element for the transfer is written as

$$T_{\beta\alpha} = \langle \phi_\beta \Phi_\beta | \Delta U | \phi_\alpha \Phi_\alpha \rangle, \tag{3.29}$$

where, following the previous notations, $\Phi_{\alpha,\beta}$ are the intrinsic wave functions of the entrance and exit channels α, β.

To be explicit, we now treat the specific case of the one neutron pickup reaction from a target nucleus A by an incident proton through the reaction (p, d) to populate a state i in the residual nucleus B. The transfer potential ΔU can be approximated by the potential between the proton and the neutron to be picked up with $\mathbf{r} = \mathbf{r}_p - \mathbf{r}_n$ being the relative coordinates between the proton and the picked-up neutron.

In the Born approximation, when the incoming and outgoing waves are treated as unperturbed by the potential (i.e. assuming $U = 0$ in (3.28), the initial and final wave functions are

$$|\Psi_\alpha\rangle = e^{i\mathbf{k}_p.\mathbf{r}_p}\Phi_A|\Psi_\beta\rangle = e^{i\mathbf{k}_d.\mathbf{r}_d}\Phi_B\Phi_d. \tag{3.30}$$

In the case of a pure single particle neutron state with quantum numbers $n\ell j$, the wave function of the final state can be written as $\Phi_A = \Phi_B\phi_{n\ell j}(\mathbf{r}_n)$ and the transition matrix element is then given by

$$T^{n\ell j}_{\beta\alpha} = \int e^{-i\mathbf{k}_d.\mathbf{r}_d}\Phi^*_d(\mathbf{r})\Phi^*_B\Delta U(\mathbf{r})e^{i\mathbf{k}_p.\mathbf{r}_p}\phi_{n\ell j}\Phi_B\mathbf{dr}_B\mathbf{dr}_n\mathbf{dr}_p. \tag{3.31}$$

Considering the relations

$$\mathbf{r}_d = \frac{1}{2}(\mathbf{r}_n - \mathbf{r}_p) \text{ and } \mathbf{r}_p = \frac{A}{A+1}\mathbf{r}_n - \mathbf{r}, \tag{3.32}$$

one gets

$$\mathbf{k}_p.\mathbf{r}_p - \mathbf{k}_d.\mathbf{r}_d = -\left(\mathbf{k}_d - \frac{A}{A+1}\mathbf{k}_p\right).\mathbf{r}_n - \left(\mathbf{k}_p - \frac{1}{2}\mathbf{k}_d\right).\mathbf{r} = -\mathbf{q}.\mathbf{r}_n - \mathbf{K}.\mathbf{r}, \tag{3.33}$$

where $\mathbf{q}$ is the momentum carried by the picked-up neutron and $\mathbf{K} = \mathbf{k}_p - \mathbf{k}_d/2$. The transition matrix element of (3.31) can then be formulated as

$$T^{n\ell j}_{\beta\alpha} = \int e^{-i\mathbf{K}.\mathbf{r}}\Phi^*_d(\mathbf{r})\Delta U(\mathbf{r})\mathbf{dr} \times \int_R^\infty e^{-i\mathbf{q}.\mathbf{r}_n}\phi_{n\ell j}\mathbf{dr}_n, \tag{3.34}$$

which can be intuitively interpreted as the product of two Fourier transforms: the first one of the deuteron wave function and the potential and the second one of the picked-up neutron wave function. Under the Born approximation, it appears explicitly that the transition matrix element is a product of a reaction term and a structure term. Note that the integration limit of the second integral in (3.34) has been set *by hand* to the nuclear radius R since the Born approximation does not take into account any absorption of the projectile during the reaction ($U = 0$). In the more realistic DWBA approximation, the absorption is taken care by the imaginary part of the optical potential U, neglected in the previous calculation. By assuming the initial intrinsic wave function Φ_A as a single-particle state with quantum numbers $n\ell j$ weighted by a certain spectroscopic amplitude $\sqrt{S^{n\ell j-}}$

$$|\Phi_A\rangle = \sum_{n\ell j}\sqrt{S^{n\ell j+}}|\phi_{n\ell j}\Phi_B\rangle, \tag{3.35}$$

the transition matrix element for the one neutron pick-up is given by

$$T_{\beta\alpha}^{n\ell j} = \sqrt{S^{n\ell j+}} \int \chi_d^{(-)*}(\mathbf{k}_d, \mathbf{r}_d)\Phi_d^*(\mathbf{r})\langle\Phi_B|\Delta U(\mathbf{r})|\Phi_B\rangle\phi_{n\ell j}\chi^{(+)}(\mathbf{k}_p, \mathbf{r}_n)\mathrm{d}\mathbf{r}_p\mathrm{d}\mathbf{r}_d, \tag{3.36}$$

where the implicit integrals run over the integral degrees of freedom $\mathbf{r}_B$ of nucleus B. In this equation, we note that structure and reaction information are factorized in the DWBA approach.

The population of the final state β from α may occur via the pickup from several neutron orbitals. The pickup cross section will then be a sum over all neutron orbitals $i = (n\ell j)$, following

$$\sigma_{\beta\alpha} = \sum_i S^{n\ell j}\sigma_{\beta\alpha}^{n\ell j}, \tag{3.37}$$

where $\sigma_{\beta\alpha}^{n\ell j}$ is the single-particle cross section calculated with the transition matrix element of (3.36), assuming a single-particle neutron state (with $S^{n\ell j} = 1$). The cross section from α to β is then decomposed in structure information ($S^{n\ell j}$) and terms that are calculated in the reaction formalism ($\sigma^{n\ell j}$).

As seen previously for the elastic scattering reactions, several approximations can be used to describe transfer. Some of them are introduced below:

- The *Distorted Wave Born Approximation* (DWBA) described above is the simplest (but useful) model. It assumes a direct one-step process that is weak compared to the elastic channel and may be treated by perturbation theory.
- The *adiabatic model* is a modification of the DWBA formalism for (d, p) and (p, d) that takes deuteron breakup effects into account in an approximate way.
- The *Coupled Channel Born Approximation* (CCBA) is used when the one-step approximation breaks down. Strong inelastic excitations are treated in coupled channels, while transfer is still treated with DWBA.
- The *Coupled Reaction Channel* (CRC) does not assume one-step or weak transfer process. All processes are taken into account on equal footing. It is the most complete treatment of reactions.

3.4.5 Two-Body Kinematics and the Missing-Mass Technique

One interest of nucleon transfer reactions lies in its two-body kinematics. From energy and momentum conservation, all information about the residue (4) can be obtained by measuring the momentum p_3 and using the known mass m_3 of the second outgoing particle (3). Indeed, the excitation energy $E_4^\star$ of particle (4) (its spectroscopy) is given by:

$$E_4^\star = \sqrt{E_4^2 - p_4^2c^2} - m_{4,gs}, \tag{3.38}$$

where the total energy E_4 and the momentum vector p_4 of particle (4) can be obtained from

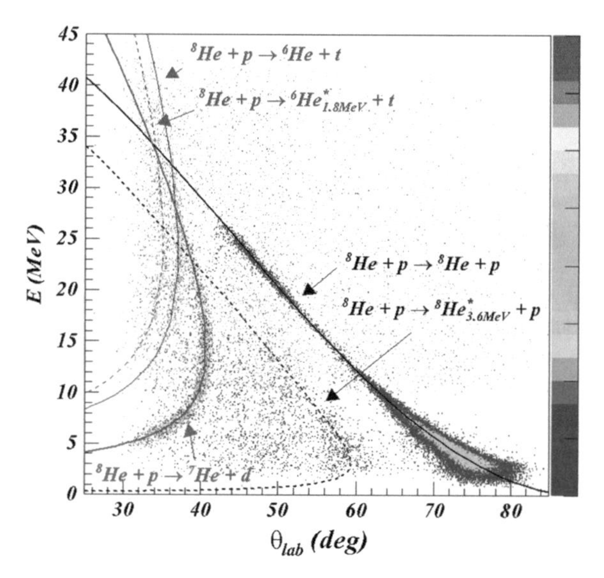

Fig. 3.17 Kinematics of ^{8}He+p at 15.7 AMeV measured at the SPIRAL facility, GANIL. Figure reprinted from [33] with permission from Springer Nature

$$E_4 = \sqrt{T_1 + m_1 + m_2 - (T_3 + m_3)} \tag{3.39}$$

$$p_4^2 = p_1^2 + p_3^2 - 2p_1p_3\cos(\theta_3), \tag{3.40}$$

where p_1 and $m_{1,2}$ are known and $T_{1,3}$, the kinetic energy of particles (1) and (3) can be obtained from their momentum following the well known identity $T = \sqrt{p(p + 2m)}$, all quantities given in MeV (with the convention $c = 1$ for the velocity of light).

This technique of reconstructing the mass, and therefore the excitation energy spectrum, of one of the two ejectiles without measuring it is denominated as *missing mass*. Since the nucleus of interest is not measured, all bound and unbound rates can be measured in the same way. The missing mass technique is one of the few possible (and most used) techniques for the spectroscopy of unbound states. The kinematics of a (p, d) reaction in inverse kinematics is shown for the case of ^{8}He+p at 15.7 MeV/nucleon measured at GANIL and illustrated in Fig. 3.17. It is clear from the data that other reaction channels take place at the same time with rather strong cross section, namely inelastic scattering and (p, t) transfer reaction. From the analysis of the data, it has been shown that the DWBA approximation is not sufficient for a proper treatment of the reaction. Only the CRC theory, by taking explicitly into account all important channels, is able to reproduce correctly the data.

3.4.6 Angular Distributions and Momentum Matching

A nucleon transfer reaction is a quantum process during which a quantized angular momentum is transferred. The differential cross section to a given final state, as a function of the scattering angle, shows an oscillatory pattern whose structure (position of maxima and minima) depends on the transferred angular momentum. In the following, we give a classical correspondence between the position of the first angular maximum after a transfer reaction and the transferred angular momentum.

For simplicity, we assume a reaction in direct kinematics where a nucleon is removed from a target whose mass is considered infinite compared to the projectile and light residue, so that all the transferred angular momentum is taken away by the light particle.

In a quantum system for which the angular momentum is a good quantum number, the operator $\hat{L}^2$ commutes with the Hamiltonian and its application to the nucleus wave function gives

$$L^2|\Phi\rangle = \ell(\ell+1)\hbar^2|\Phi\rangle. \tag{3.41}$$

In the classical limit of a transfer reaction occurring at the surface of the nucleus, the transferred momentum in the reaction is given by $L = p_\perp R$, where $p_\perp$ is the transverse momentum of the scattered particle and R is the distance between the two at grazing, leading the classical limit approximation

$$p_\perp R = \sqrt{\ell(\ell+1)}\hbar. \tag{3.42}$$

In the infinite mass target approximation, the perpendicular momentum of the light recoil can be approximated to $p_\perp = p\sin(\theta_\circ)$. Under these approximations, the scattering angle (the first maximum of the cross section) is given by

$$\theta_\circ = \arcsin\left(\frac{\sqrt{\ell(\ell+1)}\hbar}{pR}\right) \tag{3.43}$$

On can verify the validity of this classical estimate of the first maximum of the differential cross section on various examples. Consider the example ^{52}Cr(d,p)^{53}Cr at 10 MeV/nucleon, as shown in Fig. 3.18. 10 MeV/nucleon for a deuteron is equivalent to a momentum p=193 MeV/c. If one assumes the standard parameterization of the nuclear radius R $= 1.2\,\text{fm} \times A^{1/3}$, one gets R $= 4.5$ fm for ^{52}Cr. Following (3.43), for $\Delta\ell = 0, 1, 2$, one calculates $\theta_\circ = 0^\circ, 19^\circ, 34^\circ$, respectively. The $\Delta\ell = 1$ is in good agreement with the data. The values estimated for $\Delta\ell = 0, 1, 2$ are in good agreement with the microscopic calculations shown in Fig. 3.18.

Intuitively, the beam velocity will impact strongly the population of states depending on the transferred angular momentum they imply. The population with $\Delta\ell = 0$ will require lower incident energy for higher $\Delta\ell$ values for a given Q-value of the reaction. The optimum population of given states requires *momentum matching* imposed by (3.42).

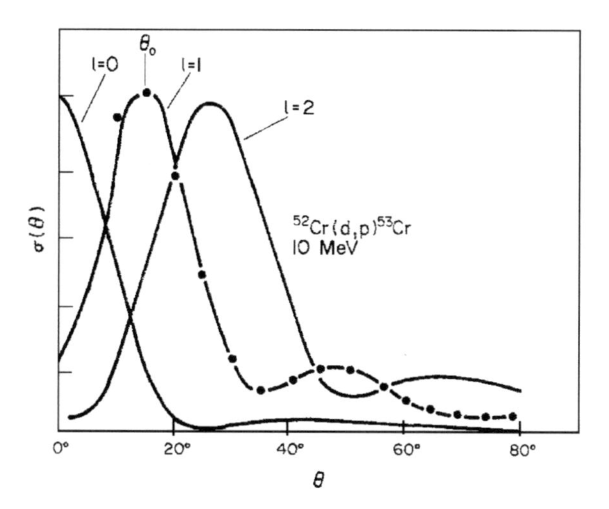

Fig. 3.18 Measured angular distribution for ^{52}Cr(d,p)^{53}Cr$_{gs}$ at 10 MeV/nucleon. The data points are compared to calculations assuming different transfer angular momenta from 0 to 2. The location of the first maximum clearly shows a $\Delta\ell = 1$ transfer. Figure from [34] with permission from Elsevier

3.4.7 One-Nucleon Transfer: The Perfect Tool for Shell Structure Studies

The shell structure evolution with the number of protons and/or neutrons is one of the most exciting questions in nuclear structure. Indeed, since the very first study of unstable nuclei, it has been understood that magic numbers of the shell model as they were known for stable nuclei are not universal across the nuclear landscape and evolve. The first historical example is the disappearance of the N = 20 shell closure in the region of ^{32}Mg [35–38].

The most cited mechanisms that play a role in shell evolution are:

- a large diffusiveness in neutron-rich nuclei that modifies the mean-field potential and the one-body spin-orbit [39],
- spin-isospin components of the two-nucleon interaction (spin-isospin term of the central part of the Hamiltonian or the tensor term of the interaction) [40–45],
- three-body forces [46, 47].

A review of observations for shell evolution interpreted in the framework of the shell model is contained in [48]. Data are necessary to support or infer theoretical pictures. The main difficulty lies in the fact that nuclei are correlated systems and their spectroscopy does not reflect necessarily the underlying single-particle nuclear structure. Correlations should then be assessed experimentally through the determination of spectroscopic strength. As examples of transfer-reaction studies for nuclear structure, one could quote the study of the reduction of the spin-orbit splitting with isospin in Sn isotopes [49], the disappearance of the N = 20 shell gap and the $N = 16$ "new" magic number [50, 51] while consistent results have been found from nucleon stripping at the NSCL [52], and the single-particle shell structure of the doubly-magic ^{132}Sn [53].

Two-body correlations are a key aspect of nuclear structure beyond the single-particle shell model picture. They can be probed via direct two-nucleon transfer reactions [54], sensitive to both momentum and spatial correlations in the nucleus. The two neutron transfer has been widely used in both direct and inverse kinematics and is known to be sensitive to configuration transitions, such as shape transitions [55], pairing in light [56–58] or heavy nuclei [59] and intruder configurations.

3.5 Knockout Reactions

The production of radioactive beams by fragmentation can reach very asymmetric nuclei (and short lived) with no chemical selectivity. By doing so, a large part of the nuclear chart can be accessed. The current and forthcoming new generation RIB facilities focus on energies from 100 MeV/nucleon to several GeV/nucleon. Nucleon knockout reactions are a key tool for spectroscopy at these facilities with the strong advantage to allow the use of thick targets to compensate the low beam intensities of the most exotic species. On the other hand knockout reactions can mostly populate hole states (i.e. nucleon removal only) while low energy transfer reaction have the advantage to populate both hole and particle (pickup reactions) states.

3.5.1 Quasi-free Scattering

Similarly to elastic scattering, high-energy electrons are in principle the ideal probe to eject protons from nuclei: the electromagnetic interaction is well known, the nuclear problem is reduced to the structure of the target nucleus (with some final state interaction corrections to consider as well) and high resolution missing mass spectroscopy can be performed by detecting the scattered electron. The formalism, initially developed by Jacob and Maris [60, 61], is detailed in dedicated review articles (see for example [62–64]). Beautiful results on nuclear structure have been obtained from $(e, e'p)$ experiments with stable nuclei [65–67]. Despite all these advantages, electron induced nucleon quasi-free scattering $(e, e'p)$ is restricted to proton knockout and can only be performed for stable nuclei today. For these reasons, proton induced quasi-free scattering (p, pN) (with $N = p, n$) at energies beyond 300 MeV/nucleon is considered today the cleanest stripping probe to investigate the structure for both stable and unstable nuclei.

The best energy for quasi-free scattering minimizes initial-state and final state interactions, i.e. the distortions of the incoming and outgoing protons in the nuclear potential of the target nucleus and the residue, respectively. This is achieved when both the incident energy and the kinetic energy of the scattered protons are close to the minimum of the nucleon-nucleon reaction cross section. Due to the equal mass of the recoil proton and ejected nucleon, when both are detected at about 45° in the laboratory frame, they approximately share the energy of the incoming proton. For example, a quasi-free scattering performed at 400 MeV/nucleon will lead to two nucleons scattered at about 200 MeV when both detected at 45°, due to momentum conservation. The minimum value of the nucleon-nucleon reaction cross sections as a function of the energy (laboratory frame) is located around 200–300 MeV/nucleon. Accordingly, the best energies for quasi-free scattering are between 300 and 800 MeV/nucleon (for both direct or inverse kinematics).

The measured momenta of the two scattered protons in a quasi-free scattering experiment give access to the intrinsic momentum $\mathbf{q} = (q_\parallel, q_\perp)$ of the removed proton inside the projectile. By use of the missing mass technique, quasi-free scattering allows to measure the separation energy E_s of the populated bound and unbound states of the residue:

$$\mathbf{q}_\perp = \mathbf{p}_{1\perp} + \mathbf{p}_{2\perp} \tag{3.44}$$

$$q_\parallel = \frac{(p_{1\parallel} + p_{2\parallel}) - \gamma\beta(M_A - M_{A-1})}{\gamma} \tag{3.45}$$

$$E_s = T_0 - \gamma(T_1 + T_2) - 2(\gamma - 1)m_p + \beta\gamma(p_{1\parallel} + p_{2\parallel}) - \frac{q^2}{2M_{A-1}}, \tag{3.46}$$

where β, γ are the velocity and Lorentz factor of the projectile in the laboratory frame, T_0 and $T_{1,2}$ the kinetic energies of the projectile and the two protons, respectively. The example of ^{12}C(p,2p)^{11}B at 392 MeV incident energy in direct kinematics measured at RCNP, Japan, represents the state of the art of what can be done in direct kinematics with stable nuclei. Very recently, quasi-free scattering ^{12}C(p,2p) was measured for the first time exclusively in complete and inverse kinematics at an energy of about 400 MeV/nucleon at GSI [68].

Quasi-free scattering is usually described theoretically based on one main assumption, the *impulse approximation* (IA), assuming that the knockout process occurs in one single step and involves only the two interacting nucleons: the incoming proton and the to-be-knocked-out nucleon. The transition matrix elements for quasi-free scattering can be expressed in impulse approximation as

$$T_{(p,pN)} = \langle \chi_{p_2}\chi_{p_1}|V_{pN}|\chi_{p_0}\Psi_{jlm}\rangle, \tag{3.47}$$

where V_{pN} is the optical potential between the incoming proton and the nucleon N to be removed, $\chi_{p_{1,2}}$ are the distorted waves for the outgoing and scattered-off protons in the potential field of the residual nucleus $A - 1$, χ_{p_0} is the incoming proton wave function distorted by the presence of the target nucleus A, and Ψ_{jlm} is the bound state wave function of the knocked out nucleon. The above relation for $T_{(p,pN)}$ is for a single-particle state. In case of correlations, spectroscopic amplitudes from a structure model have to be considered.

In the extreme limit of the plane wave approximation, the above transition matrix element reduces to the Fourier transform of the single-particle wave function $\Phi_{n\ell j}$, showing that there is a very close connection between the quasi-free scattering cross section and the wave function of the removed nucleon, similar to $(e, e'p)$ reactions.

Recent developments have been made to predict (p, pN) cross sections under the eikonal and PWIA approximations [69], DWIA [70] or within a so-called Faddeev multipole scattering reaction framework [71, 72].

Two major facilities can today produce radioactive ion beams at adequate energy for quasi-free scattering at intermediate energies: the RIBF of RIKEN in Japan and, at energies up to 1 GeV/nucleon, GSI. In the coming decade, the FAIR facility at GSI should provide the best experimental conditions to investigate nuclear structure from quasi-free scattering in inverse kinematics with the R3B setup [75].

3.5.2 Heavy-Ion Induced Inclusive Nucleon Removal Reactions

As a tool for the spectroscopy of unstable nuclei, heavy-ion induced nucleon-removal reactions have been so far more popular than quasi-free scattering. Many important results and theoretical developments have been recently accomplished with this method [76–78]. In the literature, these reactions are often referred to as "knockout" although the exact reaction process for the nucleon removal may be more complicated than a single-step nucleon removal resulting from a hard-core nucleon-nucleon collision. In most cases, the experimental technique uses in-beam gamma spectroscopy at the secondary target to tag the final state of the populated projectile-like residue. In these measurements, the final state of the target, most often ^{9}Be or ^{12}C, is not known. Note that there can be ambiguity in the use of the term "inclusive" in such experiments. Regarding the detection of reaction products, they are always inclusive since the target-like recoil and its products are not measured. On the other hand, when one focuses on the final state of the projectile-like residue, if the cross sections to individual final bound states of the residue are measured (by gamma tagging) one speaks of exclusive cross sections. The cross section to all bound states of the residue are then referred to as "inclusive". From here, the term "inclusive" will be used in the latter definition.

One asset of inverse kinematics fast nucleon-removal reactions is given by the momentum distribution of the residue. By measuring the total, parallel or perpendicular momentum distributions one has a direct access to the intrinsic momentum distribution of the removed nucleon. This quantity is highly connected to the angular moment ℓ of the removed nucleon and therefore it gives important information on the shell structure of the nucleus under study [79–83]. The technique has been pioneered and largely developed in inverse kinematics from the 90s at the NSCL at energies around or below 100 MeV/nucleon [82, 84, 85], GANIL [86] and at GSI at much higher incident energies [87]. An example of such data with gamma tagging is illustrated on Fig. 3.19.

Intermediate energy nucleon-removal cross sections are often interpreted by comparison to shell-model spectroscopic factors and single-particle cross sections calculated under the eikonal approximation. The first nuclear knockout model in the

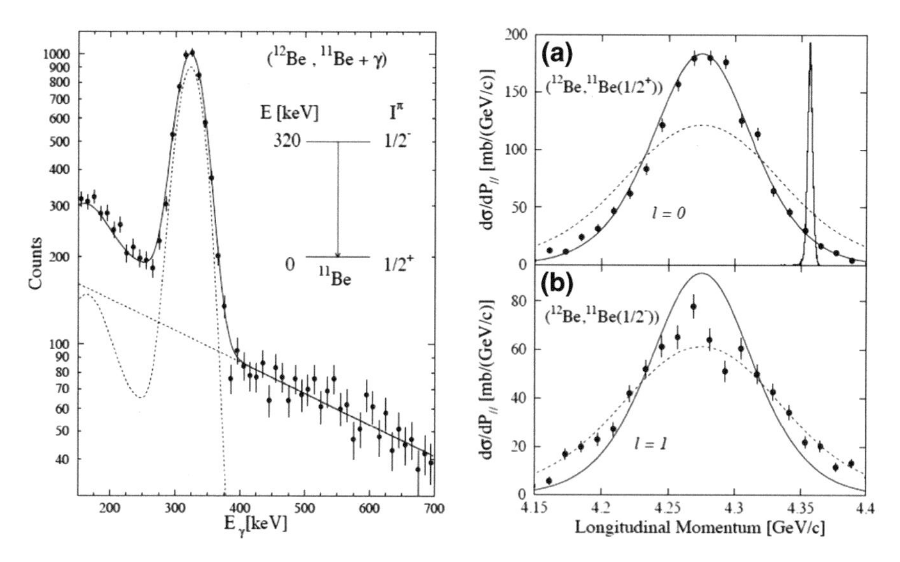

Fig. 3.19 (Left) Gamma spectrum from the de-excitation of ^{11}Be populate by one-neutron knockout ^{12}Be(^{9}Be,X)^{11}Be at 78 MeV/nucleon. (Right) Parallel momentum distributions in coincidence with the population of the first excited state of ^{11}Be and its ground state. The intrinsic angular momentum of the removed nucleon can be easily disentangled in both cases. Figures from [85] with permission from the American Physical Society

eikonal approximation was presented in [88]. The S-matrix formalism is commonly used to derive these cross sections (see for example [89], while other approaches have been developed [90, 91]).

The one-nucleon removal cross section is calculated using the eikonal formalism [92, 93] and consists of a stripping and a diffractive part

$$\sigma = \sigma_{\text{str}} + \sigma_{\text{diff}}. \tag{3.48}$$

The projectile wave function is defined as a core wave function $|\Phi_C\rangle$ complemented with the wave function of the removed nucleon $|\Phi_N\rangle$. Calculations are based on two main quantities: the S matrices for the core (S_C) and the removed nucleon (S_N), as described above.

The stripping part of the cross section corresponds to reactions where the target is excited, while the diffractive part corresponds to breakup events during which both the target and the core remain in their ground state. The diffractive part of the nucleon removal cross section can be seen as events where both the core and the nucleon are elastically scattered off the target and for which the overlap of the final core and nucleon with the incoming projectile ground state does not equal 1.

In an impact-parameter representation, the stripping part of the cross section is calculated as

$$\sigma_{\text{str}} = 2\pi \int bdb \int d\mathbf{r}|\Phi_N(\mathbf{r})|^2 |S_C(\mathbf{b_C})|^2 (1 - |S_N(\mathbf{b_N})|^2), \tag{3.49}$$

where $\mathbf{b_C}$, $\mathbf{b_N}$ are the impact parameters of the core and of the removed nucleon respectively. The S matrix for the core-target system S_C is defined from the target and core densities as well as from the in-medium NN cross section that depends on the incident energy. Equation (3.49) is intuitive: it is the sum over all possible impact parameters to preserve the core (probability $|S_C|^2$ by definition of the S matrix) and to strip off the valence nucleon (probability $1 - |S_N|^2$).

The diffractive part can be expressed as

$$\sigma_{\text{diff}} = 2\pi \int bdb \langle\Phi_0||S_C S_N|^2|\Phi_0\rangle - |\langle\Phi_0|S_C S_N|\Phi_0\rangle|^2. \tag{3.50}$$

The single-particle cross section $\sigma = \sigma_{\text{str}} + \sigma_{\text{diff}}$ combined with spectroscopic factors calculated from a nuclear-structure formalism are compared to experimental cross sections.

Although beyond the scope of the present overview, it is important to highlight that two-nucleon removal as a probe for two-body correlations inside the nucleus have been investigated both experimentally [94] and theoretically [95].

3.5.3 Recent Achievements with Unstable Nuclei

3.5.3.1 Breakdown of the $N = 28$ Shell Closure in ^{42}Si

The $N = 28$ shell closure, produced by the one-body spin-orbit term of the nuclear potential and separating the orbitals of same parity $f_{7/2}$ and $p_{3/2}$, disappears progressively below the doubly magic ^{48}Ca nucleus in ^{46}Ar [96] and ^{44}S [97, 98], after the removal of only two and four protons, respectively. This rapid disappearance of rigidity of the $N = 28$ isotones has been ascribed to the reduction of the neutron shell gap $N = 28$ combined with that of the proton subshell gap $Z = 16$, leading to increased probability of quadrupole excitations within the *fp* and *sd* shells for neutrons and protons, respectively. For the ^{44}S nucleus, its small 2^+_1 energy, large B(E2$\uparrow$) value [97], and the presence of a 0^+_2 isomer at low excitation energy [98] point to a mixed ground state configuration of spherical and deformed shapes. The energies of excited states in the ^{42}Si and the ^{41,43}P nuclei have been measured through in-beam-ray spectroscopy [99]. The low energy of the 2^+_1 state in ^{42}Si, 770(19) keV, together with the level schemes of ^{41,43}P, provide evidence for the disappearance of the $N = 28$ shell closure around ^{42}Si. It is ascribed to the combined action of proton-neutron tensor forces leading to a global compression of the proton and neutron single-particle orbitals, added to the quadrupole symmetry between the occupied and valence states which favors excitations across the $Z = 14$ and $N = 28$ shell gaps. This low 2^+_1 energy was recently confirmed by an experiment performed at the

RIBF [100]. ^{42}Si was populated from multi-nucleon removal at intermediate energy. The in-beam gamma spectroscopy of Si and S isotopes was performed by use of the DALI2 spectrometer [101]. The systematics of the ratio $R_{4/2}$ of the 4_1^+ and 2_1^+ energies in silicon isotopes from $N = 24$ to $N = 28$ shows a rapid development of deformation, further demonstrating that ^{42}Si is well deformed.

3.5.3.2 Merging of the $N = 20$ and $N = 28$ Deformation Regions

Up to very recently, the $N = 20$ island of inversion, interpreted as a deformed two-particle two-hole configuration favored for the ground state configuration [102–104], and the study of the above-mentioned collapse of the $N = 28$ shell closure were treated as independent questions. Thanks to the achievement of intense beams of light radioactive ions, the contour of the region has been established. A recent measurement performed at the RIBF from in-beam gamma spectroscopy of Mg isotopes produced from nucleon removal indicates that the island of inversion does not show any decrease of collectivity for Mg isotopes at $N > 24$ and merges with the $N = 28$ deformation region towards ^{40}Mg [105].

3.5.3.3 Search for New Shell Effects at the RIBF

A systematic search for 2^+ states in very neutron-rich even-even nuclei from Ar to Zr isotopes produced from $(p, 2p)$ reactions has been initiated at the RIBF with the so-called SEASTAR program [117] (Shell Evolution And Search for Two-plus energies in even-even nuclei At the RIBF, see Fig. 3.20). The use of the unique high primary beam intensities at the RIBF and the coupling of MINOS [118, 119] and the DALI2 array are the core of SEASTAR. The program primarily aims at the spectroscopy of ^{78}Ni and ^{110}Zr, as well as nuclei in the vicinity of the closed subshell nuclei ^{48}S and ^{60}Ca. A further study of the onset of deformation at and beyond N = 40 is also targeted along isotopic chains from $Z = 22$ to 26. This physics program was started in 2014 and the first spectroscopy of ^{52}Ar [120], ^{66}Cr, ^{70,72}Fe [121], ^{78}Ni, ^{82,84}Zn [122], $^{90-94}$Se [123], ^{98,100}Kr [124], ^{110}Zr [125] were successfully performed (Fig. 3.21).

3.5.4 *Nucleon-Stripping from Unstable Nuclei: A Comparison of Transfer and Knockout*

The distribution of spectroscopic strength in nuclei can be extracted from direct-reaction cross section measurements, assuming a modeling of the reaction mechanism. Recently, a compilation of one-nucleon removal at intermediate energies from sd-shell exotic nuclei showed that the measured cross sections for knocking out

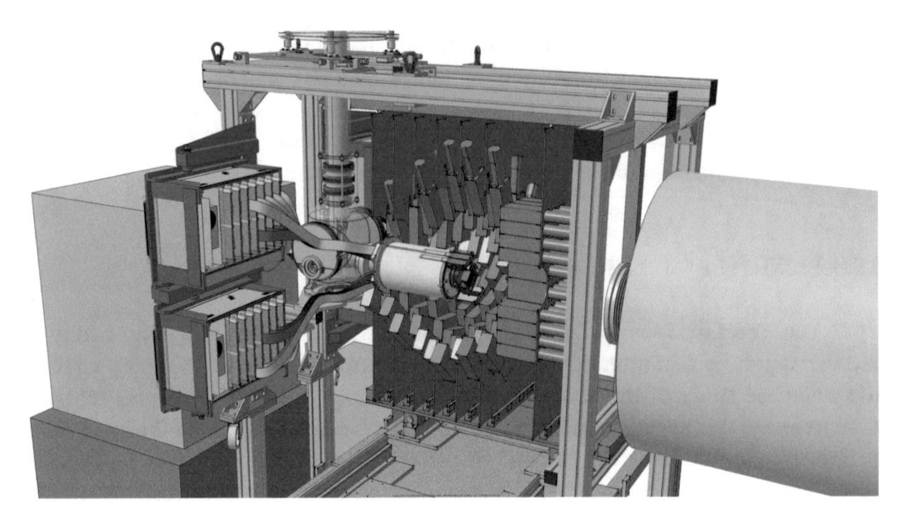

Fig. 3.20 Schematic view of the MINOS target and TPC vertex tracker in position inside the DALI2 spectrometer for in-beam gamma spectroscopy at RIBF, RIKEN

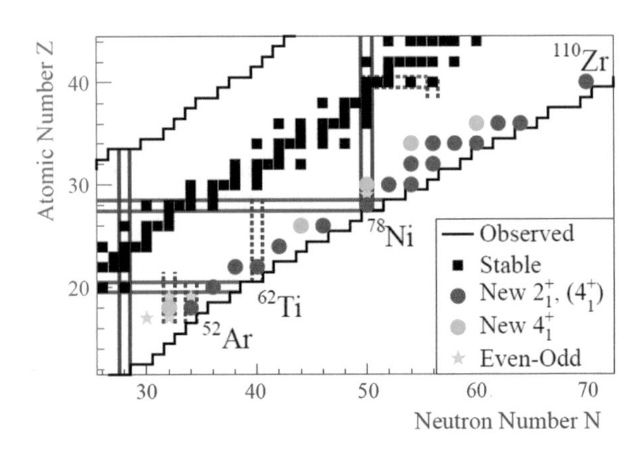

Fig. 3.21 Even-even nuclei (colored dots) to be populated via (*p*, 2*p*) knockout and studied via in-beam gamma spectroscopy at the RIBF of RIKEN with the SEASTAR setup [117, 119]

a valence nucleon in a very asymmetric nucleus (such as a neutron in ^{32}Ar, ^{28}Ar and ^{24}Si) are about four times smaller than predictions from state-of-the-art calculations [126]. On the other hand, at low energy, a study of the (*p*, *d*) neutron transfer on the proton-rich ^{34}Ar and on the neutron-rich ^{46}Ar provides experimental spectroscopic factors in agreement with large-basis shell model calculations to within 20% [127]. These findings which are in agreement with a previous systematic study of transfer reactions [128] are inconsistent with the trend deduced from the analysis of nucleon removal cross sections at about 100 MeV/nucleon.Very recently, a new systematic study over transfer reactions with stable nuclei did not evidence any dependence of the effect of short range or beyond model space correlations with the transferred angular momentum ℓ, mass of target nuclei or asymmetry ΔS [49]. Therefore, it is suggested that these two probes, transfer and knockout, lead sys-

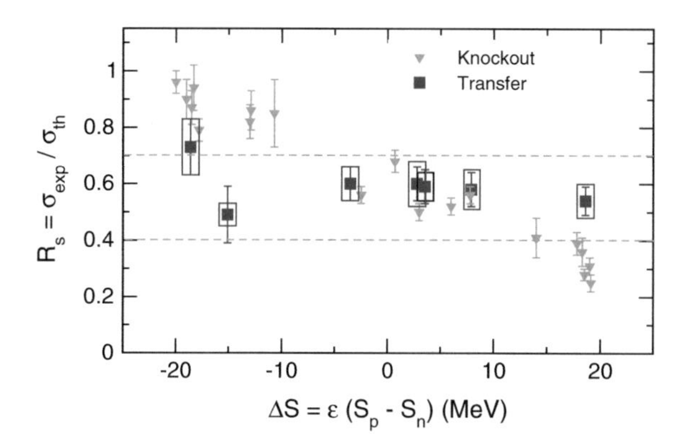

Fig. 3.22 Ratio of experimental to theoretical one-nucleon stripping cross sections. Experimental knockout data from [126, 129] are considered and their theoretical predictions are performed within the eikonal formalism under the sudden approximation. Transfer data from oxygen isotopes are taken from [131] and their theoretical predictions are performed within the Coupled Reaction Channel formalism and *traditional* (based on Woods-Saxon single-particle wave functions) form factors. Courtesy F. Flavigny

tematically to different spectroscopic factors when analyzed in the above mentioned frameworks, namely DWBA or CRC for transfer and under the sudden and eikonal approximations for knockout. The origin of this difference has to be understood.

A recent study of the nucleon removal from ^{14}O and ^{16}C ($\Delta S = |S_n - S_p|$ close to 20 MeV) at intermediate energies lower than 100 MeV/nucleon has shown that the applicability of the sudden approximation and the eikonal theory for nucleon removal depends on both the incident beam energy and the binding energy of the removed nucleon [129]. The applicability of the eikonal formalism to the previously deeply-bound nucleon removal was questioned. Indeed, one may also question the role of dissipation in deeply-bound nucleon removal [130].

The corresponding transfer stripping reactions (d, t) and $(d,^3\text{He})$ from the same ^{14}O nucleus at 18 MeV/nucleon performed at GANIL [131] and analyzed within the framework of coupled reaction channel formalism with a set of optical potentials, matter radii and spectroscopic factors did not show any systematic reduction for deeply-bound nucleon stripping, as illustrated in Fig. 3.22. This analysis is in agreement with the conclusions of [49, 127] but at variance with [126], showing that the mostly used reaction-mechanism models for transfer (DWBA, CRC) and heavy-ion induced knockout (sudden and eikonal approximation) do not lead to the same structure information in some cases. These discrepancies still need to be quantitatively understood. A more systematic study of deeply-bound nucleon removal reactions from weakly bound nuclei will definitely help in understanding the limits of current direct reaction models.

3.6 Outlook

Nuclear reactions are at the basis of the production of radioactive beams which, in turn, are essential to address the nuclear many-body problem and its most general features. Direct reactions in inverse kinematics are powerful tools to investigate nuclear structure in unstable nuclei. Elastic and inelastic scattering, transfer at low incident energy, quasi-free scattering and nucleon-removal reactions at intermediate energies offer unique capabilities to populate selectively nuclear states. The cross sections to populate final states and their dependence with momentum or scattering angle contain essential information to further our understanding of the nature of nuclear states. Direct reactions will continue to play a major role in nuclear physics in current and new-generation facilities.

The present lecture aimed at introducing nuclear reactions for non experts. The present notes should not be seen as an overview since many important facets of nuclear reactions were not covered (even not mentioned). For those who wish to go further, several books describe the formalism of direct processes. For a deeper study, one may refer to the following references [3, 132–135].

Acknowledgements The present lecture notes are based on a lecture given at La Rábida summer school in Huelva, Spain. The material of the notes is partly taken from published lecture notes [136] by the author from a lecture given at the 2015 Pisa Summer school entitled "Re-writing nuclear physics textbooks: 30 years of radioactive ion beam physics", organized by Angela Bonaccorso. I thank José-Enrique Ramos and Antonio Moro for their kind invitation and the perfect organisation of the La Rábida summer school in 2018.

References

1. E. Rutherford, Philos. Mag. **21**, 669–688 (1911)
2. G.F. Bertsch, D.J. Dean, W. Nazarewicz, SciDAC Rev. **6**, 42 (2007)
3. G.R. Satchler, *Direct Nuclear Reactions* (Oxford University Press, Oxford, 1983)
4. T. Aumann et al., Phys. Rev. Lett. **119**, 262501 (2017)
5. I. Tanihata et al., Phys. Rev. Lett. **55**, 2676 (1985)
6. H. Geissel, G. Munzinger, K. Riisager, Ann. Rev. Nucl. Part. **45**, 163 (1995)
7. D.C. Faircloth, Lecture Notes from CERN accelerator school. https://doi.org/10.5170/CERN-2013-001
8. R. Gobin et al., Rev. Sci. Instrum. **81**, 02B301 (2010)
9. O. Bajeat et al., Nucl. Instrum. Methods Phys. Res. B **317**, 411 (2013)
10. B. Frois et al., Phys. Rev. Lett. **38**, 152 (1976)
11. B. Frois, C.N. Papanicolas, Ann. Rev. Nucl. Part. Sci. **37**, 133 (1987)
12. M. Wakasugi, T. Suda, Y. Yano, Nucl. Instrum. Methods Phys. Res. **532**, 216 (2004)
13. T. Suda et al., Phys. Rev. Lett. **102**, 102501 (2009)
14. A.N. Antonov et al., Nucl. Instrum. Methods Phys. Res. A **637**, 60 (2011)
15. T. Suda, M. Wakasugi, Prog. Part. Nucl. Phys. **55**, 417 (2005)
16. A. Obertelli, T. Uesaka, Eur. Phys. J. A **47**, 105 (2011)
17. S. Ishimoto et al., Nucl. Instrum. Methods Res. A **480**, 304 (2002)
18. A. Gillibert et al., Eur. Phys. J. A **49**, 155 (2013)

19. C. Bertulani, *Encyclopedia of Physics* (Wiley-VCH, Berlin, 2009). ISBN-13: 978-3-527-40691-3
20. http://pro.ganil-spiral2.eu/spiral2/instrumentation/actar-tpc
21. C.E. Demonchy et al., Nucl. Instrum. Methods Res. Phys. A **573**, 145 (2007)
22. http://gaspard.in2p3.fr/index.html
23. http://paris.ifj.edu.pl
24. S. Akkoyun et al., Nucl. Instrum. Methods Res. Phys. A **688**, 26 (2012)
25. E.C. Pollacco et al., Nucl. Instrum. Methods Res. Phys. A **421**, 471 (1999)
26. M. Grieser et al., Eur. Phys. J. Spec. Top. **207**, 1 (2012)
27. http://www.nscl.msu.edu/exp/sr/attpc
28. A.H. Wuosmaa et al., Nucl. Instrum. Methods Res. Phys. A **580**, 1290 (2007)
29. M. Baranger, Nucl. Phys. A **149**, 225 (1970)
30. T. Duguet, G. Hagen, Phys. Rev. C **85**, 034330 (2012)
31. R.J. Furnstahl, H.-W. Hammer, Phys. Lett. B **531**, 203 (2002)
32. T. Duguet, H. Hergert, J.D. Holt, V. Soma, Phys. Rev. C **92**, 034313 (2015)
33. V. Lapoux, N. Alamanos, Eur. Phys. J. A **51**, 91 (2015)
34. L.D. Knutson, W. Haeberli, Prog. Part. Nucl. Phys. **3**, 127 (1980)
35. C. Thibault et al., Phys. Rev. C **12**, 644 (1975)
36. C. Détraz et al., Phys. Rev. C **19**, 164 (1979)
37. D. Guillemaud-Mueller et al., Nucl. Phys. A **426**, 37 (1984)
38. T. Motobayashi et al., Phys. Lett. B **346**, 9 (1995)
39. J. Dobaczewski, I. Hamamoto, W. Nazarewicz, J.A. Sheikh, Phys. Rev. Lett. **72**, 981 (1994)
40. T. Otsuka et al., Phys. Rev. Lett. **87**, 082502 (2001)
41. A.P. Zuker, Phys. Rev. Lett. **91**, 179201 (2003)
42. T. Otsuka, T. Suzuki, R. Fujimoto, H. Grawe, Y. Akaishi, Phys. Rev. Lett. **95**, 232502 (2005)
43. T. Otsuka, T. Matsuo, D. Abe, Phys. Rev. Lett. **97**, 162501 (2006)
44. T. Lesinski, M. Bender, K. Bennaceur, T. Duguet, J. Meyer, Phys. Rev. C **76**, 014312 (2007)
45. M. Bender et al., Phys. Rev. C **80**, 064302 (2009)
46. S. Bogner, R. Furnstahl, A. Schwenk, Prog. Part. Nucl. Phys. **65**, 94 (2010)
47. T. Otsuka et al., Phys. Rev. Lett. **105**, 032501 (2010)
48. O. Sorlin, M.-G. Porquet, Prog. Part. Nucl. Phys. **61**, 602 (2008)
49. J.P. Schiffer et al., Phys. Rev. Lett. **92**, 162501 (2004)
50. A. Obertelli et al., Phys. Lett. B **633**, 33 (2006)
51. S.M. Brown et al., Phys. Rev. C **85**, 011302(R) (2012)
52. J.R. Terry et al., Phys. Lett. B **640**, 86 (2006)
53. K.L. Jones et al., Nature **465**, 454 (2010)
54. G. Potel, A. Idini, F. Barranco, E. Vigezzi, R.A. Broglia, Rep. Prog. Phys. **76**, 106301 (2013)
55. P.D. Duval, D. Goutte, M. Vergnes, Phys. Lett. B **124**, 297 (1983)
56. I. Tanihata et al., Phys. Rev. Lett. **100**, 192502 (2008)
57. G. Potel, F. Barranco, E. Vigezzi, R.A. Broglia, Phys. Rev. Lett. **105**, 172502 (2010)
58. M. Assié, D. Lacroix, Phys. Rev. Lett. **102**, 202501 (2009)
59. G. Potel et al., Phys. Rev. Lett. **107**, 092501 (2011)
60. G. Jacob, ThAJ Maris, Rev. Mod. Phys. **38**, 121 (1966)
61. G. Jacob, Th.A.J. Maris, Rev. Mod. Phys. **45**, 6 (1973)
62. J. Kelly, Adv. Nucl. Phys. **23**, 75 (1996)
63. A. Dieperink et al., Annu. Rev. Nucl. Part. Sci. **40**, 239 (1990)
64. S. Boffi et al., Phys. Rep. **226**, 1 (1993)
65. J. Mougey et al., Nucl. Phys. A **262**, 461 (1976)
66. P. deWitt-Huberts, J. Phys. G **16**, 507 (1990)
67. L. Lapikas, Nucl. Phys. A **553**, 297 (1993)
68. V. Panin et al., Phys. Lett. B **753**, 204 (2015)
69. T. Aumann, C.A. Bertulani, J. Ryckebusch, Phys. Rev. C **88**, 064610 (2013)
70. K. Ogata, K. Yoshida, K. Minomo, Phys. Rev. C **92**, 034616 (2015)
71. R. Crespo, A. Deltuva, E. Cravo, Phys. Rev. C **90**, 044606 (2014)

72. R. Crespo, A. Deltuva, E. Cravo, Phys. Rev. C **93**, 054612 (2016)
73. M. Yosoi et al., Phys. Lett. B **551**, 255 (2003)
74. M. Yosoi et al., Nucl. Phys. A **738**, 451 (2004)
75. Technical Proposal, http://www-win.gsi.de/r3b/Documents/R3B-TP-Dec05.pdf
76. P.G. Hansen, J.A. Tostevin, Annu. Rev. Nucl. Part. Sci. **53**, 219 (2003)
77. A. Bonaccorso, Phys. Scr. T **152**, 014019 (2013)
78. A. Bonaccorso, JPS Conf. Proc. **6**, 010023 (2015)
79. R. Serber, Phys. Rev. **72**, 1008 (1947)
80. R. Serber, Ann. Rev. Nucl. Part. Sci. **44**, 1 (1994)
81. T. Kobayashi et al., Phys. Rev. Lett. **60**, 2599 (1988)
82. N. Orr et al., Phys. Rev. Lett. **69**, 2050 (1992)
83. C.A. Bertulani, P.G. Hansen, Phys. Rev. C **70**, 034609 (2004)
84. Aumann et al., Phys. Rev. Lett. **84**, 35 (2000)
85. A. Navin et al., Phys. Rev. Lett. **85**, 266 (2000)
86. E. Sauvan et al., Phys. Lett. B **491**, 1 (2000)
87. H. Simon et al., Phys. Rev. Lett. **83**, 496 (1999)
88. K. Hencken, G. Bertsch, H. Esbensen, Phys. Rev. C **54**, 3043 (1996)
89. J.A. Tostevin, J.S. Al Khalili, Phys. Rev. C **59**, R5(R) (1999)
90. A. Bonaccorso, D.M. Brink, Phys. Rev. C **44**, 1559 (1991)
91. A. Bonaccorso, G.F. Bertsch, Phys. Rev. C **63**, 044604 (2001)
92. J.A. Tostevin, J. Phys. G. Nucl. part. Phys. **25**, 735 (1999)
93. F. Barranco, E. Vigezzi, in *International School of Heavy-Ion Physics, 4th course*, ed. by R.A. Broglia, P.G. Hansen (World Scientific, Singapore, 1998)
94. K. Yoneda et al., Phys. Rev. C **74**, 021303(R) (2006)
95. E.C. Simpson et al., Phys. Rev. Lett. **102**, 132505 (2009)
96. L. Gaudefroy et al., Phys. Rev. Lett. **97**, 092501 (2006)
97. T. Glasmacher et al., Phys. Lett. B **395**, 163 (1997)
98. S. Grévy et al., Eur. Phys. J. A **25**, s1.111 (2005)
99. B. Bastin et al., Phys. Rev. Lett. **99**, 022503 (2007)
100. S. Takeuchi et al., Phys. Rev. Lett. **109**, 182501 (2012)
101. S. Takeuchi et al., Nucl. Instrum. Methods A **763**, 596 (2014)
102. B.H. Wildenthal, W. Chung, Phys. Rev. C **22**, 2260 (1980)
103. A. Poves, J. Retamosa, Phys. Lett. B **184**, 311 (1987)
104. E.K. Warburton, J.A. Becker, B.A. Brown, Phys. Rev. C **41**, 1147 (1990)
105. P. Doornenbal et al., Phys. Rev. Lett. **111**, 212502 (2013)
106. A. Burger et al., Phys. Lett. B **622**, 29 (2005)
107. R.V.F. Janssens et al., Phys. Lett. B **546**, 55–62 (2002)
108. D.-C. Dinca et al., Phys. Rev. C **71**, 041302(R) (2005)
109. A. Gade et al., Phys. Rev. C **74**, 021302(R) (2006)
110. F. Wienholtz et al., Nature **498**, 346–349 (2013)
111. M. Rosenbusch et al., Phys. Rev. Lett. **114**, 202501 (2015)
112. D. Steppenbeck et al., Nature **502**, 207 (2013)
113. G. Hagen et al., Phys. Rev. Lett. **109**, 032502 (2012)
114. J.D. Holt, T. Otsuka, A. Schwenk, T. Suzuki, J. Phys. G **39**, 085111 (2012)
115. V. Soma, C. Barbieri, T. Duguet, Phys. Rev. C **87**, 011303 (2013)
116. H. Hergert et al., Phys. Rev. Lett. **110**, 242501 (2013)
117. P. Doornenbal, A. Obertelli, Nucl. Phys. News **24**, 36 (2014)
118. A. Obertelli et al., Eur. Phys. J. A **50**, 8 (2014)
119. A. Obertelli, T. Uesaka, Nucl. Phys. News **25**, 17 (2015)
120. H. Liu et al. (2018), arXiv:1811.08451
121. C. Santamaria et al., Phys. Rev. Lett. **115**, 192501 (2015)
122. C.M. Shand et al., Phys. Lett. B **773**, 492–497 (2017)
123. S. Chen et al., Phys. Rev. C **95**, 041302 (2017)
124. F. Flavigny et al., Phys. Rev. Lett. **118**, 242501 (2017)

125. N. Paul et al., Phys. Rev. Lett. **118**, 032501 (2017)
126. A. Gade et al., Phys. Rev. C **77**, 044306 (2008)
127. J. Lee et al., Phys. Rev. Lett. **104**, 112701 (2010)
128. M.B. Tsang et al., Phys. Rev. Lett. **102**, 062501 (2009)
129. F. Flavigny et al., Phys. Rev. Lett. **108**, 252501 (2012)
130. C. Louchart et al., Phys. Rev. C **83**, 011601 (2011)
131. F. Flavigny et al., Phys. Rev. Lett. **110**, 122503 (2013)
132. D.F. Jackson, *Nuclear Reactions* (Methuen, London, 1970)
133. N.K. Glendenning, Direct Nuclear Reactions (World Scientific, Singapore, 2004)
134. C.A. Bertulani, P. Danielewicz, *Introduction to Nuclear Reactions*. Institute of Physics (2004)
135. I.J. Thompson, F. Nunes, *Nuclear Reactions for Astrophysics* (Cambridge University Press, Cambridge, 2009)
136. A. Obertelli, Eur. Phys. J. Plus **131**, 319 (2016)

Part II
Student's Seminars

Chapter 4
Measurement of the ^{244}Cm and ^{246}Cm Neutron-Induced Cross Sections at the n_TOF Facility

V. Alcayne, A. Kimura, E. Mendoza, D. Cano-Ott, O. Aberle, S. Amaducci, J. Andrzejewski, L. Audouin, V. Babiano-Suarez, M. Bacak, M. Barbagallo, V. Bécares, F. Bečvář, G. Bellia, E. Berthoumieux, J. Billowes, D. Bosnar, A. S. Brown, M. Busso, M. Caamaño, L. Caballero, M. Calviani, F. Calviño, A. Casanovas, F. Cerutti, Y. H. Chen, E. Chiaveri, N. Colonna, G. P. Cortés, M. A. Cortés-Giraldo, L. Cosentino, S. Cristallo, L. A. Damone, M. Diakaki, M. Dietz, C. Domingo-Pardo, R. Dressler, E. Dupont, I. Durán, Z. Eleme, B. Fernández-Domíngez, A. Ferrari, I. Ferro-Gon calves, P. Finocchiaro, V. Furman, A. Gawlik, S. Gilardoni, T. Glodariu, K. Göbel, E. González-Romero, C. Guerrero, F. Gunsing, S. Heinitz, J. Heyse, D. G. Jenkins, Y. Kadi, F. Käppeler, N. Kivel, M. Kokkoris, Y. Kopatch, M. Krtička, D. Kurtulgil, I. Ladarescu, C. Lederer-Woods, J. Lerendegui-Marco, S. Lo Meo, S.-J. Lonsdale, D. Macina, A. Manna, T. Martínez, A. Masi, C. Massimi, P. F. Mastinu, M. Mastromarco, F. Matteucci, E. Maugeri, A. Mazzone, A. Mengoni, V. Michalopoulou, P. M. Milazzo, F. Mingrone, A. Musumarra, A. Negret, R. Nolte, F. Ogállar, A. Oprea, N. Patronis, A. Pavlik, J. Perkowski, L. Piersanti, I. Porras, J. Praena, J. M. Quesada, D. Radeck, D. Ramos Doval, T. Rausher, R. Reifarth, D. Rochman, C. Rubbia, M. Sabaté-Gilarte, A. Saxena, P. Schillebeeckx, D. Schumann, A. G. Smith, N. Sosnin, A. Stamatopoulos, G. Tagliente, J. L. Tain, Z. Talip, A. E. Tarifeño-Saldivia, L. Tassan-Got, A. Tsinganis, J. Ulrich, S. Urlass, S. Valenta, G. Vannini, V. Variale, P. Vaz, A. Ventura, V. Vlachoudis, R. Vlastou, A. Wallner, P. J. Woods, T. J. Wright and P. Žugec

Abstract The neutron capture reactions of the ^{244}Cm and ^{246}Cm isotopes open the path for the formation of heavier Cm isotopes and of heavier elements such as Bk and Cf in a nuclear reactor. In addition, both isotopes belong to the minor actinides with a large contribution to the decay heat and to the neutron emission in irradiated

V. Alcayne (✉) · E. Mendoza · D. Cano-Ott · V. Bécares · E. González-Romero
T. Martínez
Centro de Investigaciones Energéticas Medioambientales y Tecnológicas (CIEMAT), Madrid, Spain
e-mail: victor.alcayne@ciemat.es

A. Kimura
Japan Atomic Energy Agency (JAEA), Tokai-mura, Japan

J.-E. García-Ramos et al. (eds.), *Basic Concepts in Nuclear Physics: Theory, Experiments and Applications*, Springer Proceedings in Physics 225,
https://doi.org/10.1007/978-3-030-22204-8_4

fuels proposed for the transmutation of nuclear waste and fast critical reactors. The available experimental data for both isotopes are very scarce. We measured the neutron capture cross section with isotopically enriched samples of ^{244}Cm and ^{246}Cm provided by JAEA. The measurement covers the range from 1 eV to 250 eV in the n_TOF Experimental Area 2 (EAR-2). In addition, a normalization measurement with the ^{244}Cm sample was performed at Experimental Area 1 (EAR-1) with the Total Absorption Calorimeter (TAC).

O. Aberle · M. Bacak · M. Barbagallo · M. Calviani · F. Cerutti · E. Chiaveri · A. Ferrari
S. Gilardoni · Y. Kadi · D. Macina · A. Masi · M. Mastromarco · F. Mingrone · C. Rubbia
M. Sabaté-Gilarte · L. Tassan-Got · A. Tsinganis · S. Urlass · V. Vlachoudis
European Organization for Nuclear Research (CERN), Meyrin, Switzerland

S. Amaducci · G. Bellia · L. Cosentino · P. Finocchiaro · A. Musumarra
INFN Laboratori Nazionali del Sud, Catania, Italy

J. Andrzejewski · A. Gawlik · J. Perkowski
University of Lodz, Lodz, Poland

L. Audouin · Y. H. Chen · D. Ramos Doval · L. Tassan-Got
IPN, CNRS-IN2P3, University of Paris-Sud, Université Paris-Saclay, 91406 Orsay Cedex, France

V. Babiano-Suarez · L. Caballero · C. Domingo-Pardo · I. Ladarescu · J. L. Tain
Instituto de Física Corpuscular, CSIC - Universidad de Valencia, Valencia, Spain

M. Bacak
Technische Universität Wien, Vienna, Austria

M. Barbagallo · N. Colonna · L. A. Damone · A. Mazzone · G. Tagliente · V. Variale
Istituto Nazionale di Fisica Nucleare, Bari, Italy

F. Bečvář · M. Krtička · S. Valenta
Charles University, Prague, Czech Republic

G. Bellia · A. Musumarra
Dipartimento di Fisica e Astronomia, Università di Catania, Catania, Italy

E. Berthoumieux · E. Dupont · F. Gunsing
CEA Saclay, Irfu, Universit é Paris-Saclay, Gif-sur-Yvette, France

J. Billowes · E. Chiaveri · A. G. Smith · N. Sosnin · T. J. Wright
University of Manchester, Manchester, UK

D. Bosnar · P. Žugec
University of Zagreb, Zagreb, Croatia

A. S. Brown · D. G. Jenkins
University of York, York, UK

M. Busso · S. Cristallo · L. Piersanti
Istituto Nazionale di Fisica Nazionale, Perugia, Italy

M. Busso
Dipartimento di Fisica e Geologia, Università di Perugia, Perugia, Italy

4.1 Experimental Set Up and Preliminary Results

Accurate neutron capture cross section data for minor actinides (MAs) are required to estimate the production and transmutation rates of MAs in LWR reactors with a high burnup, critical fast reactors like Gen-IV systems and other innovative reactor systems such as accelerator driven systems (ADS) [1]. The ^{244}Cm ($T_{1/2}$ = 18.1 years) and ^{246}Cm ($T_{1/2}$ = 4730 years) isotopes are among the most important MAs due to the difficulties in their transmutation and their contribution to the radiotoxicity of the irradiated nuclear fuels. The first and only data available until 2012 on the ^{244}Cm and ^{246}Cm neutron capture cross sections came from an experiment [2] which used

M. Caamaño · I. Durán · B. Fernández-Domíngez
University of Santiago de Compostela, Santiago, Spain

F. Calviño · A. Casanovas · G. P. Cortés · A. E. Tarifeño-Saldivia
Universitat Politècnica de Catalunya, Barcelona, Spain

M. A. Cortés-Giraldo · C. Guerrero · J. Lerendegui-Marco · J. M. Quesada
M. Sabaté-Gilarte
Instituto de Física Corpuscular, CSIC - Universidad de Valencia, Valencia, Spain

S. Cristallo · L. Piersanti
Istituto Nazionale di Astrofisica - Osservatorio Astronomico di Teramo, Teramo, Italy

L. A. Damone
Dipartimento di Fisica, Università degli Studi di Bari, Bari, Italy

M. Diakaki · M. Kokkoris · V. Michalopoulou · A. Stamatopoulos · L. Tassan-Got
R. Vlastou
National Technical University of Athens, Athens, Greece

M. Dietz · C. Lederer-Woods · S.-J. Lonsdale · P. J. Woods
School of Physics and Astronomy, University of Edinburgh, Edinburgh, UK

R. Dressler · S. Heinitz · N. Kivel · E. Maugeri · D. Rochman · D. Schumann
Z. Talip · J. Ulrich
Paul Scherrer Institut (PSI), Villigen, Switzerland

Z. Eleme · N. Patronis
University of Ioannina, Ioannina, Greece

I. Ferro-Gon Calves · P. Vaz
Instituto Superior Técnico, Lisbon, Portugal

V. Furman · Y. Kopatch
Joint Institute for Nuclear Research (JINR), Dubna, Russia

T. Glodariu · A. Negret · A. Oprea
Horia Hulubei National Institute of Physics and Nuclear Engineering (IFIN-HH), Bucharest, Romania

K. Göbel · D. Kurtulgil · R. Reifarth
Goethe University Frankfurt, Frankfurt, Germany

J. Heyse · P. Schillebeeckx
European Commission, Joint Research Centre, Geel, Retieseweg 111, 2440, Geel, Belgium

the neutrons produced in an under-ground nuclear explosion. Recently, a second measurement of both (n, γ) cross sections has been performed with a large coverage Ge array in the Accurate Neutron Nucleus Reaction Measurement Instrument ANNRI at J-PARC [3].

We have used the same samples as in J-PARC to measure both ^{244}Cm and ^{246}Cm neutron capture cross sections at n_TOF. Our results, obtained in a different facility and using a different detection system, will allow to validate the results obtained at J-PARC, thus reducing the systematic uncertainties. In addition, we expect to increase the neutron energy range of the J-PARC measurement from 100 eV up to at least 250 eV.

F. Käppeler
Karlsruhe Institute of Technology, Campus North, IKP, 76021 Karlsruhe, Germany

S. Lo Meo · A. Mengoni
Agenzia nazionale per le nuove tecnologie, l'energia e lo sviluppo economico sostenibile (ENEA), Bologna, Italy

S. Lo Meo · A. Manna · C. Massimi · A. Mengoni · G. Vannini · A. Ventura
Istituto Nazionale di Fisica Nucleare, Sezione di Bologna, Bologna, Italy

A. Manna · C. Massimi · G. Vannini
Dipartimento di Fisica e Astronomia, Università di Bologna, Bologna, Italy

P. F. Mastinu
Istituto Nazionale di Fisica Nucleare, Sezione di Legnaro, Legnaro, Italy

F. Matteucci · P. M. Milazzo
Istituto Nazionale di Fisica Nazionale, Trieste, Italy

F. Matteucci
Dipartimento di Fisica, Università di Trieste, Trieste, Italy

A. Mazzone
Consiglio Nazionale delle Ricerche, Bari, Italy

R. Nolte · D. Radeck
Physikalisch-Technische Bundesanstalt (PTB), Bundesallee 100,
38116 Braunschweig, Germany

F. Ogállar · I. Porras · J. Praena
University of Granada, Granada, Spain

A. Pavlik
Faculty of Physics, University of Vienna, Vienna, Austria

T. Rausher
Department of Physics, University of Basel, Basel, Switzerland

T. Rausher
School of Physics, Astronomy and Mathematics, University of Hertfordshire, Hertfordshire, UK

A. Saxena
Bhabha Atomic Research Centre (BARC), Mumbai, India

S. Urlass
Helmholtz-Zentrum Dresden-Rossendorf, Dresden, Germany

A. Wallner
Australian National University, Canberra, Australia

Both cross sections have been measured at the n_TOF EAR-2 [4] with three C_6D_6 detectors [5]. The ^{244}Cm samples have been also measured with the TAC [6] in EAR-1 [7], which has a significant lower neutron fluence and a larger flight path, in order to validate the results obtained in the EAR-2 and to obtain a more accurate normalization. In addition, with this second measurement we will obtain spectroscopic information about the γ-ray cascades following the ^{244}Cm(n,γ) and ^{240}Pu(n,γ) reactions (the samples contain a significant amount of ^{240}Pu originated by the ^{244}Cm alpha-decay).

At the n_TOF facility neutrons are produced by spallation reactions in a Pb target induced by 20 GeV/c proton pulses. Neutrons travel through 20 and 185 m in vacuum until reaching the EAR-2 and EAR-1 experimental areas, respectively. There the Cm samples and the detection systems are located. In the case of the measurement performed in the EAR-2 a dedicated set up was prepared for the Cm campaign. The γ-ray cascades following neutron capture were detected with 3 BICRON C_6D_6 scintillator detectors placed at 5 cm of the sample. In Fig. 4.1 we show a picture of the experimental set up together with the geometry implemented in the Geant4 code [8], needed to perform high accurate Monte Carlo simulations for the Total Energy Detection technique.

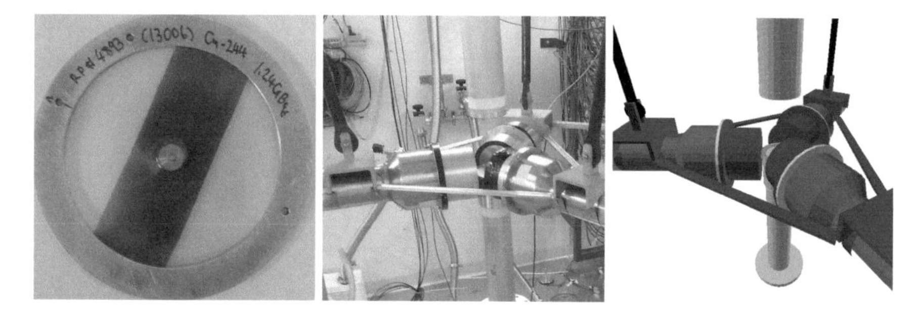

Fig. 4.1 Sample and ring used during the experiment (*left*). Picture of the set up in the EAR-2 (*center*). Geometry implemented in Geant4 to perform Monte Carlo simulations (*right*)

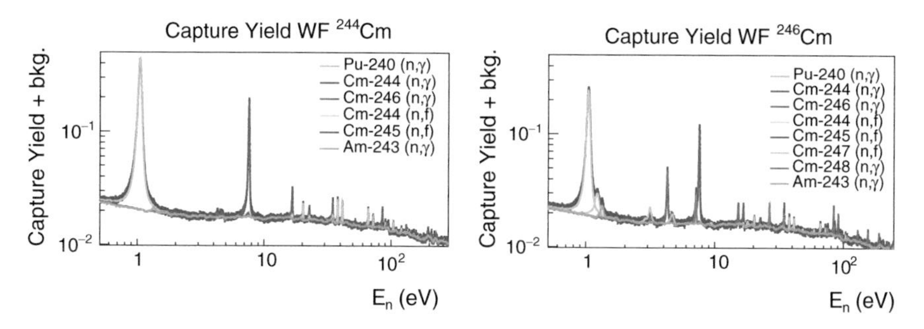

Fig. 4.2 Preliminary experimental results compared with JEFF-3.2

Preliminary results of the measured capture yields are presented in Fig. 4.2. In both panels we show the experimental capture yields (unnormalized and no background subtracted) together with the different capture yields calculated from the cross sections available in JEFF-3.2 [9] normalized to the experimental data.

References

1. G. Aliberti et al., Nuclear data sensitivity, uncertainty and target accuracy assessment for future nuclear systems. Ann. Nucl. Energy **33**, 700–733 (2006)
2. M.S. Moore, G.A. Keyworth, Phys. Rev. C **3**, 1656–1667 (1971)
3. A. Kimura et al., J. Nucl. Sci. Technol. **49**, 708–724 (2012)
4. C. Weiss et al., Nucl. Instrum. Methods A **799**, 90 (2015)
5. U. Abondano et al., Nucl. Instrum. Methods A **521**, 454–467 (2004)
6. C. Guerrero et al., Nucl. Instrum. Methods A **608**, 424–433 (2009)
7. C. Guerrero et al., Eur. Phys. J. A **49**, 27 (2013)
8. S. Agostinelli et al., Nucl. Instrum. Methods A **506**, 250 (2003)
9. A. Koning et al., Status of the JEFF nuclear data library. J. Korean Phys. Soc. **59**(2), 1057–1062 (2011). The JEFF library is available from https://www.oecd-nea.org/dbdata/jeff

Chapter 5
First Steps Towards An Understanding of the Relation Between Heavy Ion Double Charge Exchange Nuclear Reactions and Double Beta Decays

Jessica I. Bellone, S. Burrello, Maria Colonna, Horst Lenske and José A. Lay Valera

Abstract Theoretical studies are performed, providing a relation between heavy ion double charge exchange (DCE) cross section and $\beta\beta$ strength, up to momentum transfers of 25–30 MeV. DCE reactions can be interpreted in terms of two correlated or uncorrelated single charge exchange processes, thus mimicking $0\nu\beta\beta$ or $2\nu\beta\beta$ decay, respectively. The dominance of the former mechanism would allow to gain information on $0\nu\beta\beta$ strength, thus helping in improving neutrino effective mass evaluation.

Jessica I. Bellone, S. Burrello, Maria Colonna, Horst Lenske, José A. Lay Valera—for the NUMEN collaboration

---

J. I. Bellone (✉) · S. Burrello · M. Colonna
INFN - LNS, via Santa Sofia 62, 95125 Catania, Italy
e-mail: bellone@lns.infn.it

S. Burrello
e-mail: burrello@lns.infn.it

M. Colonna
e-mail: colonna@lns.infn.it

S. Burrello · J. A. L. Valera
Departamento de FAMN, Universidad de Sevilla, Apartado 1065, 41080 Sevilla, Spain
e-mail: lay@us.es

H. Lenske
Institut für Theoretische Physik, Justus-Liebig-Universitat Giessen, 35392 Giessen, Germany
e-mail: horst.lenske@theo.physik.uni-giessen.de

J.-E. García-Ramos et al. (eds.), *Basic Concepts in Nuclear Physics: Theory, Experiments and Applications*, Springer Proceedings in Physics 225,
https://doi.org/10.1007/978-3-030-22204-8_5

5.1 Introduction

The study of heavy ion charge exchange reactions has been receiving a lot of interest in the last decades, because of its multidisciplinarity. The present work focuses on the study of double charge exchange (DCE) reactions, described as a sequence of two uncorrelated single charge exchange (SCE) processes; in this way, DCE strength is analogous to that of $2\nu\beta\beta$ decay, representing the main mechanism competing with $0\nu\beta\beta$ (whose relation with DCE reactions is still under study [3], in the hope to improve Majorana mass evaluation, if $0\nu\beta\beta$ decay were observed).

5.2 Heavy Ion Charge Exchange Cross Section Factorization

Once studied the role of the optical potentials and nuclear structure terms, heavy ion SCE [4] and DCE cross section factorization into the product of a nuclear structure ($K_{\alpha\beta}$) and a reaction term (N^D) is provided, by assuming a gaussian shape for the former term

$$\frac{d^2\sigma}{dEd\Omega} = \frac{E_\alpha E_\beta}{4\pi^2(\hbar c)^4}\frac{k_\beta}{k_\alpha}\frac{1}{(2J_a+1)}\frac{1}{(2J_A+1)}\sum_{\substack{m_a,m_A\\ m_b,m_B}}|\sum_{\substack{\tau=C,Tn\\ SL}}\sum_{S,T}K^{\tau,ST}_{\alpha\beta}(\mathbf{q}_{\alpha\beta})\int d^3q\, h^{ST}_{\alpha\beta}(\mathbf{q},\mathbf{q}_{\alpha\beta})N^D(\mathbf{q})|^2 \tag{5.1}$$

DCE reactions described as two independent SCE processes are simulated, through a new code, developed by the authors, assuming pole approximation and single state dominance. Numerical simulations [2], performed for heavy nuclei studied within the NUMEN collaboration [1], provide that the factorized expression in (5.1), which is exact for $q_{\alpha\beta} = 0$, works up to a momentum transfer $q_{\alpha\beta} \simeq 25-30$ MeV.

DCE simulations also show that partial compensation occur between optical potentials in the intermediate channel, thus leaving a SCE-like diffraction pattern in DCE angular distribution and a DCE N^D value larger than the product of two SCE ones.

5.3 Conclusions

Cross section factorization is obtained for $q_{\alpha\beta} \leq 25-30$ MeV, for SCE and DCE processes. Simulations performed for DCE reactions, described in terms of two independent SCE processes, allow to give information on the main mechanism competing with the one resembling $0\nu\beta\beta$ decay.

References

1. F. Cappuzzello et al., The NUMEN project: NUclear Matrix Elements for Neutrinoless double beta decay. Eur. Phys. J. A **54**(72) (2018)
2. F. Cappuzzello, H. Lenske et al., Analysis of the $^{11}B(^7Li,^7Be)^{11}Be$ reaction at 57 MeV in a microscopic approach. Nucl. Phys. A **739**, 30–56 (2004)
3. H. Lenske, Probing double beta-decay by heavy ion charge exchange reactions. J. Phys.: Conf. Ser. **1056**, 012030 (2018)
4. H. Lenske, J.I. Bellone, M. Colonna, J.A. Lay, Theory of single charge exchange heavy ion reactions. Phys. Rev. C **98**, 044620 (2018)

Chapter 6
Study on the Decay of $^{46}Ti^{*}$

M. Cicerchia, F. Gramegna, D. Fabris, T. Marchi, M. Cinausero, G. Mantovani, A. Caciolli, G. Collazzuol, D. Mengoni, M. Degerlier, L. Morelli, M. Bruno, M. D'Agostino, C. Frosin, S. Barlini, S. Piantelli, M. Bini, G. Pasquali, P. Ottanelli, G. Casini, G. Pastore, D. Gruyer, A. Camaiani, S. Valdré, N. Gelli, A. Olmi, G. Poggi, I. Lombardo, D. Dell'Aquila, S. Leoni, N. Cieplicka-Orynczak and B. Fornal

Abstract The pre-formation of α-clusters in α-conjugate nuclei or their dynamical condensation during nuclear reactions was largely debated; among the methods to probe a possible cluster structure in nuclei, it has been suggested to study

M. Cicerchia (✉) · F. Gramegna · T. Marchi · M. Cinausero · G. Mantovani
INFN Laboratori Nazionali di Legnaro, Legnaro (PD), Italy
e-mail: cicerchia@lnl.infn.it

M. Cicerchia · G. Mantovani · A. Caciolli · G. Collazzuol · D. Mengoni
Dipartimento di Fisica e Astronomia dell'Università di Padova, Padua, Italy

D. Fabris · A. Caciolli · G. Collazzuol · D. Mengoni
INFN Sezione di Padova, Padua, Italy

M. Degerlier
Science and Art Faculty, Physics Department, Nevsehir Haci Bektas Veli Univ, Nevsehir, Turkey

L. Morelli · M. Bruno · M. D'Agostino · C. Frosin
INFN Sezione di Bologna e Dipartimento di Fisica e Astronomia, Università Di Bologna, Bologna, Italy

S. Barlini · S. Piantelli · M. Bini · G. Pasquali · P. Ottanelli · G. Casini · G. Pastore · A. Camaiani · S. Valdré · N. Gelli · A. Olmi · G. Poggi
INFN Sezione di Firenze e Dipartimento di Fisica e Astronomia, Università Di Firenze, Florence, Italy

D. Gruyer
Grand Accélérateur National d'Ions Lourds, 14076 Caen, France

I. Lombardo · D. Dell'Aquila
INFN Sezione di Napoli e Dipartimento di Fisica, Università Federico II Napoli, Naples, Italy

D. Dell'Aquila
Institut de Physique Nuclèaire (IPN) Université Paris-Sud 11, Orsay, Île-de-France, France

S. Leoni · N. Cieplicka-Orynczak
INFN Sezione di Milano e Dipartimento di Fisica, Università Di Milano, Milan, Italy

N. Cieplicka-Orynczak · B. Fornal
Institute of Nuclear Physics, Polish Academy of Sciences Krakow, Krakow, Poland

J.-E. García-Ramos et al. (eds.), *Basic Concepts in Nuclear Physics: Theory, Experiments and Applications*, Springer Proceedings in Physics 225,
https://doi.org/10.1007/978-3-030-22204-8_6

pre-equilibrium emitted particles and clusters following their emission during the dynamical part of the reaction, before a full thermalization takes place.

The pre-formation of α-clusters in α-conjugate nuclei or their dynamical condensation during nuclear reactions was largely debated; among the methods to probe a possible cluster structure in nuclei, it has been suggested to study pre-equilibrium emitted particles and clusters following their emission during the dynamical part of the reaction, before a full thermalization takes place. Indeed, a strong correlation between nuclear structure and reaction dynamics arises when some nucleons or clusters of nucleons are emitted or captured [1]. The NUCL-EX collaboration (INFN, Italy) is carrying out an extensive research campaign on pre-equilibrium emission of light charged particles from hot nuclei with the ultimate goal to study how possible cluster structures may nuclear reactions [2]. For this purpose, the emission of light charged particles from hot ^{46}Ti nuclei formed in the reactions ^{16}O+^{30}Si, ^{18}O+^{28}Si and ^{19}F+^{27}Al, was investigated, using the GARFIELD+RCo 4π array, fully equipped with digital electronics [3], at Legnaro National Laboratories. For central impact parameters, the systems form the same compound nucleus, namely ^{46}Ti*. Since the abundance of pre-equilibrium particles is demonstrated to be dependent by the beam velocity [4], it was kept constant (7 A MeV) for the three reactions (^{16}O+^{30}Si, ^{18}O+^{28}Si, ^{19}F+^{27}Al) for sake of comparison: in such a way, the non–equilibrium processes are expected to be almost the same. The reaction ^{16}O+^{30}Si has been also measured at a beam energy of 8 AMeV to populate the ^{46}Ti at the same excitation energy of the ^{18}O+^{28}Si at 7 AMeV reaction to obtain the same statistical component.

The experimental observables of selected events have been studied and compared with those simulated by the statistical code GEMINI++ [5], which has been used, as a starting point, with standard input parameters and has been filtered with a software replica of the experimental array to take into account the finite size of the detecting device. The analysis has been performed on an event by event basis. As expected at these bombarding energies, for each studied reaction, the prediction from GEMINI++ accounts for the major part of the cross section both looking at the angular distributions and at the different light charged particle energy spectra, demonstrating that complete fusion is the main mechanism occurring between the colliding partners. See [6] for more details.

The observed differences between the experimental and the predicted observables, among which the α-particle angular distribution, shown in the right panel of Fig. 6.1 as an example, put into evidence effects related to both the different entrance channels and to a small contribution from fast emission: an overproduction of forward angular emitted a-particles is observed, that may represent the onset of pre-equilibrium emission, probably facilitated by the α-clustering structure of reacting partners. To understand if the pre-equilibrium process is well accounted for by theory, a more quantitative analysis is needed. A comparison to predictions by dynamical code, like AMD [7] and HIPSE [8] are also under study and will be used in forthcoming more exclusive analysis.

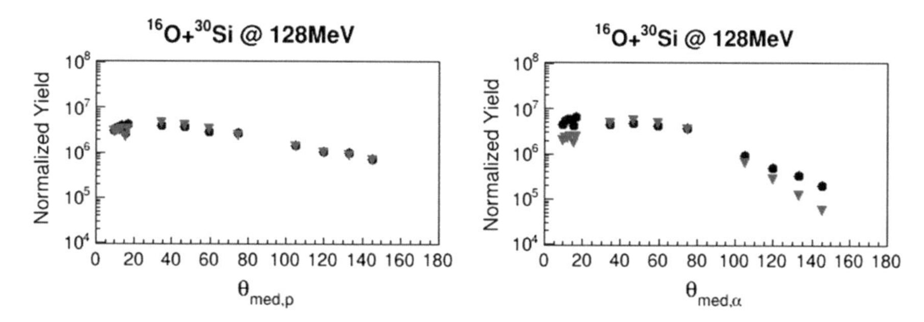

Fig. 6.1 Comparison between experimental (black dots) and GEMINI++ simulated (red triangles) angular distribution of proton (left panel) and α-particles (right panel) in coincidence with evaporation residue for the reaction ^{16}O+^{30}Si at 128 MeV. Experimental and simulated distributions are normalized to the residues number

References

1. P.E. Hodgson, E. Běták, Phys. Rep. **374**, 1–89 (2003)
2. L. Morelli et al., J. Phys. G **41**(2014) 075107; L. Morelli et al., J. Phys. G **41**(2014) 075108; D. Fabris et al., PoS (X LASNPA), 2013, p. 061.D; V.L. Kravchuk, et al. EPJ WoCs, **2**(2010) 10006; O.V. Fotina et al., Int. J. Mod. Phys. E **19**(2010) 1134
3. F. Gramegna et al., Proceedings of IEEE Nuclear Symposium, 2004, Roma, Italy, 0-7803-8701-5/04/; M. Bruno et al., M. Eur. Phys. J. A **49** (2013), 128
4. J. Cabrera et al., Phys. Rev. C **68**, 034613 (2003)
5. R.J. Charity et al., Phys. Rev. C **82**, 00016 (2017)
6. M. Cicerchia, Nuovo Cimento C, **41** (2018)
7. A. Ono, EPJ WoC **31**, 012011 (2012)
8. D. Lacroix et al., Phys. Rev. C **69**, 054604 (2004)

Chapter 7
Bayesian Reconstruction of Axial Dose Maps Using the Measurements of a Novel Detection System for Verification of Advanced Radiotherapy Treatments

A. D. Domínguez-Muñoz, M. C. Battaglia, J. M. Espino, R. Arráns, M. A. Cortés-Giraldo and M. I. Gallardo

Abstract In this work, a reconstruction algorithm is used to obtain axial dose map distributions for the verification of advanced photon radiotherapy treatments. The experimental data is obtained with a detection system designed, developed and constructed specifically for this purpose. This system is basically composed by two perpendicular single sided silicon strip detectors (SSSSD) placed inside a rotating polyethylene phantom. Measured data consist on mean absorbed dose in each strip at different angular positions. Dividing the dose map into pixels, statistical bayesian methods can be applied in order to estimate pixel data from measured one. These methods are applied to a hypothetical treatment plan and the results converge to a solution of a dose map distribution that agrees with treatment planning system (TPS) calculation.

7.1 Introduction

The new features of modern photon radiotherapy treatments force the treatment planning system (TPS) calculations to be far from reference conditions. Then, calculated dose map distribution may differ from the real one [1]. These treatments generally include dose escalation in the vicinity of critical organs that should be spared, so it is understood the importance of dose calculation accuracy. Thus, additional test are required to verify this.

A. D. Domínguez-Muñoz (✉) · M. C. Battaglia · J. M. Espino · M. A. Cortés-Giraldo
M. I. Gallardo
Departamento de Física Atómica, Molecular y Nuclear, Universidad de Sevilla,
41013 Seville, Spain
e-mail: adominguez18@us.es

R. Arráns
Hospital Universitario Virgen Macarena (HUVM), 41071 Seville, Spain

J.-E. García-Ramos et al. (eds.), *Basic Concepts in Nuclear Physics: Theory, Experiments and Applications*, Springer Proceedings in Physics 225,
https://doi.org/10.1007/978-3-030-22204-8_7

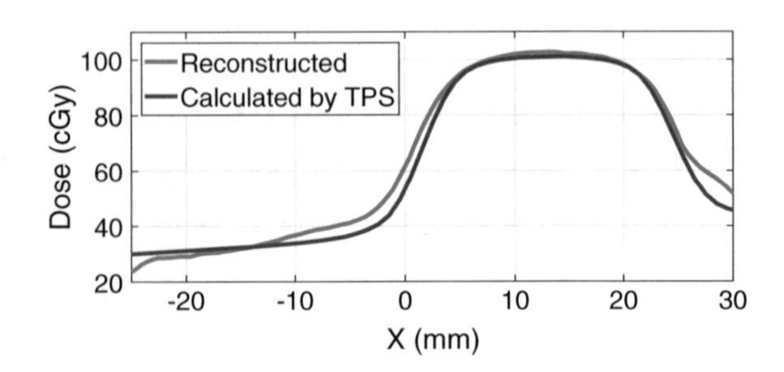

Fig. 7.1 Dose profile of 2D treatment reconstructed map and comparison with the TPS calculation

7.2 Material and method

The system [2] consists of a rotating cylindrical polyethylene phantom that houses the dual single-sided silicon strip detector (DSSSSD), that is two singled-sided silicon strip detectors positioned so that the strips of each one are perpendicular to the other. The active area of the DSSSSD is placed centred in an axial plane of the phantom, parallel in general to the beam incidence direction. Each strip works as an independent detector that is previously calibrated in dose to water. The experimental data set contains statistical uncertainties, that are taken into account by means of using reconstruction methods based on statistical estimation techniques. The issue consists on obtaining the mean absorbed dose in a grid of pixels, treating them like parameters included in the probability density function of the experimental data. Bayesian methods, as *maximum a posteriori*, provide an iterative solution in which prior information can be added to improve the results.

7.3 Results

The proposed method is applied to a theoretical treatment plan that consists of three fields pointing to a position outside the cylinder center. Reconstructed dose map of treatments show a good agreement with TPS calculated ones (Fig. 7.1).

References

1. J. Dyk et al., Commissioning and quality assurance of treatment planning computers. Int. J. Radiat. Oncol. Biol. Phys. **26**, 261–273 (1993)
2. M.I. Gallardo et al., System and method of radiotherapy treatment verification. Patent at the OEMP—Number ES 2,409,760, B1 (2014)

Chapter 8
Be-10 Measurements in Atmospheric Filters Using the AMS Technique: The Data Analysis

K. De Los Ríos, C. Méndez-García, S. Padilla, C. Solís, E. Chávez, A. Huerta and L. Acosta

Abstract For measurements with the AMS technique, data set analysis is carry out to determine the measure of central tendency and then to obtain the ^{10}Be concentrations from atmospheric samples. We proposed to compare the traditional statistical method with an alternative exploratory data analysis in order to optimize de data series extracting outlier values. Our preliminary results are here discussed.

8.1 Experimental Procedure

At LEMA,[1] the possibility to measure the concentrations of radioisotopes, such as ^{10}Be, ^{26}Al and actinides in environmental samples is been explored. To test for the first time these nuclei we used atmospheric filters. The ^{10}Be was the most recently radioisotopes measured [2]. To make this by using AMS technique, it is necessary to extract the radioisotope from the atmospheric filters in the form of BeO molecule. We optimized a known radiochemical procedure [3, 4] and applied it in test samples.[2] The processed sample is introduced into an Al cathode and later inserted in the carousel of AMS system to be measured. The sample is ionized with a Cs ion source. The ion beam generated pass through two different optical stages and an acceleration

[1]In 2013 LEMA (Laboratorio Nacional de Espectrometria de Masas con Aceleradores) was installed in Mexico. It is a Laboratory specialized in the radiocarbon measurements [1].
[2]Quartz filters PM10, samples taken at the Instituto Mexicano del Petróleo, Mexico City, the chosen here is labelled as TO4 corresponding to November 27 to 29, 2012 sampling campaign.

K. De Los Ríos · C. Méndez-García · S. Padilla · C. Solís · E. Chávez
A. Huerta · L. Acosta (✉)
Instituto de Física, UNAM, Mexico City, Mexico
e-mail: acosta@fisica.unam.mx

C. Méndez-García
CONACyT, Mexico City, Mexico

J.-E. García-Ramos et al. (eds.), *Basic Concepts in Nuclear Physics: Theory, Experiments and Applications*, Springer Proceedings in Physics 225,
https://doi.org/10.1007/978-3-030-22204-8_8

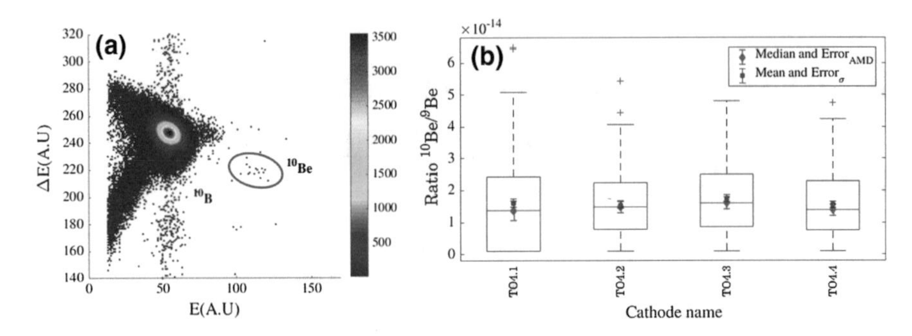

Fig. 8.1 **a** Energy spectrum of the cathodes with TO4 test sample. **b** Schematic diagram, measures of central tendency with the two proposed methods and the associated statistical errors

system (1 MV). The beam produced (^{10}Be or ^{9}Be handled by a bouncer system) is focused to be measured by Faraday cups (^{9}Be) and finally, the ^{10}Be events by using the ΔE-E technique in a gas detector[3] (See Fig. 8.1a).

To develop the analysis, the ^{10}Be/^{9}Be ratio is used. Traditionally, the weighted mean and the standard deviation are used, for which it is necessary a stable behaviour of the data taking. Such condition is not presented in all the cases for this kind of data [5]. For such a reason, we propose another way to develop the statistical analysis called Exploratory Data Analysis (EDA), applied to the test sample TO4. EDA is a robust and resistant statistical method [5]. There are different ways to determine the measure of central tendency but we use the median. It is possible to show the behaviour of the data set with schematic plots, the box of this graphical method contains the 75% of the data set and show the median (red lines in Fig. 8.1b), the minimum and maximum data of the distribution (whiskers) and the outliers, i.e. data that are not inside the box+whiskers (red crosses in Fig. 8.1b).

The results observed in Fig. 8.1b indicate that the EDA method is equivalent to the traditional method, even considering that EDA excludes the atypical data. This result is verifiable taking into account two aspects: the mean and the median (and their errors) are comparable. Applying the Kruskalwallis test [5], a value of $p = 0.3312$ obtained indicates that both distributions come from the same data-set with a 1% of significance level. However this just happens when stability exists, in other cases, EDA will rise big differences, helping to identify such data series with high instability.

Acknowledgements This work was partially supported by CONACYT 51600, 82692, 123655, DGAPA-PAPIIT IA101616, IA103218 and PIIF-2018 Projects.

[3]Before gas detector, a passive absorbent of Ni_3Si_4 75 nm of diameter is used to optimize the separation between ^{10}Be and its isobar interference ^{10}B.

References

1. C. Solís et al., Nucl. Inst. Meth. Phys. Res. B **331** (2014)
2. K. De Los Ríos, J. Phys.: Conf. Ser. **1078** (2018) 012009
3. M. Auer et al., Earth Plan. Sci. Lett. A **287** (2009)
4. S. Padilla et al., J. Env. Rad. A **189** (2018)
5. D. Wilks, 3rd edn. (Academic Press, 2011), pp. 15–60

Chapter 9
Fission Studies in Inverse Kinematics

M. Feijoo, J. Benlliure, J. L. Rodríguez-Sanchéz and J. Taieb

Abstract In this work, we present the results obtained from the data analysis of the SOFIA experiment, performed in 2014 at the GSI, whose characteristics allowed us to achieve the complete identification (charge and mass) of both fission fragments for the first time. In particular, the nuclear-collision fission induced reaction ^{236}U+Al at 720 A MeV is being matter of work. Our goal is to study the fission dynamics through the study of its dissipative effects. The work done so far have showed us a good resolution in charge of the fission fragments, setting a solid base to go further. In the future, we plan to improve the experimental technique inducing fission using quasifree (p,2p) reactions, to better control the initial conditions of the process.

Fission is a nuclear process where an excited heavy nucleus deforms into a transient state called saddle point. If the excitation continues beyond this point, fission will irrevocably occur, and the nucleus will be splitted in two different fragments. In spite of its apparent simplicity, the process presents many degrees of freedom (mass and charge of the fissioning nuclei, excitation energy, mass asymmetry of the fragments...), making its theoretical description really complicated [1].

In order to improve our understanding of the process, multiples experiments have been performed through the years. Nowadays, the inverse kinematics experimental technique is the most useful approach to study the process. By accelerating the projectile into relativistic energies, high velocity fission fragments will be produced and emitted in forward direction, making their detection easier and more efficient.

This technique was used in the SOFIA experiment [2]. It represented a real breakthrough in the study of fission, achieving for the first time the complete identification

M. Feijoo, J. Benlliure, J. L. Rodríguez-Sanchéz and J. Taieb on behalf the SOFIA collaboration.

M. Feijoo (✉) · J. Benlliure · J. L. Rodríguez-Sanchéz
Department of Particle Physics, IGFAE, University of Santiago de Compostela, 15782 Santiago de Compostela, Spain
e-mail: manuelfeijoo.rodriguez@usc.es

J. Taieb
CEA DAM Bruyères-le-Châtel, 91297 Arpajon, France

J.-E. García-Ramos et al. (eds.), *Basic Concepts in Nuclear Physics: Theory, Experiments and Applications*, Springer Proceedings in Physics 225,
https://doi.org/10.1007/978-3-030-22204-8_9

of both fission fragments. We are interested on the study of fission induced by nuclear collision using the ^{236}U+Al reaction. To obtain the fissioning ^{236}U, a primary beam of ^{238}U at 1 A GeV will collide against a Be target creating a large chain of isotopes. At the entrance of SOFIA setup, the ^{236}U will be fully identified, arriving with an average kinematic energy of 720 A MeV and the fission reaction will be produced on the Active Target, where the Al layers are settled down. Thanks to the powerful SOFIA setup, we can measure the fragments energy lost, using two MUlti-Sampling Ionization Chambers (MUSIC), track the beam and fragments positions with 3 MWPC and know the velocity of the fragments with a ToF Wall. Thereby, the identification in charge (Fig. 9.1) and mass (currently on work) of the fission fragments are achieved. We are interested on the study of the dissipative effects of the reaction and on modeling their behavior. By measuring the total charge of the fission fragments, we will able to know the excitation energy of the process [3].

Finally, we plan to go a step forward and use (p,2p) reactions to induce fission, in order to improve our understanding of the fission process by controlling the initial stage of the reaction, from the ground state to the saddle point. The R3B/CALIFA [2] detector will be coupled to SOFIA (Fig. 9.2) to detect the energy lost of both emitted protons and obtain the excitation energy of the fission process. This will give us a more reliable value of the excitation energy, as it will not be model dependent but a direct measurement.

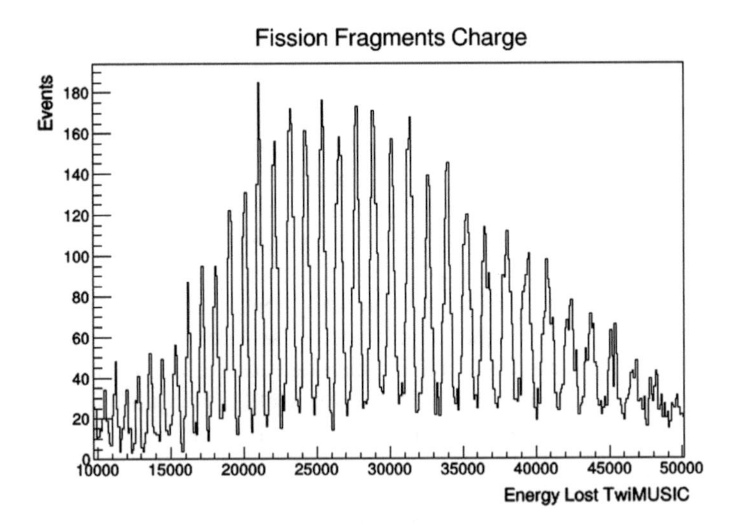

Fig. 9.1 Charge of the fission fragments. Each peak represents a different atomic number

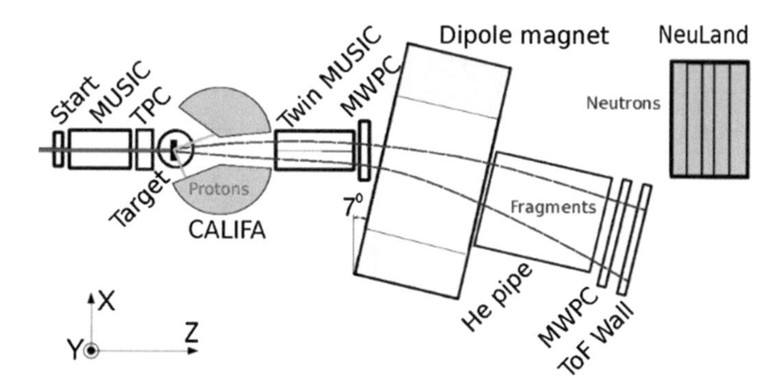

Fig. 9.2 CALIFA detector (red circle) coupled to the SOFIA setup

References

1. J. Benlliure, *Spallation Reactions in Applied and Fundamental research in The Euroschool Lectures on Physics with Exotic Beams*, vol. II (Springer, Berlin, Heidelberg, 2006)
2. J.-F. Martin, Studies on fission with ALADIN. Eur. Phys. J. A **51**, 174 (2015)
3. J.L. Rodríguez-Sánchez, Presaddle and Postsaddle dissipative effects in fission using complete kinematics measurements. Phys. Rev. C **94**, 061601 (2016)

Chapter 10
Iterative Algorithm for Optimal Super Resolution Sampling

P. Galve, A. López-Montes, J. M. Udías and J. López Herraiz

Abstract The resolution recovery in PET iterative methods such as OSEM, including all the physical effects involved in the system response matrix, is often limited by the reduced sampling in the projection space. In this work, we propose a method to further improve resolution recovery in the PET image reconstruction process by iteratively refining the measurements with data-driven increased sampling. In this method we first reconstruct the image by standard OSEM methods. After that, we define four subLORs around each initially measured LOR, and estimate the subLOR contribution spreading the original number of counts with maximum-likelihood based weights, computed with the relative value of the projections in each subLOR with respect to the four subLORs. Now we reconstruct the image with the standard OSEM algorithm using the subLORs set of data. We call this step a superiteration, which may be repeated 2–3 times until convergence is achieved. We have evaluated the improvements in image quality obtained using data acquired in the Argus PET/CT scanner. The method shows promising results increasing the recovery coefficients without noise raise measured in an Image Quality phantom, and also remarkable image quality improvements for live patients reconstructions.

The limited resolution of Positron Emission Tomography (PET) is one of its main drawbacks. It is caused by a combination of different factors intrinsic to the technique [1]. Physical factors can be modeled in the System Response Matrix (SRM) [2, 3], whereas the geometry and configuration of modern state-of-the-art scanners are already optimized to get the best performance [4, 5].

In this work we try to improve the image resolution by using a novel method to iteratively increase the object sampling. It is based on a Maximum Likelihood algorithm developed by [6] to recover triple coincidences in PET. The algorithm was also used for demultiplexing multiplexed data in SPECT [7] and identification of inter-crystal scatter events [8]. In our implementation, we virtually subdivide each

P. Galve (✉) · A. López-Montes · J. M. Udías · J. López Herraiz
Grupo de Física Nuclear and Iparcos, Facultad de Ciencias Físicas, Universidad Complutense de Madrid, CEI Moncloa, 28040 Madrid, Spain
e-mail: pgalve@nuclear.fis.ucm.es

J.-E. García-Ramos et al. (eds.), *Basic Concepts in Nuclear Physics: Theory, Experiments and Applications*, Springer Proceedings in Physics 225,
https://doi.org/10.1007/978-3-030-22204-8_10

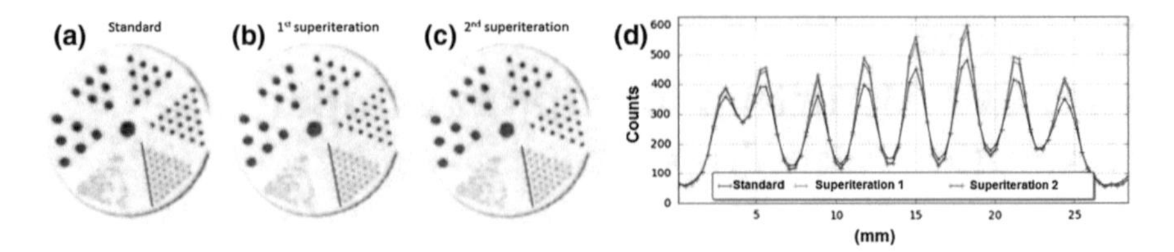

Fig. 10.1 **a** Standard reconstruction of a Derenzo phantom acquired with the SuperArgus scanner with OSEM (20 iterations, 5 subsets). **b** and **c** are the images 1 and 2 superiterations. The line profile along the darker line is shown in figure **d**

crystal into two halves, and therefore each line-of-response (LOR) into 4 subLORs. The whole process can be summarized as follows:

1. Standard image reconstruction.
2. Projection of the previously reconstructed image along the new subLORs. We use the relative subLOR projections to proportionally spread the original number of LOR counts among the different subLORs.
3. Standard reconstruction considering each subLOR is an independent LOR.

In Fig. 10.1 we can observe a Derenzo phantom acquired with the Super Argus scanner, newer version of the Argus scanner [3]. A peak-to-valley ratio increase of 41 $\pm$ 3% in 1.5 mm rods with respect to the normal OSEM method has been achieved after first superiteration, and 54 $\pm$ 3% after the second one. These results show promising future for the algorithm presented. It is important to note that the computation time per iteration increases ~4 times (the number of data is four times the original data size). Therefore the full reconstruction time increases several times, but it is mitigated with the use of acceleration GPUs.

Acknowledgements This work was supported by Comunidad de Madrid (S2013/MIT-3024 TOPUS-CM), Spanish Ministry of Science and Innovation, Spanish Government (FPA2015-65035-P, RTC-2015-3772-1). This is a contribution for the Moncloa Campus of International Excellence. Grupo de Física Nuclear-UCM, Ref.: 910059. This work acknowledges support by EU's H2020 under MediNet a Networking Activity of ENSAR-2 (grant agreement 654002). J. L. Herraiz is also funded by the EU Cofund Fellowship Marie Curie Actions, 7th Frame Program. P. Galve is supported by a Universidad Complutense de Madrid and Banco Santander predoctoral grant, CT27/16-CT28/16.

References

1. W.W. Moses, NIM-A **648**, S236–S240 (2011)
2. J.L. Herraiz et al., Phys. Med. Biol. **51**(18), 4547–4565 (2006)
3. K. Gong et al.," *IEEE TMI*, v. 36, n. 10, pp. 2179–2188, 2017
4. Y. Wang et al., J. Nucl. Med. **47**(11), 1891–1900 (2006)
5. S. Krishnamoorthy et al., Phys. Med. Biol. **63**(15), 155013 (2018)

6. E. Lage et al., Med. Phys. Med. Phys. Med. Phys **42**(24), 1398–102502 (2015)
7. S. Moore et al., in *Fully3D conference* (2015), pp. 515–517
8. M.S. Lee et al., Phys. Med. Biol. (2018)

Chapter 11
Adiabatic Correction to the Eikonal Approximation

C. Hebborn, D. Baye and Pierre Capel

Abstract This work focuses on the development of an adiabatic correction to the eikonal model. The preliminary results computed for the Coulomb-dominated breakup of the one-neutron halo nucleus ^{11}Be at $69A$ MeV are encouraging. Further analyses on the accuracy of the correction have still to be performed.

For the last three decades, the development of Radioactive-Ion Beams (RIBs) has enabled the study of nuclei away from stability. Near the neutron dripline, halo nuclei have been observed [1]. These light nuclei display a very peculiar structure: one or two of their valence neutrons are located far from the bulk of the nucleus and form a diffuse halo around it. They have challenged the usual description of nuclei, seeing all nucleons piling up into well-defined orbitals and forming a compact object. Accordingly, halo nuclei are modelled as two- or three-body objects: a compact core to which one or two neutrons are loosely bound.

As halo nuclei are short-lived, they cannot be studied through the usual spectroscopic techniques but can be probed through indirect methods, such as reaction processes. In the present paper, we focus on breakup reactions, which describe the dissociation of the projectile into its more fundamental constituents. Breakup measurements can thus reveal the cluster structure of the nucleus. To obtain reliable information about this structure, one needs an accurate reaction model coupled to a realistic description of the projectile.

The eikonal model [2] is very efficient from a computational point of view and provides a straightforward interpretation of the collision. It simplifies the many-body

C. Hebborn (✉) · D. Baye · P. Capel
Physique Nucléaire et Physique Quantique (CP 229), Université libre de Bruxelles (ULB), 1050 Brussels, Belgium
e-mail: chebborn@ulb.ac.be

D. Baye
e-mail: dbaye@ulb.ac.be

P. Capel
Institut für Kernphysik, Johannes Gutenberg-Universität Mainz, 55099 Mainz, Germany
e-mail: pcapel@uni-mainz.de

J.-E. García-Ramos et al. (eds.), *Basic Concepts in Nuclear Physics: Theory, Experiments and Applications*, Springer Proceedings in Physics 225,
https://doi.org/10.1007/978-3-030-22204-8_11

Schrödinger equation by assuming that the projectile-target (P-T) wavefunction does not differ much from an initial plane wave. In its usual form, it also considers the adiabatic approximation, which sees the internal coordinates of the projectile as frozen during the collision. The asymptotic wavefunction is given by a plane wave shifted by the so-called eikonal phase $\chi(\mathbf{b}, \mathbf{r}) \propto \int_{-\infty}^{+\infty} V_{PT}(\mathbf{R}, \mathbf{r})\mathrm{d}Z$ computed from the optical potential V_{PT} simulating both the Coulomb and the nuclear interactions between the projectile clusters and the target. This potential depends on the P-T coordinate $\mathbf{R} = (\mathbf{b}, Z)$ and the internal coordinate of the projectile $\mathbf{r}$.

Unfortunately, the adiabatic approximation is valid only when the collision is brief enough, and thus is not compatible with long-range interactions. Therefore, for Coulomb-dominated reactions, the usual eikonal model diverges. The aim of this study is to develop an adiabatic correction which naturally removes the divergence and accounts for a part of dynamics of the projectile. This correction is derived from the Dynamical Eikonal Approximation (DEA) [3], which does not rely on an adiabatic approach. With a unitary change of the wavefunction, one finds an eikonal-like model where the phase χ is replaced by the first-order term χ^{FO}

$$\chi(\mathbf{b}, \mathbf{r}) \rightarrow \chi^{FO}(\mathbf{b}, \mathbf{r}, E) \propto \int_{-\infty}^{+\infty} e^{i\frac{(E-E_0)Z}{\hbar v}} V_{PT}(\mathbf{R}, \mathbf{r})\mathrm{d}Z. \tag{11.1}$$

In this equation E_0 denotes the energy of the ground state of the projectile and v the asymptotic P-T velocity. Contrary to the usual eikonal model, the first-order term depends on the excitation energy E in the continnum of the projectile.

In Fig. 11.1, we plot the energy distribution of the breakup cross sections of ^{11}Be with ^{208}Pb at 69A MeV. The solid red line is obtained by the DEA, which is very accurate at these energies and that we take as reference. The usual eikonal model (dashed green line), computed with a cutoff in impact parameter $b_{max} = 71$ fm to avoid the divergence, reproduce neither the shape nor the magnitude of the cross section. The adiabatic correction (ACE, dotted blue line) does not need a cutoff in

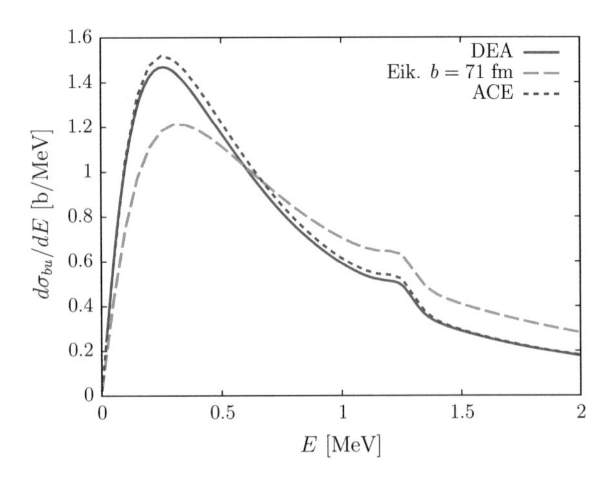

Fig. 11.1 Breakup cross sections of ^{11}Be impinging on ^{208}Pb at 69A MeV

impact parameter and provides results close to the DEA. As these preliminary results are very encouraging, we will explore in the future its efficiency to reproduce other observables such as parallel-momentum distributions.

References

1. I. Tanihata, J. Phys. G **22**, 157 (1996)
2. R.J. Glauber, in *Lecture in Theoretical Physics*, vol. 1, ed. by W.E. Brittin, L.G. Dunham (Interscience, New York, 1959), p. 315
3. D. Baye, P. Capel, G. Goldstein, Phys. Rev. Lett. **95**, 082502 (2005)

Chapter 12
Modeling Neutrino-Nucleus Interactions for Neutrino Oscillation Experiments

G. D. Megias, S. Dolan and S. Bolognesi

Abstract We present our recent progress on the relativistic modeling of neutrino-nucleus reactions for their implementation in MonteCarlo event generators (GENIE, NEUT) employed in neutrino oscillation experiments. We compare charged-current neutrino (ν) and antineutrino ($\bar{\nu}$) cross sections obtained within the SuSAv2 model, which is based on the Relativistic Mean Field theory and on the analysis of the superscaling behavior exhibited by (e, e') data. We also evaluate and discuss the impact of multi-nucleon excitations arising from 2p–2h states excited by the action of weak forces in a fully relativistic framework, showing for the first time their implementation in GENIE and their comparison with recent T2K data.

Current efforts in long-baseline ν experiments are aimed at improving knowledge of ν oscillations, where the development and implementation of realistic ν-nucleus interaction models are essential to constrain experimental uncertainties. The current state of the art for experimental systematics is in the region of 5–10% [1] and are mostly related to flux and cross section predictions (3–4%). A decrease of 2–3% on these uncertainties would allow to shorten running time and experimental costs (reducing by half either the experimental exposure or the detector volume) while increasing the sensitivity to determine ν mass hierarchy or CP violation in the neutrino sector. Such a reduction of systematics will therefore represent an essential step toward understanding the matter-antimatter asymmetry in the Universe, whilst also aiding in other areas of fundamental physics, such as the analysis of supernovae

G. D. Megias (✉)
Departamento de Física Atómica, Molecular y Nuclear, Universidad de Sevilla, 41080 Sevilla, Spain
e-mail: megias@us.es

S. Dolan
Laboratoire Leprince-Ringuet, IN2P3-CNRS, 91120 Palaiseau, France
e-mail: stephen.dolan@llr.in2p3.fr

S. Bolognesi
DPhP, IRFU, CEA Saclay, 91191 Gif-sur-Yvette, France
e-mail: sara.bolognesi@cea.fr

J.-E. García-Ramos et al. (eds.), *Basic Concepts in Nuclear Physics: Theory, Experiments and Applications*, Springer Proceedings in Physics 225,
https://doi.org/10.1007/978-3-030-22204-8_12

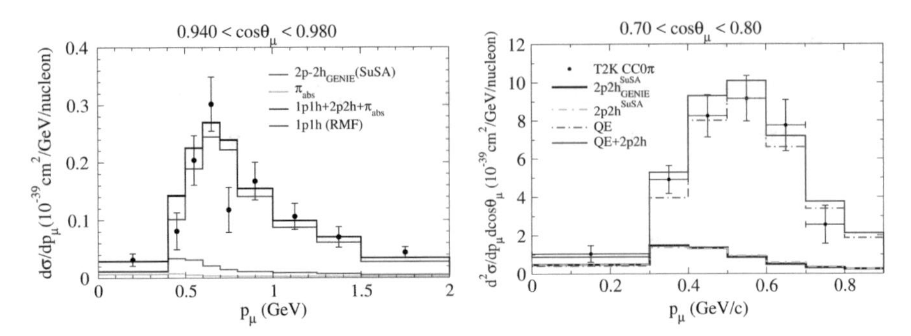

Fig. 12.1 Left panel: Comparison of T2K CC0πNp data on ^{12}C [12] for 0 protons above 500 MeV/c with the SuSAv2-MEC model and the 2p2h GENIE implementation ($2p2h^{SuSA}_{GENIE}$). Pion absorption effects are also included. Right panel: Comparison of T2K CC0π data on ^{12}C [15] with the SuSAv2-MEC model (QE+2p2h). Comparison between 2p2h GENIE implementation ($2p2h^{SuSA}_{GENIE}$) and the microscopic calculation ($2p2h^{SuSA}$) is also shown

explosions and the search for both sterile ν and proton decay. Accordingly, an accurate understanding of ν interaction physics is essential for current and upcoming experiments. Thus, the SuSAv2-MEC approach [2–4] is applied to the analysis of data from ν oscillation experiments with the aim of shedding light on the systematics arising from nuclear effects in both initial and final states. For practical purposes, the SuSAv2 model and the 2p–2h MEC contributions can be described in a simple way for different kinematics and nuclei [5–9], translating sophisticated and demanding microscopic calculations into a relatively straightforward formalism hence easing its implementation in event generators. In Fig. 12.1, we show the comparison of the SuSAv2-MEC model, which is based on relativistic, microscopic calculations [10, 11], with T2K CC0πNp data [12, 13] for 0 protons above 500 MeV/c (left panel). The 1p1h channel corresponds to RMF-based calculations and the effect of π emission followed by re-absorption in the nuclear medium is provided by the GENIE ν-nucleus event generator [14]. The 2p2h channel is generated for the first time by new implementation of the SuSAv2-MEC model within GENIE. The accurate modeling of 2p2h microscopic calculations (thick dot-dashed lines) within GENIE (solid maroon line) can be observed in the right panel, where a comparison of the full SuSAv2-MEC model with CC0π data [15] is also shown. Its capability to describe data in a wide energy range and its ease to be implemented in event generators makes the SuSAv2-MEC model a promising candidate to reduce experimental systematics in current and future ν experiments.

Acknowledgements This work was partially supported by the Spanish Ministerio de Economia y Competitividad and ERDF (European Regional Development Fund) under contracts FIS2017-88410-P, and by the Junta de Andalucia (grant No. FQM160). GDM acknowledges support from a Junta de Andalucia fellowship (FQM7632, Proyectos de Excelencia 2011). We acknowledge the support of CEA, CNRS/IN2P3 and P2IO, France; and the MSCA-RISE project JENNIFER, funded by EU grant n.644294, for supporting the EU-Japan researchers mobility.

References

1. K. Abe et al., [T2K Collaboration] (2016), arXiv:1607.08004 [hep-ex]
2. G.D. Megias et al., Phys. Rev. D **94**, 013012 (2016)
3. G.D. Megias et al., Phys. Rev. D **94**, 093004 (2016)
4. M.V. Ivanov et al., Phys. Rev. C **89**, 014607 (2014)
5. G.D. Megias et al. (2017), arXiv:1711.00771 [nucl-th]
6. G.D. Megias et al. (2018), arXiv:1807.10532 [nucl-th]
7. M.B. Barbaro et al., Phys. Rev. C **98**, 035501 (2018)
8. I. Ruiz-Simo et al., Phys. Lett. B **762**, 124 (2016)
9. J.E. Amaro et al., Phys. Rev. C **95**, 065502 (2017)
10. I. Ruiz-Simo et al., J. Phys. G: Nucl. Part. Phys. **44**, 065105 (2017)
11. G.D. Megias et al., Phys. Rev. D **91**, 073004 (2015)
12. K. Abe et al., [T2K Collaboration]. Phys. Rev. D **98**, 032003 (2018)
13. S. Dolan et al. (2018), arXiv:1804.09488 [hep-ex]
14. C. Andreopoulos et al., Nucl. Instrum. Meth. A **614**, 87–104 (2010)
15. K. Abe et al., [T2K Collaboration]. Phys. Rev. D **93**, 112012 (2016)

Chapter 13
Neutron Radiography at CNA

M. A. Millán-Callado, C. Guerrero, B. Fernández, A. M. Franconetti, J. Lerendegui-Marco, M. Macías, T. Rodríguez-González and J. M. Quesada

Abstract Neutron radiography is a non-invasive imaging technique that uses the attenuation of a neutron beam as a probe to characterize an object [1]. At Centro Nacional de Aceleradores (CNA) in Seville, we want to make the best use of our facilities, particularly in terms of the beams characteristics and neutron production, to obtain the best contrast and spatial resolution possible with a neutron camera. For this purpose, we have tested a comercial camera with a fast-neutrons converter on different configurations of the beam and the neutron production target. This has resulted in a first assessment of the limitations, future requirements and viability of introducing neutron radiography as one of the available analysis tools at CNA.

13.1 Trials and Results

At CNA, we have a 3 MV tandem accelerator type pelletron that allows us accelerating protons until the 3 MeV chosen for this experiment [2]. A solid lithium target is installed at the end of the beam line and a water cooling system allows us to reach a proton current over the target up to 10 μA [3]. The fast neutrons are then produced by means of a lithium-berilium nuclear (p, n) reaction.

The neutron camera (mini-iCam 36 mm of NeutronOptics Grenoble) is located at a determinated distance from the source with the sample placed as close as possible from the camera's aperture. The neutron camera has just in the entrance a filter composed by a combination of a rich hydrogen plastic converter for the neutrons and a scintillator material (ZnS). The incident neutrons interact with the converter, exciting the scintillator. The scintillation photons (in the visible range) are reflected

M. A. Millán-Callado (✉) · C. Guerrero · J. Lerendegui-Marco · T. Rodríguez-González · J. M. Quesada
Department FAMN, Facultad de Física, Universidad de Sevilla, Seville, Spain
e-mail: mmillan5@us.es

M. A. Millán-Callado · C. Guerrero · B. Fernández · A. M. Franconetti · M. Macías
Centro Nacional de Aceleradores, Universidad de Sevilla, CSIC, Junta de Andalucía, Seville, Spain

J.-E. García-Ramos et al. (eds.), *Basic Concepts in Nuclear Physics: Theory, Experiments and Applications*, Springer Proceedings in Physics 225,
https://doi.org/10.1007/978-3-030-22204-8_13

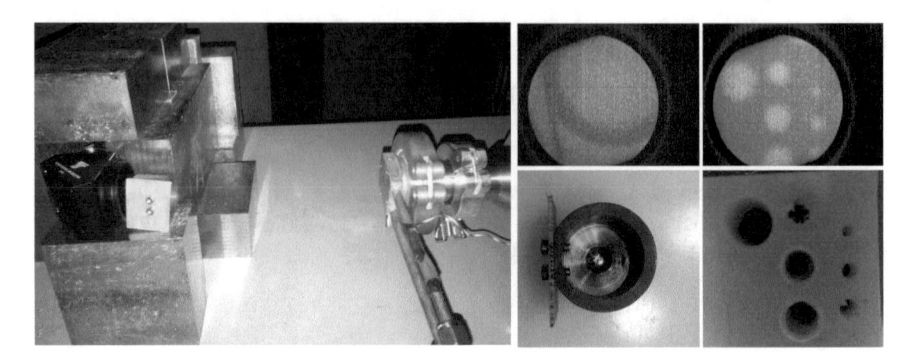

Fig. 13.1 Big panel: Picture of the experimental setup. Small panels: Object and image of some of the tests realized in the campaign. Left: Copper and plastic Faraday Cup. Right: Milimetric holes in a polyethylene piece

at a mirror and illuminate a CCD to produce the image. The CCD is shielded with lead blocks to protect it from the gamma radiation emitted by the target (Fig. 13.1, big panel).

We did a series of tests in order to quantify the viability of the mentioned setup (Fig. 13.1, small panels). These tests supposed the first neutron radiography in a Spanish research center. The conclusions are that we can obtain acceptable images in exposures of tens of minutes, identifying different materials and detecting different thicknesses of the materials, see under metallic shielding and resolve milimetric structures. Despite this promising results, we have to deal with different limitations as the low neutron fluence or the upper limit in the proton current that we can use over the target because of the high activation that we produce on it. In addition, despite the thick shield (5 cm of lead blocks), the long exposure of the camera to the gamma radiation that comes from the source ended up damaging the CCD, increasing the number of dead pixels from a 3 to a 55%.

The future work, in order to reduce this limitations, should be focused on preventing the CCD damage by means of new shieldings and cooling systems, optimizing the image treatment and improving the neutron fluence investigating other production reactions, and studying the viability of collimating the neutron source.

References

1. M. Strobl et al., Advances in neutron radiography and tomography. J. Phys. D: Appl. Phys. **42**, 243001 (2009)
2. J. García López et al., CNA: The first accelerator-based IBA facility in Spain. Nucl. Inst. Methods B **161–163**, 1137–1142 (2000)
3. J. Praena et al., Measurement of the MACS of $^{181}Ta(n,\gamma)$ at $kT = 30\,keV$ as a test of a method for Maxwellian neutron spectra generation. Nucl. Inst. Methods A **727**, 1–6 (2013)

Chapter 14
Modeling Nuclear Effects for Neutrino-Nucleus Scattering in the Few-GeV Region

K. Niewczas, R. González-Jiménez, N. Jachowicz, A. Nikolakopoulos and J. T. Sobczyk

Abstract Accelerator-based neutrino oscillation experiments rely on the description of neutrino interactions with bound nucleons inside atomic nuclei. Neutrino fluxes used in modern experiments (T2K, NOvA) are peaked in the 0.5–5 GeV energy region where one can identify contributions from multiple interaction channels and various nuclear effects. The neutrino-nucleus cross sections in this region are known with a precision not exceeding 20% and have to be investigated further in pursue to reduce systematic errors in oscillation measurements. The concept of Monte Carlo neutrino event generators, which provide essential cross section expectations for oscillation experiments, is examplified by NuWro, the generator developed at the University of Wroclaw. We discuss various implementations of nuclear effects on top of the factorization framework used to describe neutrino-nucleus scattering, focusing specifically on possible generator development using more sophisticated microscopic models.

Theoretical description of the neutrino-nucleus scattering process comes as one of the crucial systematic uncertainties in modern accelerator-based neutrino oscillation experiments [1]. As monoenergetic neutrino beams are not available, models must provide predictions for all possible reaction channels in the wide energy region covered by the neutrino spectra. It has been shown, that it is important to include contributions of two-body currents in addition to the standard one-body current operators for the analysis of oscillation experiments [2].

An HF model developed by the Ghent group is a nonrelativistic calculation of the exclusive two-nucleon knockout via meson-exchange currents (MEC) that uses Hartree-Fock wave functions for both bound and emitted nucleons [3]. The model

K. Niewczas (✉) · N. Jachowicz · A. Nikolakopoulos
Ghent University, Proeftuinstraat 86, B-9000 Gent, Belgium
e-mail: kajetan.niewczas@ugent.be

K. Niewczas · J. T. Sobczyk
University of Wrocław, Plac Maxa Borna 9, 50-204 Wrocław, Poland

R. González-Jiménez
Universidad Complutense de Madrid, CEI Moncloa, 28040 Madrid, Spain

J.-E. García-Ramos et al. (eds.), *Basic Concepts in Nuclear Physics: Theory, Experiments and Applications*, Springer Proceedings in Physics 225,
https://doi.org/10.1007/978-3-030-22204-8_14

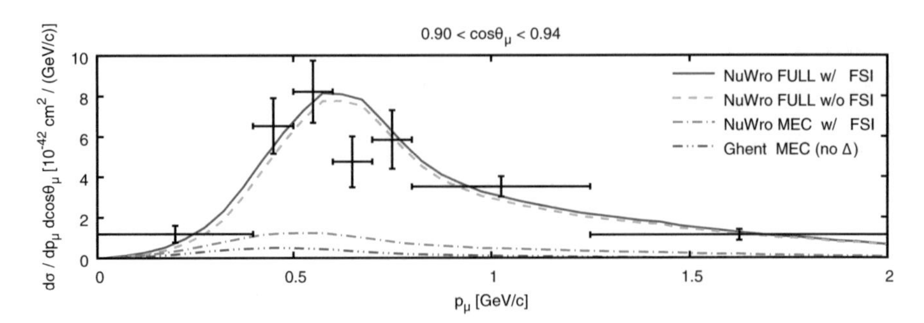

Fig. 14.1 T2K flux-folded double differential cross section per target nucleon for $^{12}C(\nu_\mu, \mu^-)$ without pions in the final state (CC0π analysis) [7]. The full prediction of NuWro compares favorably with the data. The Ghent prediction misses strength due to the lack of diagrams including the Δ-resonance and possible supply from other channels via inelastic FSI

includes both seagull and pion-in-flight diagrams, however misses contributions from any nucleon resonant excitations.

In experimental analyses, the usage of Monte Carlo neutrino event generators is essential, as they give reliable predictions for the whole available phase space. NuWro is a generator developed at the Wrocław group that is successfully used by many experimental collaborations [4]. Here, the MEC channel is modeled using an external inclusive cross section calculation [5], followed by a hadronic phase space model governing the remaining two-nucleon system [6]. On top of this calculation, the semiclassical model of intranuclear cascade is added to account for the final-state interactions (FSI) (Fig. 14.1).

Implmentation of full exclusive microscopic calculations, such as the Ghent model, in Monte Carlo neutrino event generators is essential for extending the apli-cability of those theoretical predictions in experimental analyses.

References

1. L. Alvarez-Ruso et al., Prog. Part. Nucl. Phys. **100**, 1–68 (2018)
2. M. Martini, M. Ericson, G. Chanfray, J. Marteau, Phys. Rev. C **80**, 065501 (2009)
3. T. Van Cuyck et al., Phys. Rev. C **95**, 054611 (2017)
4. T. Golan, C. Juszczak, J.T. Sobczyk, Phys. Rev. C **86**, 015505 (2012)
5. J. Nieves, I. Ruiz Simo, M. Vicente Vacas, Phys. Rev. **C83**, 045501 (2011)
6. J.T. Sobczyk, Phys. Rev. C **86**, 015504 (2012)
7. K. Abe et al., (T2K Collaboration). Phys. Rev. D **93**, 112012 (2016)

Chapter 15
Effect of Outgoing Nucleon Wave Function on Reconstructed Neutrino Energy

A. Nikolakopoulos, M. Martini, N. Van Dessel, K. Niewczas, R. González-Jiménez and N. Jachowicz

Abstract The main goal of accelerator-based neutrino experiments is the determination of the neutrino oscillation parameters, The oscillation probability depends on the ratio of the distance traveled by the neutrino to its energy, therefore the determination of the distribution of neutrino energies in a detector is crucial. The reconstructed neutrino energy is a kinematic variable which is determined by the energy and scattering angle of the final state lepton in charged current (CC) scattering off an atomic nucleus. The distribution of reconstructed energies around the true incoming energy depends on the nuclear model used to describe the ν-nucleus cross section. We show the effect of distortion of the outgoing nucleon wave function on these distributions.

In a detector one observes the CC scattering of a neutrino off a nucleus where a single final-state lepton, with energy E_l and scattering angle $\cos\theta_l$, is detected. In the experimental analysis the reconstructed energy $\overline{E}_\nu\,(E_l, \cos\theta_l)$ is the energy of the neutrino scattering of a neutron at rest, corrected for binding [1, 2]. After binning the data in terms of $\overline{E}_\nu$, the true energy distribution has to be recovered. This requires a nuclear model for the interaction. In the experimental analysis a relativistic Fermi gas (RFG) model is commonly used for this task.

In the HF model for quasielastic (QE) scattering [3–5], the final-state wave function is constructed from continuum states in the same HF potential used for the bound states. In Fig. 15.1, we compare the distribution of $\overline{E}_\nu$ for a fixed real E_ν obtained with the full HF model, with the results of the HFPW model in which the final state nucleon wave function is a plane wave. In this way we directly asses the effect of the outgoing wave function.

A. Nikolakopoulos (✉) · N. Van Dessel · K. Niewczas · N. Jachowicz
Ghent University, Proeftuinstraat 86, 9000 Gent, Belgium
e-mail: alexis.nikolakopoulos@ugent.be

M. Martini
ESNT, CEA, IRFU, SPN, Université Paris-Saclay, 91191 Gif-sur-Yvette, France

R. González-Jiménez
Universidad Complutense de Madrid, CEI Moncloa, 28040 Madrid, Spain

J.-E. García-Ramos et al. (eds.), *Basic Concepts in Nuclear Physics: Theory, Experiments and Applications*, Springer Proceedings in Physics 225,
https://doi.org/10.1007/978-3-030-22204-8_15

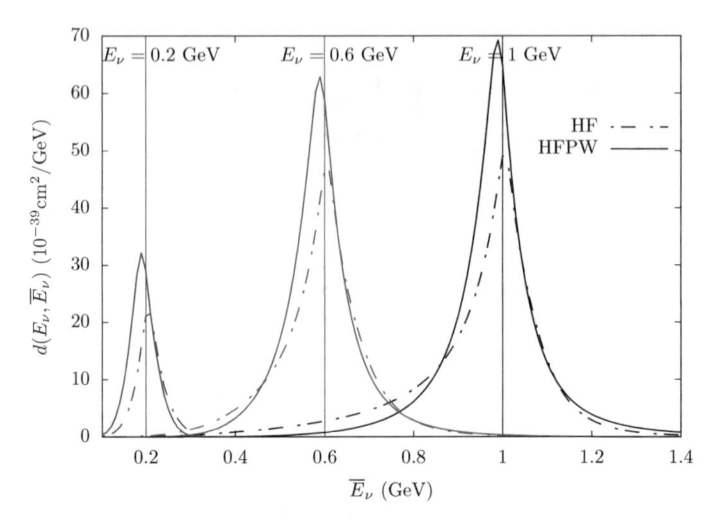

Fig. 15.1 $d(\overline{E}_\nu, E_\nu)$, defined and described in [1, 6], is proportional to the probability of a neutrino with real energy E_ν giving rise to a reconstructed energy $\overline{E}_\nu$. Calculated for CCQE scattering of electron neutrinos off ^{12}C. $\overline{E}_\nu$ is defined as in (1) of [1] with $E_B = 25$ MeV

The outgoing nucleon wave function affects the magnitude of the cross section, owing to the elimination of non-orthogonal contributions which are present in the plane wave. More important is that a reshaping occurs in the HF model, the peak of the distribution shifts to slightly larger values of $\overline{E}_\nu$, while the low $\overline{E}_\nu$ tail is strongly enhanced. This asymmetry is not reproduced with PW models or with the commonly used RFG. Not taking these more asymmetric distributions into account could introduce a significant bias in the analysis of oscillation experiments [1].

References

1. A. Nikolakopoulos et al., Phys. Rev. C **98**, 054603 (2018)
2. Aguilar-Arevalo et al., Phys. Rev. D **81**, 092005 (2010)
3. V. Pandey et al., Phys. Rev. C **92**, 024606 (2015)
4. N. Jachowicz et al., Phys. Rev. C **65**, 025501 (2002)
5. J. Ryckebusch et al., Nucl. Phys. A **503**(3), 694 (1989)
6. M. Martini et al., Phys. Rev. D **87**, 013009 (2013)

Chapter 16
Measurement of the Production Cross Sections of β^+ Emitters for Range Verification in Proton Therapy

T. Rodríguez-González, C. Guerrero, M. C. Jiménez-Ramos, J. Lerendegui-Marco, M. A. Millán-Callado, A. Parrado and J. M. Quesada

Abstract In proton therapy, there is an intensive research program aiming at in vivo range verification in order to reduce the uncertainties in the range of the proton beam, that limit the benefits of having a sharp Bragg peak [1]. In-vivo PET range verification relies on the comparison of the measured and estimated activity distributions of β^+ emitters induced on C, N, O, Ca and P by the protons along the body of the patient. As the accuracy of the estimated distribution depends on the underlying cross sections data [2], a revision of the experimental data available has been done, showing that they are not always available in the full energy range of interest (up to 250 MeV) and that there are sizeable differences in some cases [3]. The aim of this study is to develop a method for measuring the production cross sections of the β^+ emitters up to 250 MeV. For this, as starting point, the production cross sections of ^{11}C and ^{13}N in ^{nat}C, ^{nat}N and ^{nat}O have been measured at Bragg peak energies using the 18 MeV proton beam at the CNA cyclotron and a multi-layer target configuration. The activity induced in each film has been measured using the clinical PET scanner at CNA.

16.1 Experimental Set up and Results

Using the external line of the cyclotron, three targets of polyethylene (PE), PMMA and nylon-6 have been irradiated in order to obtain the production cross section of ^{11}C and ^{13}N in ^{nat}C, ^{nat}O and ^{nat}N, respectively. The energy beam is degraded using a multi-stack target configuration in order to obtain the cross section at different proton energies. A specific target holder has been designed and, attached to a motorized table, allows positioning the targets remotely (Fig. 16.1, left). In this way one can irradiate

T. Rodríguez-González (✉) · C. Guerrero · J. Lerendegui-Marco · M. A. Millán-Callado
J. M. Quesada
Departamento de Física Atómica, Molecular y Nuclear, Universidad de Sevilla, 41012 Seville, Spain
e-mail: mrodriguezg@us.es

C. Guerrero · M. C. Jiménez-Ramos · A. Parrado
Centro Nacional de Aceleradores, Av. Thomas A. Edison, Seville, Spain
e-mail: cguerrero4@us.es

J.-E. García-Ramos et al. (eds.), *Basic Concepts in Nuclear Physics: Theory, Experiments and Applications*, Springer Proceedings in Physics 225,
https://doi.org/10.1007/978-3-030-22204-8_16

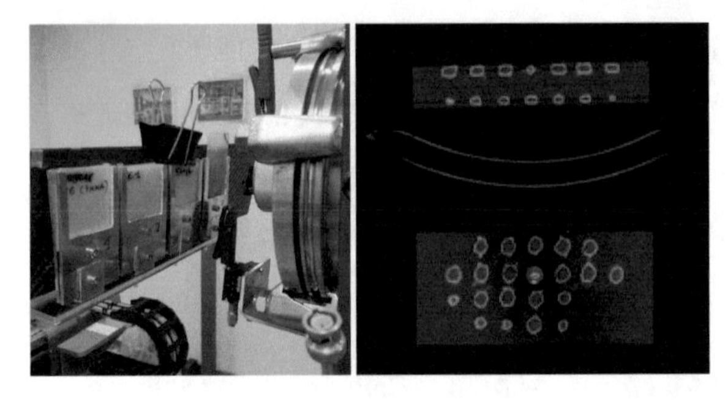

Fig. 16.1 Left: experimental set up at CNA cyclotron. Right: PET image of the activated layers placed between thick layers of PE superposed to CT image

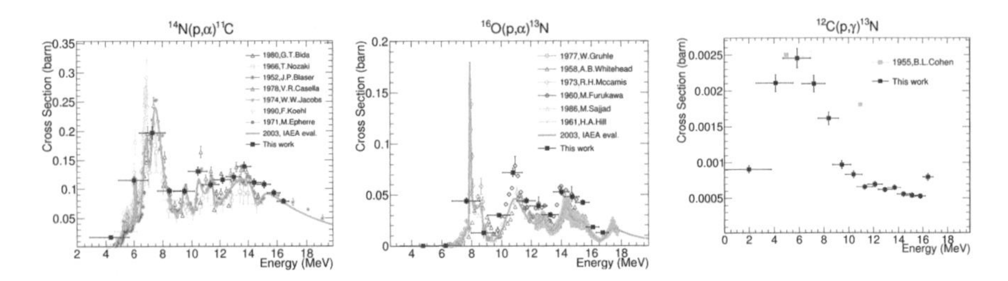

Fig. 16.2 Production cross sections for three reaction channels of interest

each stack without entering the experimental room, so that the decay of the induced activity between consecutive irradiations is minimized. The first film in each stack is a PMMA film, used to monitor the current beam and to validate the decay corrections applied in the analysis.

The activity induced in each film has been measured using the PET/CT scanner, so that the acquisition in dynamic mode (acquisitions of 1 min during 5 h) provides the activity curves as function of decay time to identify the decay of each isotope. Previously, a calibration of the PET scanner efficiency as function of the position has been done. The irradiated films were placed between thick layers of PE, acting as converters of the β^+ into photons, in two different planes (Fig. 16.1, right).

The production of ^{11}C and ^{13}N is determined by fitting the decay curves to two exponentials plus a constant and taking into account the decay during both irradiation and cooldown. The production cross sections have been obtained between 4 and 17 MeV and compared with the data available in the literature (Fig. 16.2). Given the success of the measurement with this new method, a similar experiment but at higher energies using clinical beams is being planned.

References

1. A.C. Knopf, A. Lomax, Phys. Med. Biol. **58** (2013)
2. H. Paganetti, Phys. Med. Biol. **57** (2012)
3. Experimental Nuclear Reaction Data (EXFOR). https://www-nds.iaea.org/exfor/exfor.htm

Part III
Student's Posters

Chapter 17
Calculation of Energy Level and B(E2) Values and G-Factor of Even–Even Isotopes of Sulfur Using the Shell Model Code NuShellX

Amin Attarzadeh and Saed Mohammadi

Abstract The neutron-rich isotopes with $Z \leq 20$, in particular those with neutron numbers around $N = 28$, have been the focus of a lot experimental and theoretical scrutiny during past few years. The calculation of energy levels, B (E2) values and g-factors of $^{32,34,36,38,40,42,44}S$ is divided in three types of interactions according to the sd, pf and sd-pf valance space. The model space and interaction which is used to calculate 32–36S is usdb. Because of increasing the neutron number in $^{38,40,42,44}S$ from sd to pf shell over, the sdpf and sdpfwb are selected as appropriate model space and interaction in second phase of calculations which are in good agreement with experimental data. Finally, the energy level and B (E2) values of ^{44}S and ^{36}S are calculated as isotopes with neutron number around $N = 28$ and $N = 20$ respectively. With consideration of magicity in these isotopes, one can see the higher amount of excitation energy and g-factor with respect to the others far from closed shell.

17.1 Introduction and Theory

A shell model calculation starts with a set of single particle states. The interaction between nucleons is, in general, sufficiently strong that calculations must be carried out in a Hilbert space far longer than what is practical. On the other hand, by transforming the single particle state to a Hartree-Fock basis, a large part of the nucleon-nucleon interaction may be included in the average one-body field for a nucleon. The residual interaction remaining may be weak enough that the nuclear many-body eigenvalue problem may be may be carried out in a small subset of the complete space. So must of the many body basis state are not involved in any significant way in the low-lying states of interest to us. The shell model investigations always satisfy this criterion. With transformation to such a nucleon-nucleon poten-

A. Attarzadeh (✉)
Institute for Higher Education ACECR, Khouzestan, Iran
e-mail: Attarzadeh_amin@yahoo.com

S. Mohammadi
Department of Physics, Payame Noor University (PNU), P.O. Box 19395–3697, Tehran, Iran

J.-E. García-Ramos et al. (eds.), *Basic Concepts in Nuclear Physics: Theory, Experiments and Applications*, Springer Proceedings in Physics 225,
https://doi.org/10.1007/978-3-030-22204-8_17

tial it is appropriate for the single-particle representation adopted. In addition, two further considerations must also be included. The first is the interaction of bound nucleon. The second is related to the truncation of the shell model space.

$$\mathrm{H} = \mathrm{H}_0 + \mathrm{H}_{\mathrm{res}}$$

Since many-body basis states are made of products of single-particle wave function a selection of shell model space is usually achieved by restricting the number of active single particle states. The H_0 and $\mathrm{H}_{\mathrm{res}}$ can be shown in the form of matrix as;

$$H = \begin{pmatrix} \epsilon_1 & 0 \\ 0 & \epsilon_2 \end{pmatrix} + \begin{pmatrix} <\psi_1|H_{res}|\psi_1> & <\psi_1|H_{res}|\psi_2> \\ <\psi_2|H_{res}|\psi_1> & <\psi_2|H_{res}|\psi_2> \end{pmatrix}$$

where, the $\epsilon_1\, and\, \epsilon_2$ are single particle energies given in solution to H or from experiment. Diagonal matrix elements $<\psi_x|\mathrm{H}_{\mathrm{res}}|\psi_x>$ are expectation values of $\mathrm{H}_{\mathrm{res}}$ on $|\psi_i>$. Non-diagonal matrix elements $<\psi_x|\mathrm{H}_{\mathrm{res}}|\psi_y>$ describe configuration mixing. The nuclear transition probabilities of even-even isotopes of sulphur are the other parameter that is calculated in the basis of shell model concepts. Though these probabilities are in wiesskopf unit in the result of NushellX calculation, to convert it to B (E2) up in $\mathrm{e}^2\mathrm{b}^2$ the relations

$$B_W(E\lambda) = \left(\frac{1}{4\pi}\right)\left[\frac{3}{(3+\lambda)}\right]^2 \left(1.2A^{1/3}\right)^{2\lambda} e^2\mathrm{fm}^{2\lambda}$$

and

$$B(\mathrm{E2})_{\mathrm{WU}} = 5.94 \times 10^{-6} A^{4/3}\left(e^2\mathrm{b}^2\right),$$

Are used. The other parameter which is calculated in this research is g-factor. Nuclear magnetic moments are particularly sensitive to the specific proton and neutron configurations in the wave functions due to the proton's and neutron's strikingly different spin and orbital *g*-factor values. In this article the g-factor even- even isotopes of $^{32-44}\mathrm{S}$ are calculated [1].

17.2 Result and Discussion

According to the Table 17.1, the energy level of first excited states of even—even isotopes of sulfur are calculated and compared with the experimental data [2].

Rather that energy levels, B(E2) values related to the first exited states and the g-factors are calculated in extension. The consistency in rising energy level and g-factor at closed shell in 36S and 44S is shown in Table 17.1. The g-factor for 42S and 44S nuclei are not shown because of lack of reliable empirical data in references.

Table 17.1 Experimental and calculated values of $E\left(0_1^+ \rightarrow 2_1^+\right)$, $B\left(E_2; 0_1^+ \rightarrow 2_1^+\right)$, *g-factor* of even-even isotopes of S

Nuclei	E_{exp} (KeV)	E_{cal} (KeV)	$B(E2)_{exp}$(e2b2)	$B(E2)_{cal}$(e2b2)	g-factor $_{exp}$	g-factor $_{cal}$
$^{32}_{16}S$	2230	2160	0.0288	0.0300	+0.450	+0.503
$^{34}_{16}S$	2128	2131	0.0212	0.0229	+0.500	+0.546
$^{36}_{20}\mathbf{S}$	**3290**	**3382**	**0.0089**	**0.0131**	**+1.160**	**+1.190**
$^{38}_{16}S$	1292	1530	0.0235	0.0256	+0.130	−0.005
$^{40}_{16}S$	903.6	980.0	0.3340	0.4515	−0.01	+0.070
$^{42}_{16}S$	903	1023	0.0397	0.0541	–	0.431
$^{44}_{16}\mathbf{S}$	**1329**	**1865**	**0.0310**	**0.0292**	–	**1.48**

While at this region of orbital the amount of B (E2) falls to minimum with respect to the isotopes in its neighbors. The transition of nucleons from sd to pf shell and changing the treatment of nuclei at closed shells can determine the framework of theoretical calculation in truncation and limitation consideration in NushellX [3].

References

B.A. Brown, Lecture Notes in Nuclear Structure Physics. Michigan State University
National Nuclear Data Center, Brookhaven. www.nndc.bnl.gov
"The shell model code NushellX@MSU", B.A. Brown, W.D.M. Rae, *Nuclear data sheets*, vol. 120, pp. 115–118

Chapter 18
Characterization and First Test of an i-TED Prototype at CERN n_TOF

V. Babiano-Suarez, L. Caballero, C. Domingo-Pardo, I. Ladarescu, O. Aberle, V. Alcayne, S. Amaducci, J. Andrzejewski, L. Audouin, M. Bacak, M. Barbagallo, V. Bécares, F. Bečvář, G. Bellia, E. Berthoumieux, J. Billowes, D. Bosnar, A. S. Brown, M. Busso, M.Caamaño, M. Calviani, F. Calviño, D. Cano-Ott, A. Casanovas, F. Cerutti, Y. H. Chen, E. Chiaveri, N. Colonna, G. P. Cortés, M. A. Cortés-Giraldo, L. Cosentino, S. Cristallo, L. A. Damone, M. Diakaki, M. Dietz, R. Dressler, E. Dupont, I. Durán, Z. Eleme, B. Fernández-Domíngez, A. Ferrari, I. Ferro-Gon calves, P. Finocchiaro, V. Furman, A. Gawlik, S. Gilardoni, T. Glodariu, K. Göbel, E. González-Romero, C. Guerrero, F. Gunsing, S. Heinitz, J. Heyse, D. G. Jenkins, Y. Kadi, F. Käppeler, A. Kimura, N. Kivel, M. Kokkoris, Y. Kopatch, M. Krtička, D. Kurtulgil, C. Lederer-Woods, J. Lerendegui-Marco, S. Lo Meo, S.-J. Lonsdale, D. Macina, A. Manna, T. Martínez, A. Masi, C. Massimi, P. F. Mastinu, M. Mastromarco, F. Matteucci, E. Maugeri, A. Mazzone, E. Mendoza, A. Mengoni, V. Michalopoulou, P. M. Milazzo, F. Mingrone, A. Musumarra, A. Negret, R. Nolte, F. Ogállar, A. Oprea, N. Patronis, A. Pavlik, J. Perkowski, L. Piersanti, I. Porras, J. Praena, J. M. Quesada, D. Radeck, D. Ramos Doval, T. Rausher, R. Reifarth, D. Rochman, C. Rubbia, M. Sabaté-Gilarte, A. Saxena, P. Schillebeeckx, D. Schumann, A. G. Smith, N. Sosnin, A. Stamatopoulos, G. Tagliente, J. L. Tain, Z. Talip, A. E. Tarifeño-Saldivia, L. Tassan-Got, A. Tsinganis, J. Ulrich, S. Urlass, S. Valenta, G. Vannini, V. Variale, P. Vaz, A. Ventura, V. Vlachoudis, R. Vlastou, A. Wallner, P. J. Woods, T. J. Wright and P. Žugec

Abstract Neutron capture cross section measurements are of fundamental importance for the study of the slow process of neutron capture, so called s-process. This mechanism is responsible for the formation of most elements heavier than iron in the

---

V. Babiano-Suarez (✉) · L. Caballero · C. Domingo-Pardo · I. Ladarescu · J. L. Tain
Instituto de Física Corpuscular, CSIC - Universidad de Valencia, Valencia, Spain
e-mail: vbabiano@ific.uv.es

O. Aberle · M. Bacak · M. Barbagallo · M. Calviani · F. Cerutti · E. Chiaveri · A. Ferrari
S. Gilardoni · Y. Kadi · D. Macina · A. Masi · M. Mastromarco · F. Mingrone · C. Rubbia
M. Sabaté-Gilarte · L. Tassan-Got · A. Tsinganis · S. Urlass · V. Vlachoudis
European Organization for Nuclear Research (CERN), Meyrin, Switzerland

V. Alcayne · E. Mendoza · D. Cano-Ott · V. Bécares · E. González-Romero · T. Martínez
Centro de Investigaciones Energéticas Medioambientales y Tecnológicas (CIEMAT), Madrid, Spain

J.-E. García-Ramos et al. (eds.), *Basic Concepts in Nuclear Physics: Theory, Experiments and Applications*, Springer Proceedings in Physics 225,
https://doi.org/10.1007/978-3-030-22204-8_18

Universe. To this aim, installations and detectors have been developed, as total energy radiation C_6D_6 detectors. However, these detectors can not distinguish between true capture gamma rays from the sample under study and neutron induced gamma rays produced in the surroundings of the setup. To improve this situation, we propose (Domingo Pardo in Nucl Instr Meth Phys Res A 825:78–86, 2016, [1]) the use of the Compton principle to select events produced in the sample and discard background events. This involves using detectors capable of resolving the interaction position of the gamma ray inside the detector itself, as well as a high energy resolution. These are the main features of i-TED, a total energy detector capable of gamma ray imag-

---

A. Kimura
Japan Atomic Energy Agency (JAEA), Tokai-mura, Japan

S. Amaducci · G. Bellia · L. Cosentino · P. Finocchiaro · A. Musumarra
INFN Laboratori Nazionali del Sud, Catania, Italy

J. Andrzejewski · A. Gawlik · J. Perkowski
University of Lodz, Lodz, Poland

L. Audouin · Y. H. Chen · D. Ramos Doval · L. Tassan-Got
IPN, CNRS-IN2P3, Univ. Paris-Sud, Université Paris-Saclay, 91406 Orsay Cedex, France

M. Bacak
Technische Universität Wien, Vienna, Austria

M. Barbagallo · N. Colonna · L. A. Damone · A. Mazzone · G. Tagliente · V. Variale
Istituto Nazionale di Fisica Nucleare, Bari, Italy

F. Bečvář · M. Krtička · S. Valenta
Charles University, Prague, Czech Republic

G. Bellia · A. Musumarra
Dipartimento di Fisica e Astronomia, Università di Catania, Catania, Italy

E. Berthoumieux · E. Dupont · F. Gunsing
CEA Saclay, Irfu, Universit é Paris-Saclay, Gif-sur-Yvette, France

J. Billowes · E. Chiaveri · A. G. Smith · N. Sosnin · T. J. Wright
University of Manchester, Manchester, UK

D. Bosnar · P. Žugec
University of Zagreb, Zagreb, Croatia

A. S. Brown · D. G. Jenkins
University of York, York, UK

M. Busso · S. Cristallo · L. Piersanti
Istituto Nazionale di Fisica Nazionale, Perugia, Italy

M. Busso
Dipartimento di Fisica e Geologia, Università di Perugia, Perugia, Italy

M. Caamaño · I. Durán · B. Fernández-Domíngez
University of Santiago de Compostela, Santiago de Compostela, Spain

F. Calviño · A. Casanovas · G. P. Cortés · A. E. Tarifeño-Saldivia
Universitat Politècnica de Catalunya, Barcelona, Spain

M. A. Cortés-Giraldo · C. Guerrero · J. Lerendegui-Marco · J. M. Quesada · M. Sabaté-Gilarte,
Instituto de Física Corpuscular, CSIC - Universidad de Valencia, Valencia, Spain

ing. Such system is being developed at the "Gamma Spectroscopy and Neutrons Group" at IFIC (http://webgamma.ific.uv.es/gamma/es/, [2]), in the framework of the ERC-funded project HYMNS (High sensitivitY and Measurements of key stellar Nucleo-Synthesis reactions). This work summarizes first tests with neutron beam at CERN n_TOF.

S. Cristallo · L. Piersanti
Istituto Nazionale di Astrofisica - Osservatorio Astronomico di Teramo, Teramo, Italy

L. A. Damone
Dipartimento di Fisica, Università degli Studi di Bari, Bari, Italy

M. Diakaki · M. Kokkoris · V. Michalopoulou · A. Stamatopoulos · L. Tassan-Got · R. Vlastou
National Technical University of Athens, Athens, Greece

M. Dietz · C. Lederer-Woods · S.-J. Lonsdale · P. J. Woods
School of Physics and Astronomy, University of Edinburgh, Edinburgh, UK

R. Dressler · S. Heinitz · N. Kivel · E. Maugeri · D. Rochman · D. Schumann · Z. Talip · J. Ulrich
Paul Scherrer Institut (PSI), Villigen, Switzerland

Z. Eleme · N. Patronis
University of Ioannina, Ioannina, Greece

I. Ferro-Gon calves · P. Vaz
Instituto Superior Técnico, Lisbon, Portugal

V. Furman · Y. Kopatch
Joint Institute for Nuclear Research (JINR), Dubna, Russia

T. Glodariu · A. Negret · A. Oprea
Horia Hulubei National Institute of Physics and Nuclear Engineering (IFIN-HH), Bucharest, Romania

K. Göbel · D. Kurtulgil · R. Reifarth
Goethe University Frankfurt, Frankfurt, Germany

J. Heyse · P. Schillebeeckx
European Commission, Joint Research Centre, Geel, Retieseweg 111, 2440 Geel, Belgium

F. Käppeler
Karlsruhe Institute of Technology, Campus North, IKP, 76021 Karlsruhe, Germany

S. Lo Meo · A. Mengoni
Agenzia nazionale per le nuove tecnologie, l'energia e lo sviluppo economico sostenibile (ENEA), Bologna, Italy

S. Lo Meo · A. Manna · C. Massimi · A. Mengoni · G. Vannini · A. Ventura
Istituto Nazionale di Fisica Nucleare, Sezione di Bologna, Bologna, Italy

A. Manna · C. Massimi · G. Vannini
Dipartimento di Fisica e Astronomia, Università di Bologna, Bologna, Italy

P. F. Mastinu
Istituto Nazionale di Fisica Nucleare, Sezione di Legnaro, Legnaro, Italy

18.1 i-TED Concept and First Tests at CERN n_TOF

Compton cameras are widely used in various fields such as astronomy, medicine, and the treatment of radioactive waste. In this work we explore the possibility to apply them also in the field of neutron capture experiments. The detector consists of two stages, scatter and absorber, operated in temporal coincidence. This allows us to apply the Compton principle to obtain information on the direction of origin of the gamma ray. Each stage is composed of $LaCl_3(Ce)$ scintillation crystals (thinner in the scatter than in the absorber), coupled to pixelated silicon photomultipliers (SiPM) readout by a fronted electronics from PETSyS [3].

On the left part of the Fig. 18.1, the experimental setup for the first tests of the detector in n_TOF at CERN it shown. This facility provides pulsed and intense neutron bunches over a broad every range [4]. In order to obtain the neutron energy, an external trigger input was implemented on the electronic system.

18.2 Characterization

The energy resolution is relevant because the uncertainty in this quantity leads also to an uncertainty on the Compton cone. i-TED achieves resolutions of around 5% at 662 keV [5]. On the other hand, we obtain spatial information of the gamma ray hits

F. Matteucci · P. M. Milazzo
Istituto Nazionale di Fisica Nazionale, Trieste, Italy

F. Matteucci
Dipartimento di Fisica, Università di Trieste, Trieste, Italy

A. Mazzone
Consiglio Nazionale delle Ricerche, Bari, Italy

R. Nolte · D. Radeck
Physikalisch-Technische Bundesanstalt (PTB), Bundesallee 100, 38116 Braunschweig, Germany

F. Ogállar · I. Porras · J. Praena
University of Granada, Granada, Spain

A. Pavlik
Faculty of Physics, University of Vienna, Vienna, Austria

T. Rausher
Department of Physics, University of Basel, Basel, Switzerland

T. Rausher
School of Physics, Astronomy and Mathematics, University of Hertfordshire, Hertfordshire, UK

A. Saxena
Bhabha Atomic Research Centre (BARC), Mumbai, India

S. Urlass
Helmholtz-Zentrum Dresden-Rossendorf, Dresden, Germany

A. Wallner
Australian National University, Canberra, Australia

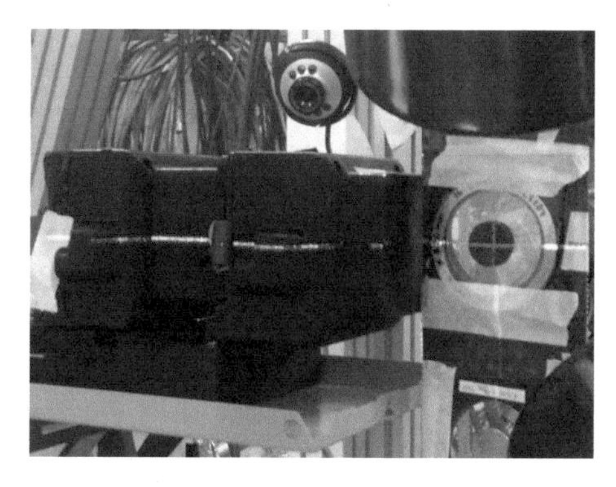

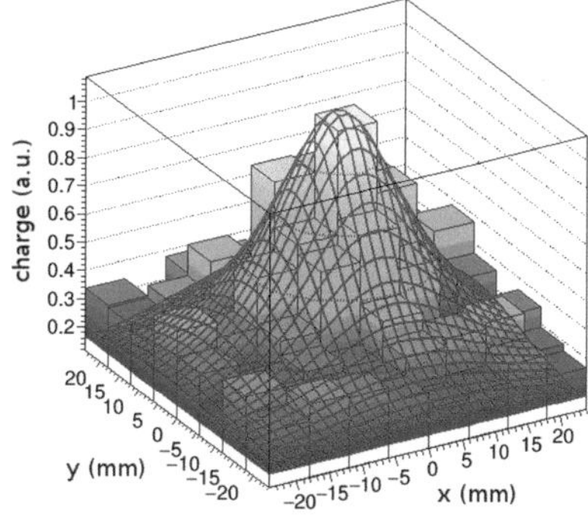

Fig. 18.1 Experimental setup with an i-TED prototype (left). Charge distribution of a gamma event fitted with one analytical formula (right)

by means of a pixelated SiPM photosensor coupled to the crystal. The information from the SiPM basically allows us to trace the vertex and the axis of the Compton cone for each detected event.

For the position reconstruction, we have investigated different algorithms in order to recover the 3D spatial coordinates of the gamma ray hit in each crystal. These characterization studies will be reported in a separate work [6]. For example, on the right part of Fig. 18.1 one can see the charge distribution of a gamma event centered in the crystal fitted by an analytical form. The accuracy in the reconstructed position ranges between 1 mm and 3 mm FWHM depending on the method used.

Acknowledgements This project has received funding from the European Research Council (ERC) under the European Union's Horizon 2020 research and innovation program (grant agreement No. 681740). We acknowledge support from the Spanish project FPA2017-83946-C2-1-P.

References

1. C. Domingo Pardo, i-TED: a novel concept for high-sensitivity (n,) cross section measurements. Nucl. Instr. Meth. Phys. Res. A **825**, 78–86 (2016)
2. http://webgamma.ific.uv.es/gamma/es/
3. http://www.petsyselectronics.com/web/
4. C. Guerrero, Performance of the neutron time-of-flight facility n_TOF at CERN. Eur. Phys. J. A **49** (2013)
5. P. Olleros, Spatial and spectroscopic characterization of high-resolution $LaCl_3$(Ce) crystals coupled to silicon photosensors (Universidad de Salamanca, 2017)
6. V. Babiano Suarez, Characterization of the 3D spatial response of large monolithic $LaCl_3$ (Ce) crystals coupled to pixelated photosensors, in preparation

Chapter 19
Development of a New Radiobiology Beam Line for the Study of Proton RBE at the 18 MeV Proton Cyclotron Facility at CNA

A. Baratto-Roldán, M. A. Cortés-Giraldo, M. C. Jiménez-Ramos, M. C. Battaglia, J. García López, M. I. Gallardo and J. M. Espino

Abstract At the National Centre of Accelerators (CNA) in Seville, Spain, a system for the irradiation of mono-layer cell cultures is under study, to be implemented at the external beam line of the cyclotron facility. This cyclotron delivers an 18 MeV proton beam, which can be used for investigating the Relative Biological Effectiveness (RBE) of protons at low energies. In the following, the characteristics of the experimental beam line are presented, together with a description of the solution proposed for the irradiation of biological samples and some preliminary results.

19.1 Introduction

Doses deposited by protons are considered to be 10% more effective for cell killing than those deposited by photons, property which is quantitatively described by the Relative Biological Effectiveness (RBE): the ratio between dose deposited by photon and proton beams determining the same biological effect. Thus, in clinical proton therapy treatments, a uniform RBE value of 1.1 is generally used [1], even if it is agreed that this quantity varies towards the distal Bragg peak region, increasing with Linear Energy Transfer (LET). Ignoring variations of RBE near the Bragg peak may have important clinical consequences [2], which makes studies of RBE at low proton energies highly relevant. To perform such studies, beam lines as the one installed at the 18 MeV proton cyclotron facility of the National Centre of Accelerators (CNA, Seville, Spain), are of great interest, since they can provide a low energy proton beam (below 18 MeV) with lower straggling as compared with those available in clinical

A. Baratto-Roldán (✉) · M. C. Jiménez-Ramos · J. García López · J. M. Espino
Centro Nacional de Aceleradores, 41092 Seville, Spain
e-mail: abaratto@us.es

M. A. Cortés-Giraldo · M. C. Battaglia · J. García López · M. I. Gallardo · J. M. Espino
Departamento de Física Atómica, Molecular y Nuclear, Universidad de Sevilla, 41012 Seville, Spain

J.-E. García-Ramos et al. (eds.), *Basic Concepts in Nuclear Physics: Theory, Experiments and Applications*, Springer Proceedings in Physics 225,
https://doi.org/10.1007/978-3-030-22204-8_19

proton therapy facilities at similar energies. In this work, the first feasibility study of a radiobiology beam line for proton RBE measurements at the CNA cyclotron facility is presented.

19.2 Methods and Results

At present, the experimental beam line installed at the cyclotron facility counts on two Faraday cups, for beam diagnostics, and on two collimators, for beam delivery and shaping, of 1.5 cm diameter and placed 1 m apart from one another, being the downstream one mounted just before a 125 μm Polyethylene terephthalate (PET) exit window. The spatial configuration of this beam line allows the irradiation of monolayer cell cultures grown in Petri dishes and mounted orthogonally with respect to the beam axis at the exit of the beam line. With this configuration, two are the major constraints when dealing with cell irradiations: low beam intensity at the sample position, of the order of some pA, to control properly the fluence within suitable irradiation time scales, and broad irradiation field, of the order of few cm, homogeneous in both energy and spatial distribution to cover the whole sample. In order to improve the homogeneity and decrease the beam intensity, the decision to use a completely defocused beam and to scatter the beam downstream the exit window has been made, by inserting tungsten scattering foils and by varying the exit-window-to-sample distance.

Experiments have been performed in different irradiation conditions, a prototype design of the sample holder has been built and measurements of the beam intensity profile have been performed using EBT3 radiochromic films. With proton beam energies of 11 MeV, obtained with a tungsten foil of 150 μm thickness and at a distance of approximately 50 cm from the exit window in air, a homogeneous irradiation field, with maximum deviations of around 8%, has been obtained in the whole sample area (3.5 cm diameter). Furthermore, preliminary dosimetric studies have been performed, using EBT3 radiochromic films and a transmission ionization chamber for dose and proton fluence evaluation. Studies of this nature are of great interest, since radiochromic films would be a handy and easy to use dosimeter solution for proton RBE studies [3].

Acknowledgements This work has received funding from the Spanish Ministry of Economy and Competitiveness under grant No. FPA2016-77689-C2-1-R and from the EU Horizon 2020 research and innovation programme under the Marie Sklodowska-Curie grant agreement No 675265, OMA—Optimization of Medical Accelerators.

References

1. M. Durante, H. Paganetti, Rep. Prog. Phys. **79**, 096702 (2016)
2. M. Wedenberg et al., Med. Phys. **41**, 091706 (2014)
3. M.C. Battaglia et al., Phys. Rev. Accel. Beams **19**, 064701 (2016)

Chapter 20
Role of Competing Transfer Channels on Charge-Exchange Reactions

S. Burrello, J. I. Bellone, Maria Colonna, José A. Lay Valera and Horst Lenske

Abstract There is a recently renewed experimental and theoretical interest in studying single and double charge exchange reactions with heavy ions. We would like to discuss a preliminary theoretical study of charge exchange reactions within the Distorted Wave Born approximation (DWBA) framework. This allows us to include the effect of the different competing transfer channels in a proper way. Evidences seems to show that, under suitable conditions, the charge exchange process could reveal safely dominant, although the competition between the two mechanisms calls for further investigation.

20.1 Introduction

Charge-exchange (CEX) reactions are well established tools for spectroscopic studies of nuclear states, where participating nuclei keep their masses constant but exchange their charge. In particular, there is a recently renewed experimental and theoretical interest in studying double charge-exchange (DCEX) reactions with heavy ions. Within this context, the NUMEN project at LNS [1] is looking at the possibility to extract from these reactions the nuclear matrix elements (NME) of interest for the double-beta decay process. Various mechanisms may contribute to the isospin transition associated to a CEX reaction, that are:

- direct conversion of nucleons, through meson exchange;
- multi-step transfer via intermediate states, feeding the same outgoing channels.

S. Burrello (✉) · J. I. Bellone · M. Colonna
INFN - LNS, Catania, Italy
e-mail: burrello@lns.infn.it

S. Burrello · J. A. L. Valera
Departamento de FAMN, Universidad de Sevilla, Seville, Spain

H. Lenske
Institut für Theoretische Physik, JLU Giessen, Giessen, Germany

J.-E. García-Ramos et al. (eds.), *Basic Concepts in Nuclear Physics: Theory, Experiments and Applications*, Springer Proceedings in Physics 225,
https://doi.org/10.1007/978-3-030-22204-8_20

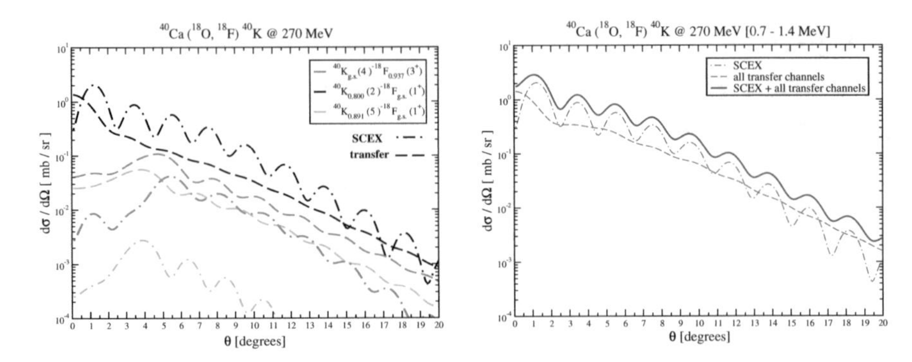

Fig. 20.1 Angular distribution of the differential cross section for the reaction ^{40}Ca (^{18}O, ^{18}F) ^{40}K @ 270 MeV for single CEX and transfer mechanism. Left panel: results for specific final states. Right panel: sum up over all contributions within the excitation energy range considered

The goal of the project is to evaluate each contribution to get a complete description of DCEX reactions, with all the mechanisms sharing the same initial and final states.

20.2 Results

One could treat these direct reactions within the DWBA framework and correspondingly evaluate the differential cross section. The projectile-target interaction is described by a complex optical potential; concerning the structure part, for the CEX, the nuclear transition form factors are microscopically deduced by performing Quasi-particle Random Phase Approximation (QRPA) calculations [2], while for the transfer, the inputs for the overlaps are given by shell model spectroscopic amplitudes. Calculations are tackled in 2nd order DWBA, employing the reaction code FRESCO [3].

The preliminary tests show that the relative importance of 2nd order transfer reactions is strongly dependent on the final state considered, as it emerges from the left panel of Fig. 20.1. Therefore, in order to isolate the CEX contribution from the experimental cross section, one should properly choose the best conditions to make the transfer practically negligible. Alternately, one has to accurately evaluate the coherent interference between the two mechanisms, as in the case of the right panel of Fig. 20.1, showing the results obtained when summing up over all the relevant intermediate transfer channels and integrating in the energy range considered. At larger excitation energies, the transfer contribution should be suppressed, while the charge-exchange is still expected to have an important strength. A comparison with the experiment is however indispensable and will be done once the data will be available.

Acknowledgements This research received funding from the Spanish Ministerio de Ciencia, Innovacion y Universidades and FEDER funds under Project No. FIS2017-88410-P and from the European Unions Horizon 2020 research and innovation programme under grant agreement N. 654002 (ENSAR2).

References

1. F. Cappuzzello et al., Eur. Phys. J. A **54**, 72 (2018)
2. H. Lenske et al., Phys. Rev. C **98**, 044620 (2018)
3. I. Thompson, Comput. Phys. Rep. **7**, 167 (1988)

Chapter 21
Two-Neutron Transfer in the $^{18}O+^{28}Si$ System

E. N. Cardozo, J. Lubian, F. Cappuzzello, R. Linares, D. Carbone, M. Cavallaro, J. L. Ferreira, B. Paes, A. Gargano and G. Santagati

Abstract In this work, we study the effect of paring correlation on the two-neutron transfer reaction $^{28}Si(^{18}O,^{16}O)^{30}Si$ at 84 MeV incident energy in the laboratory frame. For this, coupled reaction channel (CRC) and coupled channel Born approximation (CCBA) calculations were performed considering the cluster, independent coordinates and sequential transfer models for two-neutron transfer. For the calculation of the spectroscopic amplitudes the NuShellX (http://www.garsington.eclipse.co.uk/, [1]) code was used.

21.1 Theoretical Analysis

We performed coupled channel Born approximation (CCBA) calculation for sequential transfer model and coupled reaction channel (CRC) calculation for cluster and independent coordinates transfer models. The two-neutron transfer cross section was determined using the FRESCO code [2]. For the optical potential, in these calculations, the Sao Paulo double folding potential [3] was used for both parts, real and imaginary. In the entrance partition a scaling factor of 0.6 in the imaginary part of the optical potential was used for considering missing couplings which are not explicitly considered and for other dissipative processes [4]. On the other hand, the imaginary part of outgoing partition was multiplied by factor 0.78 for the reason no coupling was considered. This parameter has been demonstrated to be appropriate for describing the elastic scattering cross section for many systems at energies above the Coulomb barrier [5]. Details about interactions and model spaces, as well as, tables of the spectroscopic amplitudes for the projectile and target overlaps used can

E. N. Cardozo (✉) · J. Lubian · R. Linares · J. L. Ferreira · B. Paes
Universidade Federal Fluminense, Niteroi, Brazil
e-mail: ericanunes@fisica.if.uff.br

F. Cappuzzello · D. Carbone · M. Cavallaro · G. Santagati
Laboratori Nazionali del Sud, Istituto Nazionale di Fisica Nucleare, Catania, Italy

A. Gargano
Istituto Nazionale di Fisica Nucleare, Sezione di Napoli, Catania, Italy

J.-E. García-Ramos et al. (eds.), *Basic Concepts in Nuclear Physics: Theory, Experiments and Applications*, Springer Proceedings in Physics 225,
https://doi.org/10.1007/978-3-030-22204-8_21

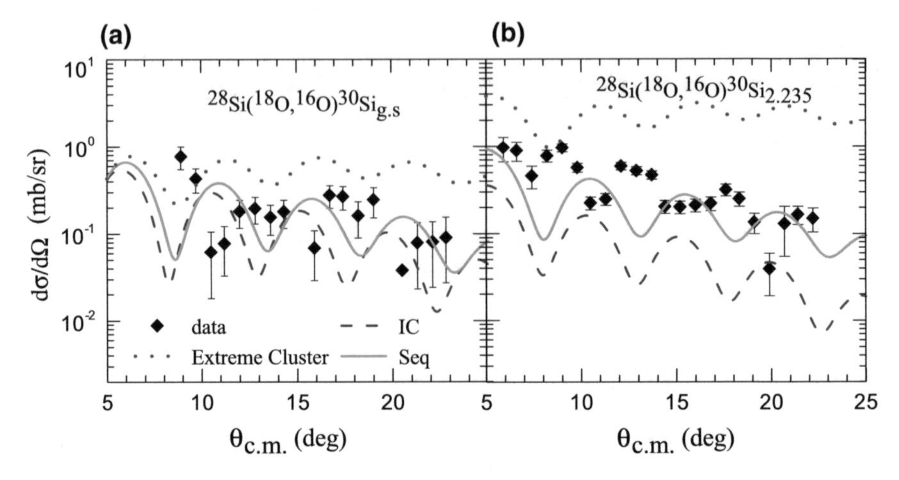

Fig. 21.1 Comparison between experimental angular distributions [6] and the theoretical transfer cross section for the ^{28}Si(^{18}O,^{16}O)^{30}Si reaction leading to the population of **a** the ground state and **b** first excited state (2^+) of ^{30}Si

be found in Cardozo et al. [6]. The comparison of the theoretical differential cross section using the extreme cluster (using spectroscoping amplitudes 1.0 and spin 0 for the two-neutron cluster) and independent coordinates (IC) models, for the direct two-neutron transfer and the two-step CCBA (Seq) results with the experimental data for the ^{28}Si(^{18}O,^{16}O)^{30}Si reaction are shown in Fig. 21.1.

One can observe that the extreme cluster model is above the experimental angular distributions. For the ground state of ^{30}Si the direct and sequential mechanisms have the same order of magnitude. For the 2.235 MeV excited state, the sequential mechanism is dominant. These dominance can be related to deformation of ^{30}Si ground state which accentuate the long-range correlations weakening the short-range correlations (paring).

21.2 Conclusion

In the present work, angular distributions obtained for the two-neutron transfer in the ^{18}O+^{30}Si system at bombarding energy of 84 MeV in the laboratory frame are analyzed. The short-range correlation among the two neutrons suffer a interference due to collective nature of ^{30}Si making the sequential transfer mechanism become relevant. The same conclusion has been obtained for the first excited state (2_1^+) in ^{66}Ni nucleus for two-neutron transfer in the identical bombarding energy [7].

References

1. W. D. M. Rae, http://www.garsington.eclipse.co.uk/
2. J.I.Thompson. http://www.fresco.org.uk
3. L.C. Chamon et al., Phys. Rev. C **66**, 014610 (2002)
4. D. Pereira et al., Phys. Lett. B **670**, 330 (2009)
5. L.R. Gasques et al., Nucl. Phys. A **764**, 135 (2006)
6. E.N. Cardozo et al., Phys. Rev. C **97**, 064611 (2018)
7. B. Paes et al., Phys. Rev C **96**, 044612 (2017)

Chapter 22
Study of the Neutron-Rich Region in Vicinity of ^{208}Pb via Multinucleon Transfer Reactions

P. Čolović, A. Illana, S. Szilner, J. J. Valiente-Dobón and PRISMA GALILEO MINIBALL Collaborations

Abstract We exploited a multinucleon transfer reactions with the neutron-rich radioactive beams in order to populate the region around the heaviest doubly-magic nucleus (^{208}Pb), the region of the nuclear landscape which is rather difficult to reach experimentally. We, thus, recently performed an experiment at HIE-ISOLDE, CERN by using the ^{94}Rb beam onto a thick ^{208}Pb target, and by detecting the reaction products in the high-resolution MINIBALL spectrometer coupled to position sensitive CD detector. With the used method we were able to successfully select the transfer channels of interest and their associated γ-rays. Preliminary analysis shows prevalence of the transfer flux in the direction of the neutron rich Pb isotopes, revealing the mechanism of transfer reactions as a competitive tool for the production of heavy neutron-rich nuclei.

22.1 Objectives and Experimental Method

Multinucleon transfer (MNT) reactions with the stable beams were extensively used for the nuclear structure and reaction dynamics studies of the neutron-rich nuclei, mostly in vicinity of the light partner [1, 2]. With the use of the unstable neutron-rich beams, nuclear reaction models [1, 4] predict large primary transfer cross sections for the neutron-rich target-like nuclei. These predicted cross sections are comparable, or even larger, than those of other competitive experimental methods. Our main objective is to exploit MNT mechanism with heavy neutron-rich unstable beam, on the heavy ^{208}Pb target, and to evaluate the production cross sections of neutron-rich heavy nuclei close to ^{208}Pb.

To enrich the existing spectroscopic data of this heavy neutron-rich region, we made a first measurements of MNT reactions at HIE-ISOLDE by accelerating a

P. Čolović (✉) · S. Szilner
Ruder Bošković Institute, Zagreb, Croatia
e-mail: petra.colovic@irb.hr

A. Illana · J. J. Valiente-Dobón
INFN, Laboratori Nazionali di Legnaro, Legnaro, Italy

J.-E. García-Ramos et al. (eds.), *Basic Concepts in Nuclear Physics: Theory, Experiments and Applications*, Springer Proceedings in Physics 225,
https://doi.org/10.1007/978-3-030-22204-8_22

^{94}Rb beam to 6.2 MeV/A onto ^{208}Pb targets of 1 and 13 mg/cm^2 thickness. The high-resolution MINIBALL spectrometer, coupled to a position sensitive silicon detector, allowed the identification of reaction products via their associated γ rays. By constructing the matrix of energy vs. scattering angle measured in the range of $\theta = 24°–63°$ a clear separation between the beam-like and target-like fragments was achieved. The selection of the beam-like and target-like fragments improves Doppler correction for the γ-rays emitted in flight. The obtained resolution after Doppler correction is 1.2% at 1.5 MeV.

22.2 Results and Outlook

The preliminary analysis (see Fig. 22.1) illustrates that the dominant transfer flux is in the neutron transfer channels, as expected. The very exciting result is the prevalence of this neutron transfer flux in the direction of the more-neutron rich Pb isotopes. Construction of full level-schemes of nuclei around ^{208}Pb will be obtained from the γ-γ analysis, where we expect to enhance present knowledge in this region [3]. The measured cross section will be compared with the state-of-the-art models [4, 5]. In conclusion, the preliminary results show that MNT is an efficient process to populate neutron-rich heavy binary partners and that it is a very competitive method compared with cold fragmentation for the population of these difficult-to-reach species.

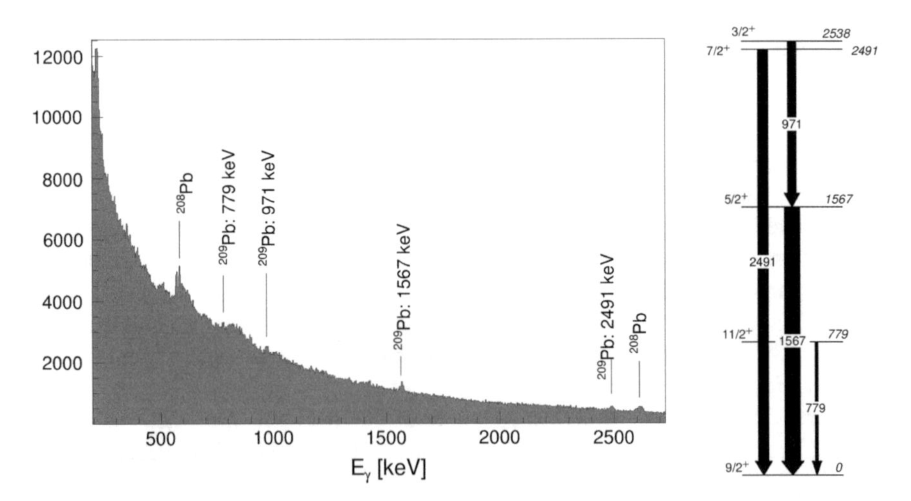

Fig. 22.1 Left: γ-spectrum obtained with the 1 mg/cm^2 ^{208}Pb target (Doppler corrected for target-like products). The dominant γ-transitions associated to the ^{209}Pb are plotted in the level scheme (right)

References

1. L. Corradi, G. Pollarolo, S. Szilner, J. of Phys. G **36**, 113101 (2009)
2. D. Montanari, L. Corradi, S. Szilner, G. Pollarolo et al., Phys. Rev. Lett. **113**, 052501 (2014)
3. A. Gottardo, J.J. Valiente-Dobón, G. Benzoni et al., Phys. Rev. Lett. **109**, 162502 (2012)
4. T. Mijatović, S. Szilner, L. Corradi, D. Montanari et al., Phys. Rev. C **94**, 064616 (2016)
5. F. Galtarossa, L. Corradi, S. Szilner et al., Phys. Rev. C **97**, 054606 (2018)

Chapter 23
The QClam-Spectrometer at the S-DALINAC

Antonio D'Alessio, Peter von Neumann-Cosel, N. Pietralla, Maxim Singer and V. Werner

Abstract The multi-wire drift chamber detectors of the QCLAM (Quadrupole CLAM shell) spectrometer at S-DALINAC were refurbished. This includes the replacement of the cathode foils, all seals and a cleaning of the inside. The detectors have, thus, been restored to operational condition. Stable gas conditions and low oxygen levels were achieved. In September 2018 the refurbished detectors were used for data acquisition for the first time.

The Technische Universität Darmstadt operates the superconducting linear electron accelerator S-DALINAC [1]. It provides electron beams with energies of up to 130 MeV and currents of 20 μA for experimental investigations in nuclear structure physics, nuclear astrophysics and detector development. Inelastic electron scattering off atomic nuclei represents a major experimental tool. For this purpose two electron spectrometers are available. The LINTOTT spectrometer [2] provides a very high energy resolution of down to $\approx$8 keV, with a comparatively small acceptance.

The QCLAM spectrometer has been developed as a large-acceptance spectrometer for (e,e') and (e,e'x) experiments with an angular acceptance of 35 msr and an energy resolution around $8 \cdot 10^{-4}$ [3–5]. It also offers the possibility of electron scattering experiments at 180° [6] for measurements of transversal nuclear modes. Measurements will be performed in the scientific context of the Collaborative Research Centre 1245 [7].

The data acquisition [8], as well as the detector system consisting of three multi-wire drift chambers (MWDCs) [9] were rebuild. The MWDCs were completely renewed at the beginning of 2018. Since no stable gas conditions and thus no constant drift time distributions could be achieved, all seals were also replaced and the grooves renewed.

A mixture of 80% Argon as counting gas and 20% carbondioxide as quenching gas with a total flow of 100 sccm is used. Recent measurements showed, that an oxygen level below 200 ppm in the gas could be achieved. A stable drift time distribution

A. D'Alessio (✉) · P. von Neumann-Cosel · N. Pietralla · M. Singer · V. Werner
Institut für Kernphysik, TU Darmstadt, Schlossgartenstraße 9, 64289 Darmstadt, Germany
e-mail: adalessio@ikp.tu-darmstadt.de

J.-E. García-Ramos et al. (eds.), *Basic Concepts in Nuclear Physics: Theory, Experiments and Applications*, Springer Proceedings in Physics 225,
https://doi.org/10.1007/978-3-030-22204-8_23

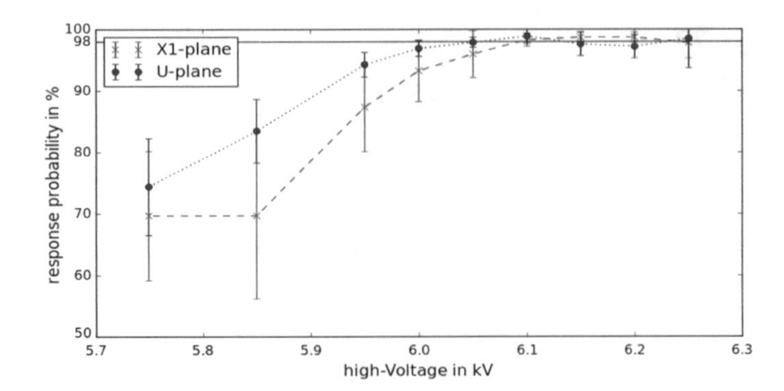

Fig. 23.1 If more than 6 kV are applied to the high-voltage foils, a plateau forms in the single wire response probability. In this region the probability is about 98%

can be expected for values below 400 ppm oxygen. The high-voltage foils that had visible corrosion damages were replaced by new aluminum-coated mylar foils with a thickness of 2 μm. With the renewed high voltage foils a broad working plateau can be achieved. In the region of 6–6.25 kV the response probability of a single wire in the MWDC is almost constant at 98% (Fig.23.1). With that and the requirement of at least 4 drift time events for detecting the position and therefore the energy of the scattered electron the detector reaches a detection probability of over 99%. The experimental program at the refurbished QCLAM-spectrometer at the S-DALINAC has started in September 2018.

The next steps are the new construction of the MWDCs with a suitable high-frequency shielding of the entire detector structure including wires and preamplifiers. As well as the development of a new housing, so that all wire planes are in one gas volume.

Acknowledgements We thank Jörg Hehner and the staff of the GSI detector laboratory for their help. This work was supported by the Deutsche Forschungsgemeinschaft under the Grant No. GRK 2128, the Grant No. SFB 1245 and by the Helmholtz Graduate School for Hadron and Ion Research.

References

1. N. Pietralla, The institute of nuclear physics at the TU Darmstadt. Nucl. Phys. News **28**, 4–11 (2018). https://doi.org/10.1080/10619127.2018.1463013
2. T. Walcher et al., Nucl. Inst. Meth. **153**, 17 (1978)
3. B. Reitz, Dissertation, TU Darmstadt (2000)
4. K.-D. Hummel, Dissertation, TH Darmstadt (1992)
5. M. Knirsch, Dissertation, TH Darmstadt (1991)
6. N. Ryezayeva et al., Phys. Rev. Lett. **100** (2008)
7. Summary of the Collaborative Research Centre-1245: http://www.sfb1245.tu-darmstadt.de/. Accessed 13 Sept 2018
8. M. Singer, Dissertation in preparation, TU Darmstadt
9. A. D'Alessio, TU Darmstadt (2016)

Chapter 24
Kaonic Atoms Measurement at DAΦNE: SIDDHARTA and SIDDHARTA-2

L. De Paolis, D. Sirghi, A. Amirkhani, A. Baniahmad, M. Bazzi, G. Bellotti, C. Berucci, D. Bosnar, M. Bragadireanu, M. Cargnelli, C. Curceanu, A. Dawood Butt, R. Del Grande, L. Fabbietti, C. Fiorini, F. Ghio, C. Guaraldo, M. Iliescu, M. Iwasaki, P. Levi Sandri, J. Marton, M. Miliucci, P. Moskal, S. Niedźwiecki, S. Okada, D. Pietreanu, K. Piscicchia, H. Shi, M. Silarski, F. Sirghi, M. Skurzok, A. Spallone, H. Tatsuno, O. Vazquez Doce, E. Widmann and J. Zmeskal

Abstract Light kaonic atoms studies provide the unique opportunity to perform experiments equivalent to scattering at threshold, being their atomic binding energies in the keV range. High precision atomic X-rays spectroscopy ensures that the energy shift and broadening of the lowest-lying states of the kaonic atoms, induced by the strong interaction between the kaon and nucleus, can be detected. Kaonic hydrogen and kaonic deuterium are the lightest atomic systems and their study deliver

L. De Paolis (✉) · D. Sirghi · M. Bazzi · C. Berucci · C. Curceanu · R. Del Grande · C. Guaraldo · M. Iliescu · P. Levi Sandri · M. Miliucci · D. Pietreanu · K. Piscicchia · F. Sirghi · A. Spallone · O. Vazquez Doce · J. Zmeskal
INFN, Laboratori Nazionali di Frascati, Frascati (Roma), Italy
e-mail: Luca.DePaolis@lnf.infn.it

L. De Paolis
Department of Physics, Faculty of Science MM.FF.NN., University of Rome 2 (Tor Vergata), Rome, Italy

D. Sirghi · M. Bragadireanu · D. Pietreanu · F. Sirghi
Horia Hulubei National Institute of Physics and Nuclear Engineering (IFIN-HH), Magurele, Romania

A. Amirkhani · A. Baniahmad · G. Bellotti · A. Dawood Butt · C. Fiorini
Politecnico di Milano, Dipartimento di Elettronica, Informazione e Bioingegneria and INFN Sezione di Milano, Milano, Italy

C. Berucci · M. Cargnelli · J. Marton · E. Widmann · J. Zmeskal
Stefan-Meyer-Institut für Subatomare Physik, Vienna, Austria

D. Bosnar
Department of Physics, Faculty of Science, University of Zagreb, Zagreb, Croatia

L. Fabbietti · O. Vazquez Doce
Excellence Cluster Universe, Technische Universiät München, Garching, Germany

F. Ghio
INFN Sez. di Roma I and Inst. Superiore di Sanita, Roma, Italy

J.-E. García-Ramos et al. (eds.), *Basic Concepts in Nuclear Physics: Theory, Experiments and Applications*, Springer Proceedings in Physics 225,
https://doi.org/10.1007/978-3-030-22204-8_24

the isospin-dependent kaon-nucleon scattering lengths. The SIDDHARTA collaboration was able to perform the most precise kaonic hydrogen measurement to date, together with an exploratory measurement of kaonic deuterium. The measurement of the kaonic deuterium will be realized in the near future by a major upgrade of SIDDHARTA: SIDDHARTA-2. In this paper an overview of the main results obtained by SIDDHARTA together with the future plans are presented.

24.1 The SIDDHARTA Experiment

The DAΦNE (Double Annular Φ Factory for Nice Experiments) accelerator is an electron-positron collider [1, 2] at the National Laboratory Frascati (LNF) in Italy. It is a unique low-energy kaons source via the decay of ϕ-mesons produced almost at rest, with a probability of about 48.9% in K^+K^-. The charged kaons are produced with a momentum of 127 MeV/c, and a momentum spread $\Delta p/p < 0.1\%$.

The SIDDHARTA (Silicon Drift Detector for Hadronic Atom Research by Timing Application) experiment measured various kaonic exotic atoms using the kaons delivered by DAΦNE.

The kaonic atoms are produced efficiently by stopping the low-energy monochromatic charged kaons inside a cryogenic gaseous target. The kaons are captured by target atoms forming kaonic atoms in highly excited orbits. The system decays by emitting radiation, which is measured.

The charged kaon trigger is a crucial feature of the experiment and it is based on the coincidence of two plastic scintillation counters mounted top and bottom of the interaction point of e^+e^-. This trigger system takes advantage of the back-to-back topology of the produced low-energy kaons: $\Phi \rightarrow K^+K^-$ and its use drastically increases the signal-to-background ratio, because most of the background is generated by e^+ and e^- particles lost from the beams, in asynchronous timing with collisions. Another fundamental element of the apparatus is the gas-target system, because the yields of kaonic-atoms X-rays decrease at sensitively higher densities due to collisions and Stark mixing.

M. Iwasaki · S. Okada
RIKEN, Tokyo, Japan

P. Moskal · S. Niedźwiecki · M. Silarski · M. Skurzok
The M. Smoluchowski Institute of Physics, Jagiellonian University, Kraków, Poland

K. Piscicchia
Museo Storico della Fisica e Centro Studi e Ricerche Enrico Fermi, Rome, Italy

H. Shi
Institute for High Energy Physics of Austrian Academy of Science,
HEPHY - Institut für Hochenergiephysik der ÖAW, Nikolsdorfergasse 18, 1050 Vienna, Austria

H. Tatsuno
Lund Univeristy, Lund, Sweden

The most precise kaonic hydrogen measurement existing in literature was realized by the use of new triggerable X-ray detectors, the Silicon Drift Detectors, characterized by excellent energy and timing resolutions, essential for the background suppression. A detailed description of the experimental setup is given in [3].

With SIDDHARTA the following measurements were performed:

- kaonic hydrogen X-ray transitions to the $1s$ level [3].
- kaonic helium4 transitions to the $2p$ level, the first measurement using a gaseous target [4, 5].
- kaonic helium3 transitions to the $2p$ level, the first measurement [5, 6].
- kaonic deuterium X-ray transitions to the $1s$ level—as exploratory measurement [7].

The $1s$—level shift ε_{1s} and width Γ_{1s} of kaonic hydrogen measured by SIDDHARTA are:

$$\varepsilon_{1s} = -283 \pm 36\ (stat) \pm 6\ (syst)\ \ eV \tag{24.1}$$

$$\Gamma_{1s} = 541 \pm 89\ (stat) \pm 22\ (syst)\ \ eV. \tag{24.2}$$

The precise determination of the K-series X-rays for kaonic hydrogen atoms provides new constraints on theories, having reached a quality which demands refined low-energy $\overline{K}N$ interaction calculations [8, 9].

24.2 SIDDHARTA-2 Experiment

SIDDHARTA-2 is a new experiment, which will be installed on DAΦNE collider in spring 2019 and will take advantage of the experience gained in the preceding SIDDHARTA experiment [3–6]. The goal of the new apparatus is to increase drastically the signal-to-background ratio, by gaining in solid angle, taking advantage of new SDDs with improved timing resolution, and by implementing additional veto systems. Figure 24.1 shows the schematic of the SIDDHARTA-2 apparatus.

A detailed Monte Carlo simulation was performed within GEANT4 framework to optimise the critical parameters of the setup, like target size, gas density, detector configuration and shielding geometry. The Monte Carlo simulation took into account all the improvements with the following assumptions: the values of shift and width of the 1s ground state of kaonic deuterium are -800 eV and 750 eV, respectively; yields ratios $K_{\alpha} : K_{\beta} : K_{total}$ are those of kaonic hydrogen, with an assumed K_{α} yield of 10^{-3}. Figure 24.2 shows the expected spectrum for an integrated luminosity of 800 pb^{-1} delivered by DAΦNE in similar machine background conditions as in the SIDDHARTA runs. The shift and width for kaonic deuterium 1s level can be determined with precisions of about 30 eV and 80 eV, respectively. These values are of the same order as the SIDDHARTA results for kaonic hydrogen.

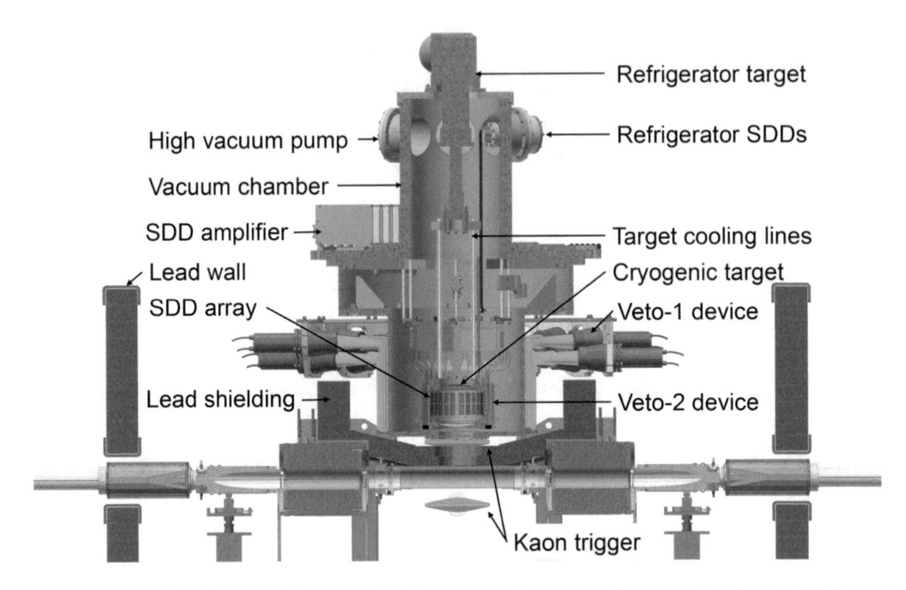

Fig. 24.1 The SIDDHARTA-2 setup with the cryogenic target cell surrounded by the SDDs and the Veto-2 system within the vacuum chamber, while the Veto-1 device is surrounding the vacuum chamber on the outside

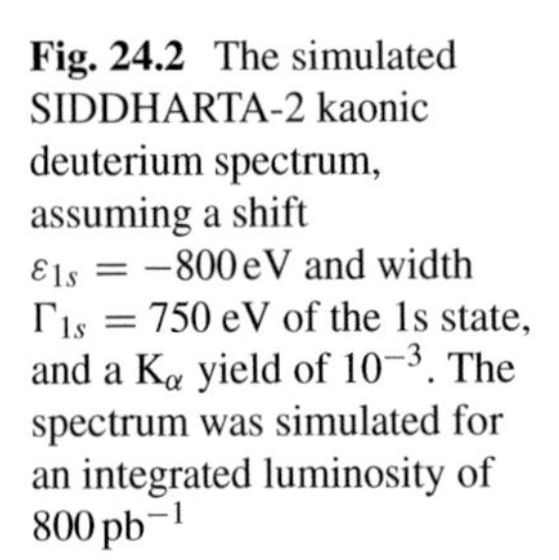

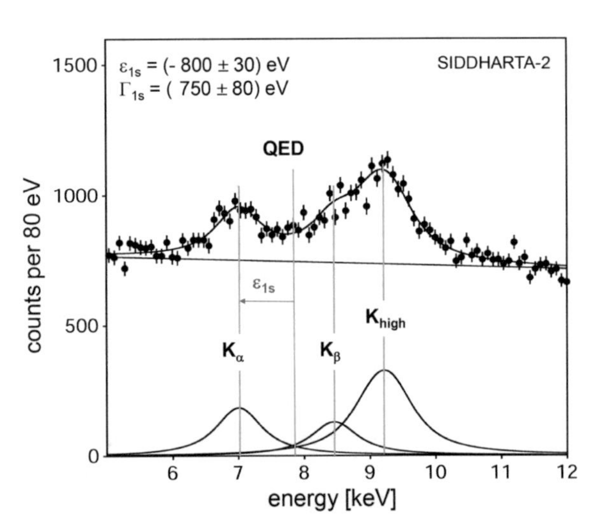

Fig. 24.2 The simulated SIDDHARTA-2 kaonic deuterium spectrum, assuming a shift $\varepsilon_{1s} = -800\,\text{eV}$ and width $\Gamma_{1s} = 750$ eV of the 1s state, and a K_α yield of 10^{-3}. The spectrum was simulated for an integrated luminosity of $800\,\text{pb}^{-1}$

Acknowledgements We thank C. Capoccia and G. Corradi, from LNF-INFN; and H. Schneider, L. Stohwasser, and D. Stuekler from Stefan-Meyer-Institut, for their fundamental contribution in designing and building the SIDDHARTA setup. We thank as well the DAΦNE staff for the excellent working conditions and permanent support. Part of this work was supported by the Austrian Science Fund (FWF): [P24756-N20]; Austrian Federal Ministry of Science and Research BMBWK 650962/0001 VI/2/2009; the Grantin-Aid for Specially Promoted Research (20002003), MEXT, Japan; the Croatian Science Foundation, under project 1680; Minstero degli Affari Esteri e della Cooperazione Internazionale, Direzione Generale per la Promozione del Sistema Paese

(MAECI), Strange Matter project; Polish National Science Center through grant No. UMO-2016/21/D/ST2/01155; Ministry of Science and Higher Education of Poland grant no 7150/E-338/M/2018.

References

1. C. Milardi et al., Int. J. Mod. Phys. A **24**, 360 (2009)
2. M. Zobov et al., Phys. Rev. Lett. **104**, 174801 (2010)
3. M. Bazzi et al., Phys. Lett. B **704**, 113 (2011)
4. M. Bazzi et al., Phys. Lett. B **681**, 310 (2009)
5. M. Bazzi et al., Phys. Lett. B **714**, 40 (2012)
6. M. Bazzi et al., Phys. Lett. B **697**, 199 (2011)
7. M. Bazzi et al., Nucl. Phys. A **907**, 69 (2013)
8. Y. Ikeda et al., Phys. Lett. B **706**, 63 (2011)
9. Y. Ikeda et al., Nucl. Phys. A **881**, 98 (2012)
10. D. Sirghi et al., Experiment with low-energy kaons at the DAΦNE Collider, Beach (2018)
11. C. Curceanu et al., The kaonic atoms research program at DAΦNE: overview and perspectives. J. Phys.: Conf. Ser. (2018)
12. C. Curceanu et al., Unlocking the secrets of the kaon-nucleon/nuclei interactions at low-energies: The SIDDHARTA(-2) and the AMADEUS experiments at the DAΦNE collider. Nucl. Phys. A (2013)

Chapter 25
PIGE Technique Within the EnsarRoot Framework

E. Galiana, D. Galaviz, H. Alvarez-Pol, P. Teubig and P. Cabanelas

Abstract The PIGE (Particle Induced Gamma Ray Emission) technique is a non-destructive, isotopically sensitive and quantitative technique used in the determination of the elemental composition of a material or sample. The method is based on the detection of gamma-rays, an excited nuclear state is induced by a proton beam of a few MeV impinging on a target. Simulations of PIGE reactions will be carried out using the EnsarRoot framework; this will be further enhance the applications by including additional analysis tools. The main goal is to provide a common framework for experimental data analysis and simulation. We aim to benchmark this tool with the measurement of the ^{35}Cl(p,p'γ)^{35}Cl and ^{37}Cl(p,p'γ)^{37}Cl reactions. The work presented here provides an overview of the status of the PIGE technique, as well as, the plans to investigate a scarcely explored region of the nuclear chart.

25.1 PIGE Technique

PIGE is an Ion Beam Analysis (IBA) analytical technique, that has a wide application scope ranging from environmental science to cultural heritage analysis. All IBA methods are non-destructive, highly sensitive and allow the detection of elements in depths ranging up to hundred of micrometers [1]. The technique uses energetic ion beams to probe the surface of materials in order to determine their composition and the determination of absolute concentration. Specifically, proton, deuteron or α-beams at low energy are used, traditionally up to 4 MeV, where only low energy nuclear forces are involved. The basic mechanism is the formation of a highly excited compound nucleus, that de-excites by the emission of gamma rays [2]. These gamma emissions conform to a unique gamma level scheme for each nucleus, allowing for

E. Galiana (✉) · D. Galaviz · P. Teubig
LIP and UL, Campo Grande 016, 1749-016 Lisboa, Portugal
e-mail: eligaliana@lip.pt

E. Galiana · H. Alvarez-Pol · P. Cabanelas
IGFAE, Instituto Gallego de Física de Altas Energías and Dpto. de Física de Partículas, Universidad de Santiago de Compostela, 15705 Santiago de Compostela, Spain

J.-E. García-Ramos et al. (eds.), *Basic Concepts in Nuclear Physics: Theory, Experiments and Applications*, Springer Proceedings in Physics 225,
https://doi.org/10.1007/978-3-030-22204-8_25

isotopical differentiation. Strong resonances populated in (p,γ) reactions allow depth profiling studies on the samples.

The fact that the repulsive Coulomb barrier has to be overcame results in a limitation of this method, as only nuclei with $Z<20$ are accessible. This light element differentiation is specially important for biomedical and organic samples.

25.2 EnsarRoot Developments to Include PIGE

The PIGE simulations and analysis will be performed within the EnsarRoot code [3, 4]. It is based on FAIRRoot [5] package which is fully based on the ROOT code and Virtual Monte Carlo that supports Geant4 transport engine. The EnsarRoot framework offers the possibility to perform the simulation of a particular experimental setup, as well as the direct comparison of simulated and measured data on equal footing using the same analysis tools. EnsarRoot will be further developed to include a subroutine based on the photopeak analysis for gamma ray spectra. Afterwards, the code will be expanded to contain the quantitative and qualitative analysis of light elements in thick samples necessary to analyse PIGE spectra.

25.3 Experimental Analysis of Chlorine

In addition to the benchmark of the PIGE analysis tool (Sect. 25.2), we will investigate and characterize PIGE reactions on the two stable chlorine isotopes, ^{35}Cl and ^{37}Cl. A quick literature review has revealed the information on both of these stable nuclei still remains to be elucidated. Proton induced reactions on ^{37}Cl play an determining role in the ultimate abundances of ^{34}S, ^{37}Cl and ^{38}Ar for explosive oxygen and silicon burning in type II supernovae. Furthermore, from the nuclear structure standpoint, these reactions are of particular interest as the target ^{37}Cl nucleus has a 20 neutrons closed shell [6].

References

1. H.R.Verma, *Atomic and Nuclear Analytical Methods* (Springer, Berlin, 2007), pp. 269–295
2. IAEA TECDOC SERIES-18822 (2017)
3. https://github.com/EnsarRootGroup/EnsarRoot, http://igfae.usc.es/satnurse/ensarroot.html
4. P. Cabanelas et al., J. Phys.: Conf. Ser. **1024**, 012038 (2018)
5. https://github.com/FairRootGroup/FairRoot, https://fairroot.gsi.de/
6. Webber et al., Nucl. Phys. A **439**, 176–188 (1985)

Chapter 26
Gogny Force Useful for Neutron Star Calculations

C. Gonzalez-Boquera, M. Centelles, X. Viñas and L. M. Robledo

Abstract We propose a new Gogny parametrization, which we call D1M*, aimed to be successful in the study of the properties of neutron stars and, at the same time, to describe finite nuclei with a similar quality as the D1M Gogny interaction. D1M* is able to reach up to 2 solar masses ($M_\odot$) for a neutron star and give good global properties for them [1].

The core of neutron stars (NSs) is the region responsible for almost all the mass and size of the star, implying that several global properties of the star are determined mostly by the core. The study of NSs with Gogny interactions, which are finite-range effective forces which describe at the same time the mean field and pairing field, works less well than when they are used for studying finite nuclei [2, 3]. That is, the most successful Gogny forces used in the description of finite nuclei (D1S, D1N and D1M) are unable to reach the maximum observed NS masses, around $2M_\odot$ [4]. Recently we have introduced a new Gogny parametrization, dubbed D1M*, aimed to preserve the quality of D1M for finite nuclei and that also may be used to study neutron-rich matter physics, such as in the context of NSs. The adjustment of the force, as well as the values of its parameters, can be found in [1]. In Fig. 26.1a we plot the symmetry energy ($E_{sym}(\rho)$) as a function of the density from a few Gogny interactions and from the well-known SLy4 Skyrme force. Generally, Gogny forces tend to have too soft $E_{sym}(\rho)$, presenting isospin instabilities at densities a few times the value of the saturation density, which gives, in turn, an equation of state (EoS) that cannot support massive NSs. On the contrary, the symmetry energy calculated

C. Gonzalez-Boquera (✉) · M. Centelles · X. Viñas
Departament de Física Quàntica i Astrofísica and Institut de Ciències del Cosmos (ICCUB), Facultat de Física, Universitat de Barcelona, Martí i Franquès 1, 08028 Barcelona, Spain
e-mail: cgonzalezboquera@ub.edu

L. M. Robledo
Departamento de Física Teórica, Facultad de Física, Universidad Autónoma de Madrid, 28049 Madrid, Spain

Center for Computational imulation and Center for Computational Simulation, Universidad Politécnica de Madrid, Campus de Montegancedo, Boadilla del Monte, 28660 Madrid, Spain

J.-E. García-Ramos et al. (eds.), *Basic Concepts in Nuclear Physics: Theory, Experiments and Applications*, Springer Proceedings in Physics 225,
https://doi.org/10.1007/978-3-030-22204-8_26

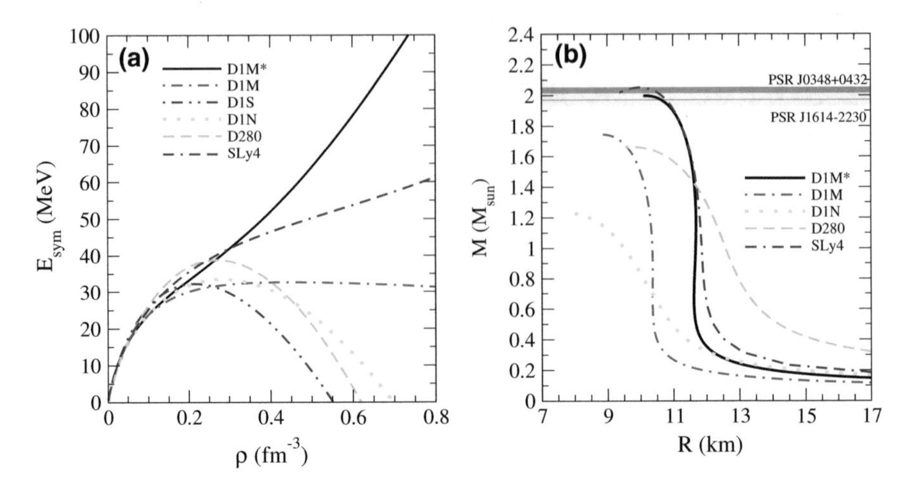

Fig. 26.1 **a** Symmetry energy against the density from the D1M*, D1M, D1S, D1N, and D280 Gogny forces and from the SLy4 Skyrme force. **b** Mass-radius relation in neutron stars from the D1M*, D1M, D1N and D280 Gogny forces and the SLy4 Skyrme force. Horizontal bands from [4]

using the D1M* interaction increases with the density, providing a stiffer EoS than the other Gogny interactions. In Fig. 26.1b we present the mass-radius relation for neutron stars for the same interactions as in Fig. 26.1a, as long as they converge in the NS calculation. This relation is determined by the behaviour of the beta-stable nuclear matter EoS. The D1M* force predicts a maximum mass of $2M_\odot$ with a radius of 10.2 km, and a canonical star of $1.4M_\odot$ with a radius of 11.6 km.

On the other hand, we have also studied the ability of this new D1M* interaction for describing finite nuclei, and the results are compared to the ones obtained with D1M. These calculations are performed for 620 even-even nuclei of the 2012AME compilation and have been carried out with the HFBaxial code in a harmonic oscillator basis [5]. We obtain for D1M a value of the rms deviations of the binding energies of $\sigma_E = 1.36$ MeV, and for the D1M* we obtain $\sigma_E = 1.34$ MeV, which points out a similar description of binding energies of finite nuclei on the average. Moreover, the differences $\Delta B = B_{th} - B_{exp}$ are scattered around the zero value, and do not present drifts with increasing neutron number.

Further studies for finite nuclei, such as studies of fission barriers and excitation energies, using the D1M* Gogny interaction are in progress.

Acknowledgements We acknowledge Grants FIS2017-87534-P from MINECO and FEDER, FPA2015-65929-P, FIS2015-63770-P and BES-2015-074210 from MINECO and MDM-2014-0369 of ICCUB (Unidad de Excelencia María de Maeztu) from MINECO.

References

1. C. Gonzalez-Boquera, M. Centelles, X. Viñas, L.M. Robledo, Phys. Lett. B **779**, 195 (2018)
2. R. Sellahewa, A. Rios, Phys. Rev. C **90**, 054327 (2014)
3. C. Gonzalez-Boquera, M. Centelles, X. Viñas, A. Rios, Phys. Rev. C **96**, 065806 (2017)
4. P.B. Demorest, et al., Nature **467**, 10810 (2010); J. Antoniadis, et al., Science **340**, 448 (2013)
5. L.M. Robledo, G.F. Bertsch, Phys. Rev. C **84**, 014312 (2011); L.M. Robledo, HFBaxial computer code (2002)

Chapter 27
Nuclear Structure and β^- Decay of A = 90 Isobars

Nadjet Laouet and Fatima Benrachi

Abstract In the aim of estimating the β^- decay properties of few nuclei in the A ~ 90 region with few hole protons and particle neutrons in addition to ^{100}Sn doubly magic core, some spectroscopic calculations have been realized. The using interaction is derived from snet one taking into account the three body interaction effect in the studied mass region, and using recent single particle and hole energies. The calculations are carried out in the framework of the nuclear shell model by means of NuShellX@MSU code. The getting results have been compared to the available experimental data.

27.1 Theoretical Framework

The approximation that consider the nucleus as an inert core without any interactions with the valence particles, fail to reproduce nuclear properties of some isotopic chains [1]. It is important to consider the effect resulting from the interactions between the core and the valence particles in the aim of reproducing these missing nuclear properties [2, 3]:

$$V_{st}^{T} = \frac{\sum_J (2J+1)\langle j_s j_t | V_{st}\, | j_s j_t \rangle_J^T \left[1 - (-1)^{J+T}\delta_{st}\right]}{\sum_J (2J+1)\left[1 - (-1)^{J+T}\delta_{st}\right]}$$

The two body matrix elements (TBME) of the using interaction snetm are modified taking in consideration the proton–neutron monopole and mass effects for odd-odd nuclei in the ^{100}Sn region, basing on the original interaction snet [4].

N. Laouet (✉) · F. Benrachi (✉)
Laboratoire de Physique Mathematique et de Physique Subatomique (LPMS), Frères Mentouri Constantine-1 University, Constantine, Algeria
e-mail: nadjet.laouet@umc.edu.dz

F. Benrachi
e-mail: fatima.benrachi@umc.edu.dz

J.-E. García-Ramos et al. (eds.), *Basic Concepts in Nuclear Physics: Theory, Experiments and Applications*, Springer Proceedings in Physics 225,
https://doi.org/10.1007/978-3-030-22204-8_27

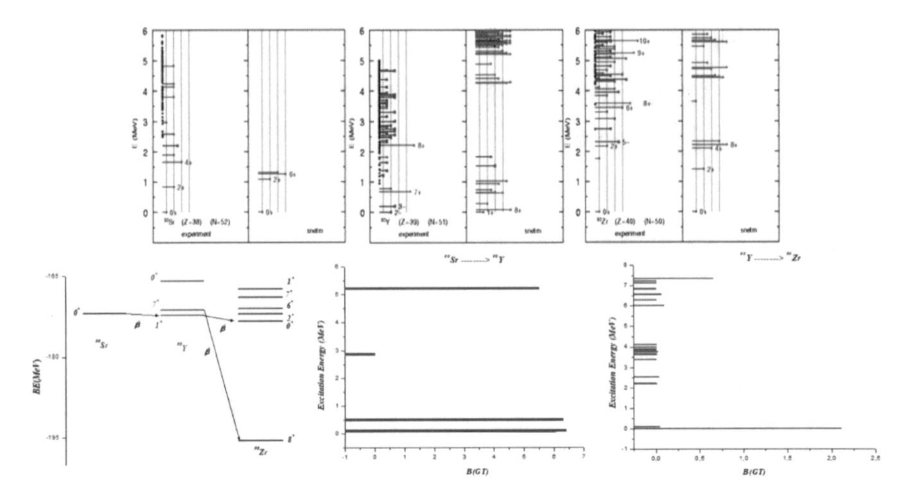

Fig. 27.1 Spectra, BE(MeV) and *B(GT) strengths as a function of excitation energy*

27.2 Calculation Results and Discussion

Using the single particle energy SPE values taken from the available experimental data and those given by Grawe et al. 2007 [5], some calculations are released by means of NuShellX@MSU [6] nuclear structure code. The obtained results using snetm interaction are presented in Fig. 27.1.

27.3 Conclusion

The calculations are realized by means of NushellX@MSU code. The modifications are based on the mass effect and proton–neutron monopole interaction of snet interaction to get snetm one. The calculation results are in agreement with the experimental data for even-even isobars ^{90}Sr and ^{90}Zr, however the calculations can not reproduce the spin and parity of the odd-odd ^{90}Y ground state. The ^{90}Sr neutron in the $\nu g_{9/2}$ shell populate the ^{90}Y proton in $\pi p_{1/2}$ one. Moreover, the ^{90}Y neutron in $\nu g_{9/2}$ shell populate the ^{90}Zr proton in $\pi g_{9/2}$ one. Most of the ^{90}Sr GT$^-$ transition strength (blue graph) is located in two peaks concentrated at about 0.5 and 5 in MeV. For the ^{90}Y (red graph) GT$^-$ transitions, most of the strength is located in four peaks concentrated at about 0.02, 2.5, 4.0 and 0.5 in MeV.

Acknowledgements Authors of this article thank the organizers for the organization and the quality of the scientific subjects presenting in the conference. Special thanks are owed to B. A. Brown for the rich discussion during the conference.

References

1. T. Otsuka et al., Phys. Rev. Lett. **105**, 1032501 (2010)
2. A. Poves, A.P. Zuker, Phys. Rep. **70**, 235 (1981)
3. O. Sorlin, M.G. Porquet, Prog. Part. Nucl. Phys. **61**, 602 (2008)
4. Z. Hu et al., Phy. Rev. C **60**, 024315 (1999)
5. H. Grawe, K. Langanke, G. Martinez-Pinedo, Rep. Prog. Phys. **70**, 1525 (2007)
6. B.A. Brown, W.D.M. Rae, Nucl. Data Sheets **120**, 115 (2014)

Chapter 28
Real-Time Tomographic Image Reconstruction in PET Using the Pseudoinverse of the System Response Matrix

A. López-Montes, P. Galve, J. M. Udías and J. López Herraiz

Abstract Positron Emission Tomography (PET), is one of the most recent medical imaging techniques, and it is commonly used in oncology, cardiology and neurology. PET is based on the detection in coincidence of the two 511 keV photons produced in the annihilation of a positron with an electron of the media. Positrons are emitted by radionuclides linked to a molecule of interest (tracer) administered to the patient. The information of the coincidences of the 511 keV photons is used within a reconstruction algorithm to generate an image of the biodistribution of the tracer in the body. The images are usually obtained after the end of the acquisition, but in some cases it would be useful to be able to obtain PET images in real-time. Iterative image reconstruction methods usually provide better resolution and less statistical fluctuations than analytical methods, but they are too slow to be used in real time applications. On the other hand, analytical methods are fast but, they produce images with artifacts when the acquired data are incomplete, they are more sensitive to noise, and their resolution is not optimal as they are based on pure mathematical approaches and assumptions. In this work, we propose to combine the best features of both analytical and iterative methods. We incorporate the most relevant physical processes involved in the emission and detection of the radiation into a System Response Matrix (SRM), and then the reconstruction is obtained using the pseudoinverse (PINV) of the SRM. This method present an improvement in image quality comparing with common real time analytical methods while it is fast enough to be used in real-time applications.

28.1 Introduction

PINV is a useful tool used to solve linear ill-posed problems that has been used in many different problems in physics [1]. In a linearized version of the reconstruction problem in PET, the activity distribution in the patient (X) and the acquired data with

A. López-Montes (✉) · P. Galve · J. M. Udías · J. López Herraiz
Grupo de Física Nuclear and Iparcos, Facultad de Ciencias Físicas, Universidad Complutense de Madrid, CEI Moncloa, 28040 Madrid, Spain
e-mail: alelopez@ucm.es

J.-E. García-Ramos et al. (eds.), *Basic Concepts in Nuclear Physics: Theory, Experiments and Applications*, Springer Proceedings in Physics 225,
https://doi.org/10.1007/978-3-030-22204-8_28

the scanner (Y) can be related using a SRM [2, 3]. Y = SRM·X The PINV of the SRM can be used to solve this linear problem [3].

28.2 Methods

As a complete SRM for a common 3D-PET acquisition is too big to be stored in current computers, we use subdivide the SRM into two parts. The PINV of an axial SRM is used to rebin 3D data into a set of 2D data slices and another PINV of a 2D-SRM is used to reconstruct each one of the different slices. Standard method for real time PET imaging [4] consists on a rebinning of the data using Single Slice ReBinning (SSRB) [5] and a further reconstruction of 2D slices using analytical methods as Filtered Backprojection (FBP) [6].

28.3 Results

Comparison of the standard method for reconstruction with our proposed method for a rat injected with FDG is shown in Fig. 28.1. The PET data was obtained in the SuperArgus preclinical scanner (SEDECAL) [7].

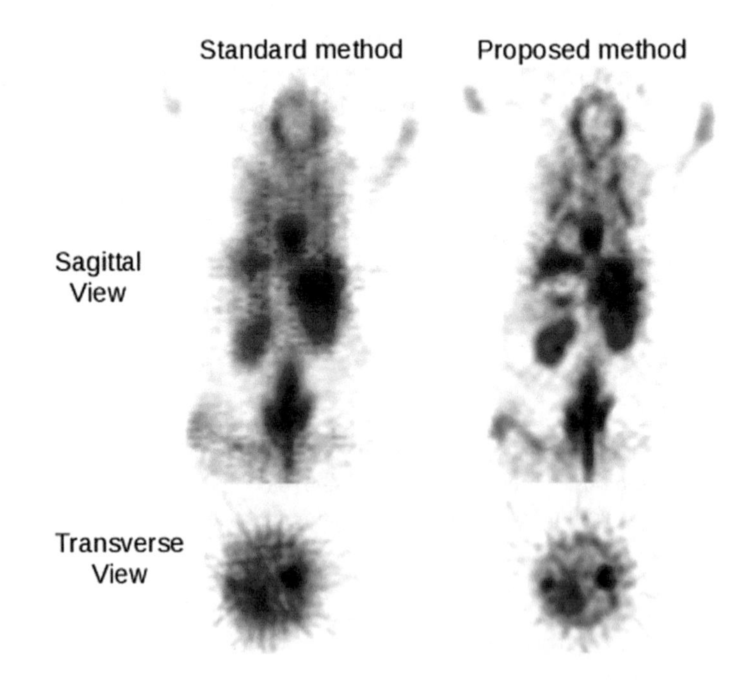

Fig. 28.1 Reconstructed images of a rat obtained with the standard analytical method and with the proposed method

Table 28.1 Approximated computing times using Fortan in a single thread of a CPU E5-2640 v4 @ 2.40 GHz processor for a 3D acquisition with data gathered in a 3 Mb sinogram and 15 Mb of reconstructed image

Method	Time (s)
Standard	4.5
Proposed	1.5

28.4 Discussion and Conclusion

According the results presented in Fig. 28.1 and Table 28.1, we can see an important improvement on the resolution recovery using PINV being even faster than analytical method SSRB+FBP. In this work we have shown that PINV methods stands out over common analytical methods used in real-time PET applications as we can include physical and geometrical information of the acquisition into the SRM.

Acknowledgements This work was supported by Comunidad de Madrid (S2013/MIT-3024 TOPUS-CM), Spanish Ministry of Science and Innovation, Spanish Government (FPA2015-65035-P, RTC-2015-3772-1). This is a contribution for the Moncloa Campus of International Excellence. Grupo de Física Nuclear-UCM, Ref.: 910059. This work acknowledges support by EU's H2020 under MediNet a Networking Activity of ENSAR-2 (grant agreement 654002). J. L. Herraiz is also funded by the EU Cofund Fellowship Marie Curie Actions, 7th Frame Program. P. Galve is supported by a Universidad Complutense de Madrid and Banco Santander predoctoral grant, CT27/16-CT28/16.

References

1. G.H. Golub, C.F. Van Loan. *Matrix Computations*
2. V.V. Selivanov, R. Lecomte, IEEE TNS 2001. https://doi.org/10.1109/23.940160
3. J. Sánchez-González, S. España, M. Abella et al., *IEEE NSS/MIC* (2005)
4. J.M. Arco, J.M. Udias, J.L. Herraiz, M. Desco, J.L. Longas, R. Matesanz, J.J. Vaquero, *WMIC* (2015)
5. F. Noo, M. Defrise, R. Clackdoyle, Phys. Med. Biol. **44**, 561 (1999)
6. J.L. Herraiz, J.J. Vaquero, J.M. Udias, *IEEE NSS/MIC* (2011)
7. http://www.sedecal.com/products/superargus-pet-ct/

Chapter 29
Photoacoustic Dose Monitoring in Radiosurgery

O. M. Giza, D. Sánchez-Parcerisa, J. Camacho, V. Sánchez-Tembleque, S. Avery and J. M. Udías

Abstract In vivo range verification in patients is determined to be crucial to utilize the full potential of proton radiotherapy Parodi et al, (Mod Phys Lett A, 30, 1540025, 2015, [1]). In this work we study the detection of acoustic waves generated by a clinical photon beam. Since the dose-acoustic effect is similar in photons and protons, the results of this work can be further extrapolated to proton beams Jones et al, (Phys Med Biol, 59, (21), 6549, 2014, [2]).

29.1 Introduction

In a radiotherapy treatment the dose absorbed by a tissue causes a small rise of the temperature (μK) which generates a pressure acoustic wave (mPa). The amplitude of the acoustic signal is proportional to the energy deposited by the radiation, the density of the material and the Grüneisen of the target material. High-density and high-Grüneisen metals have been used to amplify acoustic signals [3]: in our experiment, a lead block is submerged in the target water phantom.

O. M. Giza (✉) · D. Sánchez-Parcerisa · V. Sánchez-Tembleque · J. M. Udías
Grupo de Física Nuclear and Iparcos, Facultad de Ciencias Físicas, Universidad Complutense de Madrid, CEI Moncloa, Madrid 28040, Spain
e-mail: oliviami@ucm.es

O. M. Giza · D. Sánchez-Parcerisa · V. Sánchez-Tembleque · J. M. Udías
Instituto de Investigación Sanitaria del Hospital Clínico San Carlos (IdISSC), Madrid, Spain

J. Camacho
ITEFI, Spanish National Research Council (CSIC), Madrid, Spain

S. Avery
Department of Radiation Oncology, Hospital of the University of Pennsylvania, Philadelphia, PA 19104, USA

J.-E. García-Ramos et al. (eds.), *Basic Concepts in Nuclear Physics: Theory, Experiments and Applications*, Springer Proceedings in Physics 225,
https://doi.org/10.1007/978-3-030-22204-8_29

29.2 Methods and Materials

The experimental setup consists of several ultrasonic transducers of different resonance frequencies and bandwidths, a combination of preamplifiers and differential amplifiers and a digital oscilloscope. The setup records and amplifies the acoustic wave generated by two clinical beams (a Cyberknife™ radiosurgery beam and a clinical LINAC) irradiating a block of lead submerged in a water tank.

Simulations of the experimental setup were carried out to study the optimal measurement geometry and choice of transducer. The dose distributions were calculated with Monte Carlo code *FLUKA*, while the acoustic simulations were performed with analytical wave transport code *k-Wave*.

The temporal profiles of the dose pulses were measured with a scintillating crystal coupled with a PMT [4] and used as an input to the acoustic simulation.

29.3 Results

Gammaacoustic signals were detected in both clinical beams, with an amplitude, arrival time and pulse shape in agreement with simulations (Fig. 29.1).

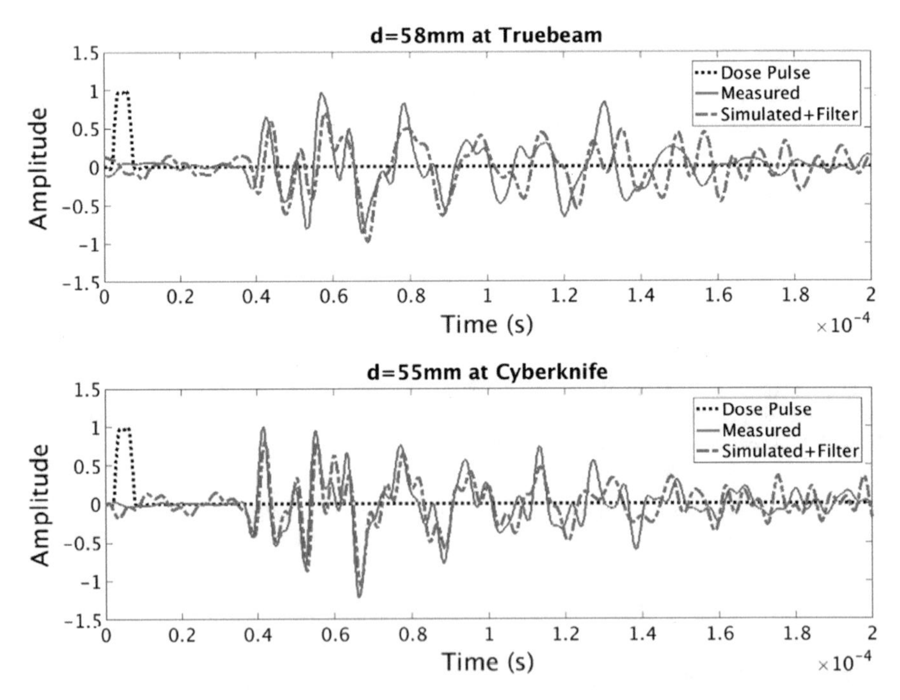

Fig. 29.1 Simulated (blue) and measured (red) gammaacoustic signals from Truebeam (top) and Cyberknife (bottom) clinical beams

29.4 Discussion and Conclusions

Photoacoustic simulation with FLUKA and *k-Wave* can reproduce the experimental response of acoustic transducers in dose monitoring applications. The proposed setup can detect photoacoustic signals originated from the penumbral areas of the treated fields and with the relevant image analysis, it could be used to monitor the position of the proton Bragg peak with mm accuracy.

Acknowledgements This work was supported by Comunidad de Madrid (S2013/MIT-3024 TOPUS-CM), Spanish Ministry of Science and Innovation, Spanish Government (FPA2015-65035-P, RTC-2015-3772-1). This is a contribution for the Moncloa Campus of International Excellence. Grupo de Física Nuclear-UCM, Ref.: 910059. This work acknowledges support by EU's H2020 under MediNet a Networking Activity of ENSAR-2 (grant agreement 654002). D.S.P is also funded by the EU Cofund Fellowship Marie Curie Actions, 7th Frame Program.

References

1. K. Parodi et al., Ionoacustics: a new direct method for range verification. Mod. Phys. Lett. A **30**, 1540025 (2015)
2. K. Jones et al., Proton beam characterization by proton-induced acoustic emission: simulation studies. Phys. Med. Biol. **59**(21), 6549 (2014)
3. S. Hickling et al., On the detectability of acoustic waves induced following irradiation by a radiotherapy linear accelerator. IEEE Trans. Ultrason. Ferroelectr. Freq. Control **63**(5), 683 (2016)
4. V. Sánchez-Tembleque et al., Simultaneous measurement of the spectral and temporal properties of a LINAC pulse from outside the treatment room. Phys. Med. Biol. **63** (2018). [submitted]

Chapter 30
Structure of Light Nuclei Studied with ^{7}Li+^{6,7}Li Reactions

D. Nurkić, M. Uroić, M. Milin, A. Di Pietro, P. Figuera, M. Fisichella, M. Lattuada, I. Martel, Đ. Miljanić, M. G. Pellegriti, L. Prepolec, A. M. Sánchez-Benítez, V. Scuderi, N. Soić, E. Strano and D. Torresi

Abstract In this contribution, a brief analysis of an experiment performed at LNS-INFN with the 30 and 52 MeV ^{7}Li beam and the ^{7}LiF and ^{6}LiF targets is given. The ^{7}Li+^{6,7}Li reactions are measured to get information on different types of structures of several light nuclei. Special attention is paid to a search for molecular states in ^{10}B and ^{10}Be. Another goal is to find information for states with different cluster configurations in ^{7}He, ^{9}Be and $^{10-12}$B. The experimental setup consists of four telescopes covering polar angles in range from 20° to 90° and providing particle identification using traditional ΔE-E techniques. The contribution includes previously published Uroić et al. (Eur Phys J A, 51, 93, 2015, [1]) procedures as well as the remaining plans for analysis.

30.1 Introduction

Detailed description and modelling of extremely deformed states in light nuclei is a task at hand for modern nuclear physics. Important prerequisites for such a task are precise information on excitation energies and reaction channels that lead to the formation of aforementioned states. The choice of projectile and target nuclei is a very important component in experiments that deal with these issues. Here, ^{6,7}Li nuclei are used because of their pronounced cluster structure (α + t and α + d).

D. Nurkić (✉) · M. Milin
Department of Physics, Faculty of Science, University of Zagreb, Bijenička cesta 32, 10000 Zagreb, Croatia
e-mail: dnurkic@phy.hr

M. Uroić · N. Soić · Đ. Miljanić · L. Prepolec
Ruđer Bošković Institute, Zagreb, Croatia

A. Di Pietro · P. Figuera · M. Fisichella · M. Lattuada · M. G. Pellegriti · V. Scuderi · E. Strano · D. Torresi
INFN-Laboratori Nazionali del Sud, Catania, Italy

I. Martel · A. M. Sánchez-Benítez
Departamento de Fisica Aplicada, Universidad de Huelva, Huelva, Spain

J.-E. García-Ramos et al. (eds.), *Basic Concepts in Nuclear Physics: Theory, Experiments and Applications*, Springer Proceedings in Physics 225,
https://doi.org/10.1007/978-3-030-22204-8_30

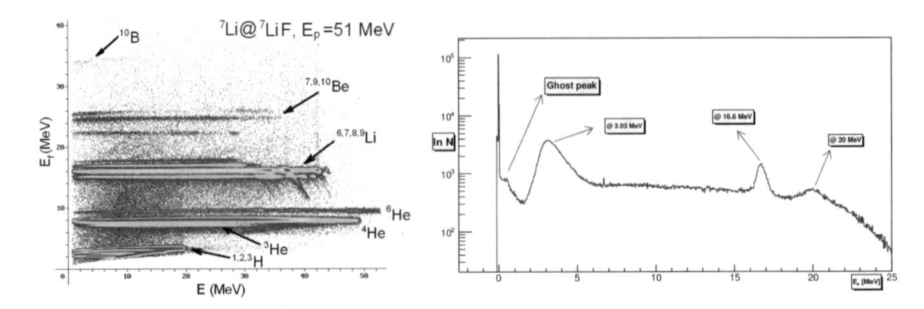

Fig. 30.1 (left) Particle identification using the slowly varying energy function E_f with the ΔE thickness non-uniformity corrections included and (right) excitation energy of ^8Be from the detection of two alpha particles. "Ghost peak"—decay of ^9Be or ^9B on the lower edge of the (wide) first excited state of ^8Be

Some of the states that are expected to be strongly populated are the three-body cluster states in ^9Be (with the α + d + t structure), ^{10}Be (α + t + t), ^{10}B ($\alpha + \alpha$ + d and α + ^3He + t) and ^{11}B ($\alpha + \alpha$ + t).

Standard ΔE-E particle identification is modified by the introduction of the slowly varying energy function E_f instead of ΔE:

$$E_f = \sqrt{\Delta E \cdot E + \alpha \Delta E^2} + \beta E \tag{30.1}$$

with the parameter values $\alpha = 0.7$ and $\beta = -0.05$. This simplifies the identification procedure and enables further ΔE thickness non-uniformity corrections. The results are shown in Fig. 30.1. More information on the procedure and other results can be found in [1].

30.2 Remaining Analysis

The remaining analysis will be focused on coincident detections. An elementary example, excitation energy of ^8Be from the detection of two alpha particles, is given in Fig. 30.1. Well behaved results encourage taking further steps in that direction.

Acknowledgements This work has been supported by the Croatian Science Foundation under Project No. 7194.

Reference

1. M. Uroić, M. Milin, A. Di Pietro et al., Eur. Phys. J. A **51**, 93 (2015). https://doi.org/10.1140/epja/i2015-15093-0

Chapter 31
ICH15: A Linac Accelerator for Proton Therapy and Radioisotope Production Using IH/CH Cavities

A. K. Orduz, I. Martel, A. C. C. Villari, J. Sánchez-Segovia, C. Bontoui, F. Manchado de Sola, R. Berjillos, J. Pérez, J. López-Morillas, J. Díaz, A. Jurado, A. M. López-Antequera, J. Vazquez, J. L. Aguado-Casas, T. Pérez, A. Pinto, D. Ablanedo, E. Hidalgo and M. Trueba

Abstract The project ICH15 aims for the design of a heavy-ion facility capable of accelerating protons up to an energy of 70 MeV and 5 MeV/u heavy-ions with A/Q = 3. The objective is to simultaneously produce radioisotopes and deliver a low energy proton beam for cancer therapy (uveal tumour). In this contribution we present an overview of the project, including a selection of medical-physics goals and technical developments: beam dynamics calculation, electromagnetic and thermo-mechanical analysis, the control system and the irradiation facilities. Studies have been carried put with Comsol Multiphysics (Comsol multiphysics 5.3. http://www.comsol.com/products/multiphysics/ [1]) and Global Tracer Particle (General particle tracer. http://www.pulsar.nl/gpt/ [2]).

A. K. Orduz (✉) · I. Martel · J. Sánchez-Segovia · C. Bontoui · F. Manchado de Sola · R. Berjillos · J. L. Aguado-Casas · A. Pinto · T. Pérez
Science and Technology Research Centre, University of Huelva, 21071 Huelva, Spain
e-mail: angie.orduz@dfa.uhu.es

A. C. C. Villari
FRIB, East Lansing, MI, USA

J. Sánchez-Segovia · C. Bontoui · F. Manchado de Sola
Hospital Juan Ramón Jiménez, Servicio Andaluz de Salud, 21005 Huelva, Spain

R. Berjillos · J. Pérez · J. López-Morillas
TTI Norte, S.L. 41300 La Rinconada (Sevilla), Spain

J. Díaz · A. M. López-Antequera · A. Jurado
E.T.S.I. Informática, University of Granada, 18071 Granada, Spain

J. Vazquez
FAYSOL S.A.L., 21810 Palos de La Frontera (Huelva), Spain

D. Ablanedo · E. Hidalgo · M. Trueba
ATI Sistemas, 15165 Bergondo (A Coruña), Spain

J.-E. García-Ramos et al. (eds.), *Basic Concepts in Nuclear Physics: Theory, Experiments and Applications*, Springer Proceedings in Physics 225,
https://doi.org/10.1007/978-3-030-22204-8_31

31.1 Introduction

Radiotherapy plays an essential role in the treatment of cancer, either by the use of radio pharmaceuticals or by the direct irradiation of the tumour using energetic radiation. An important new technique being under development is the so-called theranostics, where two isotopes of the same chemical element are used simultaneously for therapy and diagnosis. Hadron irradiation techniques (protons and other heavy ions) provide more benefits as compared to photons [3]. As the energy-deposition density along the ion path is much higher, the dose received by healthy tissues during the treatment is smaller, and the maximum deposition occurs in a narrow region at the end of the path, just over the tumour.

Design and studies for all elements have been performed. Prototypes of the solenoid, IH cavity and the RFQ were produced by the Spanish companies FAYSOL, TTI and ATI. The layout of the facility is shown in the Fig. 31.1. The prototype was integrated with an ECR source elements in the Laboratory of the University of Huelva [4] as shown in Fig. 31.2. The RFQ structure operates at 200 MHz and consists of four modulated rods and fourteen field tunning systems inside a cylindrical cavity of 2 m length. It accelerates particles with mass-to-charge ratio A/Q = 3 up to an energy of 500 keV/u.

31.2 RF and Control System

Within the framework of ICH15, it has been developed an innovative RF System, the ICH15-Solid State Power Amplifier (SSPA). The system is based on LDMOS solid state technology, and it is designed to deliver up to 300 kWs CW at a frequency of 200 MHz. The SSPA combines two HPA units, which have been integrated in a single 19”/4U rack module. The Control System consists of a set of computerized tools for on-line monitoring, communication and control of the relevant accelerator devices and diagnostic elements. The system has been structured in three levels that

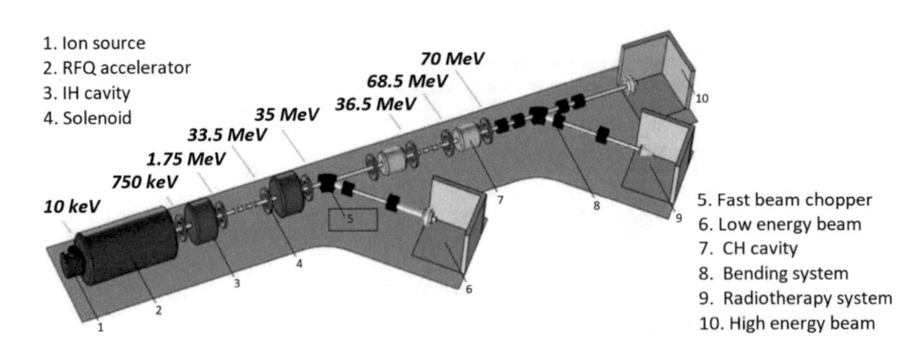

Fig. 31.1 ICH15 accelerator layout

Fig. 31.2 RFQ accelerator prototype

corresponds to the sensors, actuators, beam and RF diagnostics, security systems, PLCs and crates and finally the monitoring room, where a SCADAS system allows the remote control of the LINAC.

31.3 Conclusions

The team have carried out a design study of a medical physics facility delivering 70 MeV protons and E = 5 MeV/u heavy-ions with A/Q = 3, including the beam dynamics, electromagnetic and thermomechanical studies of the accelerator elements, RFQ, IH/CH cavities and solenoids. The work-group has also calculated the beam lines for radioisotope production and radiotherapy, including the control system, treatment and verification system. Finally, the prototypes for IH cavity, the RFQ and the solenoid has been delivered. Future work will be focused on RF test of the resonant cavities and mapping the magnetic field of the solenoid.

Acknowledgements Work partially supported by the European Regional Development Funds (FEDER), "Operational Program for Smart Growth 2014–2020" and the Ministry of Economy and Competitiveness of Spain, through the Centre for the Development of Industrial Technology (CDTI).

References

1. Comsol multiphysics 5.3. http://www.comsol.com/products/multiphysics/
2. General particle tracer. http://www.pulsar.nl/gpt/
3. W. Wieszczycka, W.H. Scharf, *Proton Radiotherapy Accelerators*, (World Scientific, Warsaw University of Technology, Poland, 2001)
4. I. Martel et al., A new RF laboratory for developing accelerator cavities at the University of Huelva, in *Proceedings, 6th International Particle Accelerator Conference* (2015). https://doi.org/10.18429/JACoW-IPAC2015-WEPMN052

Chapter 32
Near Coulomb Barrier Scattering of ^{15}C on ^{208}Pb

J. D. Ovejas

Abstract In the last years, the neutron rich carbon isotope ^{15}C has been debated as a halo nucleus ($S_n = 1215$ keV, $S_{2n} = 9395$ keV) according to reaction experiments at high energies [1]. If so, it would be the only halo nucleus exhibiting a pure s-wave configuration in its ground state. Exotic structures in light nuclei such as the n-halo of ^{11}Li and ^{11}Be or the n-skin of ^{8}He, have previously been studied via the measurement of the elastic scattering on heavy and stable targets at energies around the Coulomb barrier, see e.g. (Pesudo et al., Phys. Rev. Lett. 118:152502, 2017, [2]). Now, with the event of the HIE-ISOLDE facility at CERN, it is possible to continue these studies with ^{15}C. The IS619 experiment, carried out in August 2017, is the first dynamical study of this nucleus at near Coulomb energies and aims to prove its structure via the measurement of the angular distribution of the elastic scattering cross-section on a ^{208}Pb target.

32.1 Experimental Facility and Setup

The ^{15}C radioactive beam was produced from the impact of 1.4 GeV proton pulses on a CaO production target via nuclear spallation. The isotope of interest is purified, mass separated, and extracted to HIE-ISOLDE where it was post-accelerated to 4.3 MeV/u, i.e. just in the region of the Coulomb barrier of the system ^{15}C+^{208}Pb, and finally made to impinge on the reaction target.

Outgoing charged particles were measured with the GLORIA setup [3], consisting in six telescopes symmetrically surrounding the reaction target, each telescope

J. D. Ovejas—for the IS619 collaboration ISOLDE CERN.

J. D. Ovejas (✉)
IEM - CSIC, Serrano 113 bis, 28006 Madrid, Spain
e-mail: j.diaz@csic.es

J.-E. García-Ramos et al. (eds.), *Basic Concepts in Nuclear Physics: Theory, Experiments and Applications*, Springer Proceedings in Physics 225,
https://doi.org/10.1007/978-3-030-22204-8_32

formed by two DSSSDs: one 40 μm thick ΔE stage and a secondary 1 mm thick E detector. This configuration allows for the identification of light ions at laboratory angles comprised within 15° and 165° with a total geometric efficiency of 25% of 4π and with an angular resolution of 2.5°. The detector resolution FWHM is in the order of 30 keV.

32.2 Data Analysis and Theoretical Interpretation

The telescope ΔE–E configuration makes possible particle and ion identification and hence a gate can be imposed in the corresponding energetic range of the elastic ^{15}C events. Integration of these events in small angular intervals covered by groups of DSSSD pixels, normalized to the solid angle subtended in each interval, and finally normalized to the Rutherford scattering distribution, leads to the angular distribution of the elastic cross-section. The halo structure of ^{15}C should then be manifested as a remarkable decrease around the Coulomb rainbow.

Theoretical coupled reaction channels calculations for the elastic scattering have been performed [4] and can be seen in Fig. 32.1 (right). Dashed line corresponds to the well known Fresnel pattern, which features the Coulomb rainbow at grazing angles, enhancing the cross-section over Rutherford due to the interference between electrostatic and nuclear potentials. Solid line is the full calculation taking into account the single neutron stripping coupling, favored by the 1-n halo structure. The on-going analysis of the obtained experimental data should be able to favor one or the other of the two curves.

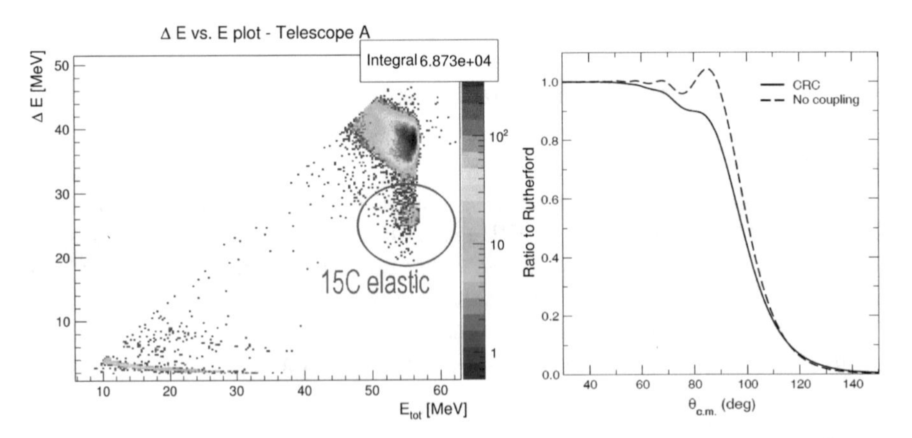

Fig. 32.1 To the left is shown a bi-dimensional plot obtained in one telescope. Here the ^{15}C elastic events can be selected for further analysis. To the right is shown the comparison of two theoretical angular distribution of the elastic cross-section depending upon taking the n-halo structure into account or not [4]

References

1. A. Ozawa et al., Nucl. Phys. A **738**, 38–44 (2004)
2. V. Pesudo et al., Phys. Rev. Lett. **118**, 152502 (2017)
3. G. Marquínez-Durán et al., Nucl. Instrum. Methods Phys. Res. **755**, 69–77 (2014)
4. N. Keeley et al., Eur. Phys. J. A **50**, 145 (2014)

Chapter 33
Preliminary Data of $^{10}Be/^{9}Be$ Ratios in Aerosol Filters in Mexico City

S. Padilla, C. G. Méndez, C. Solís, L. Acosta, M. Rodríguez-Ceja and E. Chávez

Abstract The 1 MV AMS system of Laboratorio Nacional de Espectrometría de Masas con Acelerador (LEMA) has been used to measure the meteoric ^{10}Be. The preliminary results of $^{10}Be/^{9}Be$ ratios in PM_{10} from Mexico City are presented. The aim of this work is to provide new data for meteoric $^{10}Be/^{10}Be$ ratios in this kind of samples. The radiochemical procedure to extract the ^{10}Be in aerosols collected in quartz filters has been modified and adapted in LEMA laboratory.

33.1 Introduction

The meteoric ^{10}Be ($T_{1/2} = 1.39 \times 10^6$ y) is a cosmogenic radionuclide which is produced by the interaction of cosmic rays with the isotopes of the atmosphere by spallation reactions. It is primarily produced at lower stratosphere and upper troposphere and finally, is carried and deposited on the surface joined with aerosols, in soluble form in rain or dry deposition. The ^{10}Be associated to aerosols can be used as tracer of atmospheric processes and concretely indicators of the cosmogenic interactions in lower Stratosphere, upper Troposphere, the air exchange between both and deposition processes on the Earth surface [1].

S. Padilla (✉) · C. G. Méndez · C. Solís · L. Acosta · M. Rodríguez-Ceja · E. Chávez
Laboratorio Nacional de Espectrometría de Masas Con Acelerador (LEMA). Instituto de Física, Universidad Nacional Autónoma de México (UNAM). Circuito de la Investigación Científica Ciudad Universitaria, CP 04510 Mexico, D.F, Mexico
e-mail: spadilla@fisica.unam.mx

C. G. Méndez
Cátedra CONACYT, Laboratorio Nacional de Espectrometría de Masas Con Acelerador (LEMA). Instituto de Física, Universidad Nacional Autónoma de México (UNAM), Mexico City, Mexico

J.-E. García-Ramos et al. (eds.), *Basic Concepts in Nuclear Physics: Theory, Experiments and Applications*, Springer Proceedings in Physics 225,
https://doi.org/10.1007/978-3-030-22204-8_33

33.2 Sampling and Sample Separation

The samples were taken on the roof of the Institute of Physics-UNAM (19º19′26″N 99º10′37″W), about 2300 m high. Atmospheric aerosols were collected on quartz filters (Pallflex 2500 of 20 × 25 cm, QAT-UP; Pall Sciences, Ann Arbor, MI, USA) using a high-volume sampler of PM_{10} with 1.9 $m^3 min^{-1}$ flow (Graseby Andersen SA-2000H). The sampling period for each sample was 48 h in the cold-dry season of 2012 (from November 19th to December 6th). The filters were wrapped in aluminium foil and stored at 4 °C until its processing. The radiochemical procedure consisted in a calcination, digestion with HF and the use of several ion exchange resins in order to separate and purify the ^{10}Be. Finally, the samples were calcinated to BeO and measured in the Low Energy AMS at LEMA, at the Institute of Physics of UNAM [2, 3].

33.3 Results

The measurement of ^{10}Be at LEMA were carried out at charge state +1 after of the Terminal Voltage and +1 after of absorber passive. The values obtained were a 7% from the nominal ratio of the standard. The ^{10}B interference is separated from ^{10}Be using a 75 nm absorber passive located after of ESA [4]. The characterization of the system elements such as magnets, ESA and electrostatics lens, was carried out with standards provided by K. Niishizumi and blanks mixed with Nb. The first results of $^{10}Be/^{9}Be$ obtained in the 1 MV AMS at LEMA are shown in Fig. 33.1a. The values are in a range of 1.95–4.16 × 10^{-12}. Those data are in totally agree with the ones obtained in the city of Seville by Padilla [3] (Fig. 33.1b). The samples reported by Padilla were taken on propylene filters during a week per sample, with an approximately surface of 43 × 43 cm^2 and with a mean flow of 100,000 m^3/week along 2013 [3], while in this work the flow was 3000 m^3/day. The aliquots taken in both studies correspond to 22.30–82.25 $\mu g/m^3$ in the city of Seville and 2–8.7 $\mu g/m^3$ in Mexico City, therefore a higher amount of particles were taken in the city of Seville samples. As a similar amount of ^{9}Be carrier were added in both studies, the values of ratios from the city of Seville are higher [3].

The preliminary data of $^{10}Be/^{9}Be$ ratio in aerosols collected in quartz filters from Mexico City have been obtained at LEMA. The radiochemical procedure has been succefully adapted in LEMA laboratory. The elements such as magnets, ESA and electrostatic lens of 1 MV AMS at LEMA have been optimized for the measurement of ^{10}Be in environmental samples. The data are in totally according with the ones reported in similar studies [2]. A new measurement meteoric ^{10}Be in $PM_{2.5}$ is being carried out and analyzed at LEMA. New and definitive ^{10}Be data in different campaigns and particle sizes will be reported in the future by LEMA.

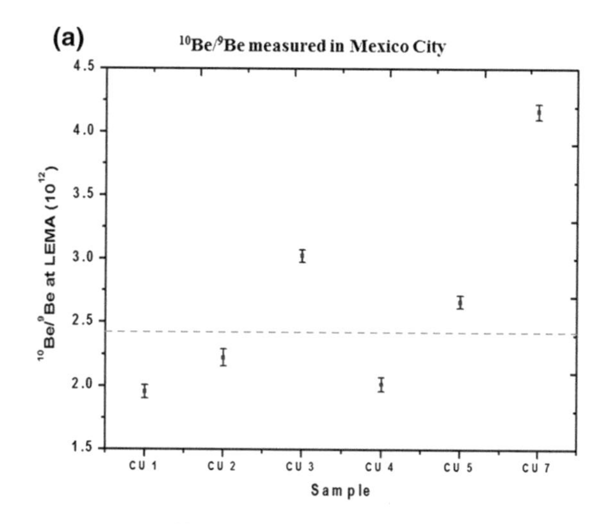

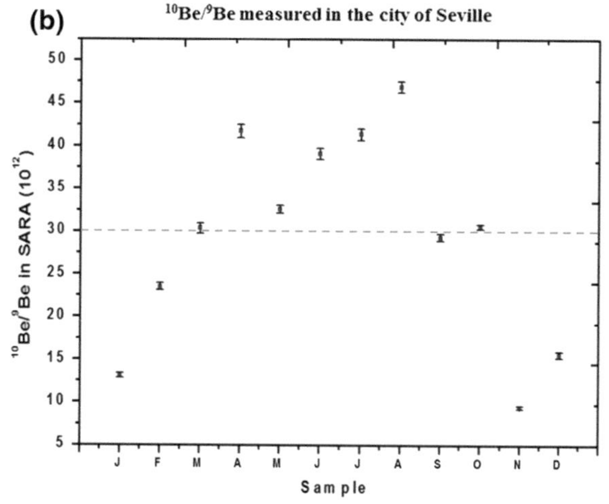

Fig. 33.1 **a** $^{10}Be/^{9}Be$ ratio obtained from quartz aerosol filters taken in Mexico City and measured in the 1 MV AMS at LEMA and **b** data of the $^{10}Be/^{9}Be$ ratio in aerosol filters in the city of Seville reported by Padilla [3]. The dashed lines correspond to the average ratio value for both studies

References

1. F. von Blanckenburg, J. Bouchez, H. Wittmann, Earth surface erosion and weathering from the 10Be (meteoric)/9Be ratio. Earth Planet. Sci. Lett. **351–352**, 295–305 (2012). https://doi.org/10.1016/J.EPSL.2012.07.022
2. M. Auer, W. Kutschera, A. Priller, D. Wagenbach, A. Wallner, E.M. Wild, Measurement of 26Al for atmospheric and climate research and the potential of 26Al/10Be ratios. Nucl. Instrum. Methods Phys. Res. Sect. B Beam Interact. Mater. At. **259**, 595–599 (2007). https://doi.org/10.1016/j.nimb.2007.01.305
3. S. Padilla Domínguez, Medidas de 10Be y 26Al en espectrometría de masas con acelerador de baja energía en el Centro Nacional de Aceleradores, PhD Thesis, University of Seville, 2015
4. C. Solís, E. Chávez-Lomelí, M.E. Ortiz, A. Huerta, E. Andrade, E. Barrios, A new AMS facility in Mexico. Nucl. Instrum. Methods Phys. Res. Sect. B Beam Interact. Mater. At. **331**, 233–237 (2014). https://doi.org/10.1016/J.NIMB.2014.02.015

Chapter 34
Measurement of Signal-to-Noise Ratio in Straw Tube Detectors for PANDA Forward Tracker

Narendra Rathod, Jerzy Smyrski and Akshay Malige

Abstract PANDA forward tracker consist of self supporting straw tube detector for reconstruction of trajectories of charged particles passes through it, particle identification. The basic properties of straw tube detector and signal-to-noise ratio with results are presented in this paper.

34.1 Self-Supporting Straw Tube Detectors

The PANDA experiment [1] will be built at the FAIR facility at Darmstadt (Germany) to conduct experimental studies of the strong interaction through pp and pA annihilation. To track charged particles emitted at the most forward angles within acceptance of the PANDA Forward Spectrometer, the Forward Tracker (FT) consisting of a set of planar straw tube layers will be used [2]. The straws for the FT consist of very thin aluminized Mylar cathodes and are made self-supporting by means of 1 bar overpressure of the working gas mixture [3]. The expected high counting rates, reaching up to 1 MHz/straw, and particle fluxes upto 25 kHz/cm^2, are the main challenges for the FT and the associated readout electronics.

The Forward Tracker (FT) is based on self-supporting straw tubes, with 10 mm in diameter. In these straws, the applied gas overpressure is of 1 bar for forward trackers which provides the mechanical stiffness and maintains the anode wire tension. The straw tubes are made of aluminized Mylar foil, with a thickness of only 27 μm, which results in a very low material budget of only 2% X_0 for the entire FT containing 24 double layers of straws. The material budget in the FT should be as low as possible

N. Rathod (✉) · J. Smyrski · A. Malige
The Marian Smoluchowski Institute of Physics, Jagiellonian University,
Łojasiewicza 11,30-348 Kraków, Poland
e-mail: nsrathore.rajput@gmail.com

J.-E. García-Ramos et al. (eds.), *Basic Concepts in Nuclear Physics: Theory, Experiments and Applications*, Springer Proceedings in Physics 225,
https://doi.org/10.1007/978-3-030-22204-8_34

(less than 5% X_0) in order to achieve the required momentum resolution $\Delta p/p$ 1.5%. Central anode is gold-plated tungsten-rhenium wire, with 20 μm diameter, is used.

34.2 Signal-to-Noise Ratio

The signal-to-noise ratio was measured for the straw tube pulses at the analog output of the Front End Electronic card (FEE). Amplitude of pulses was measured for 5.9 keV X-rays from ^{55}Fe source (see Fig. 34.1) at the anode voltage of 1700 V, the measurements was done for the gain parameter in the FEE set to 1 mV/fC and three different settings of the peaking time parameter: 15, 20 and 35 ns and there respective amplitudes are 162, 168 and 174 mV. The pulses were registered by using CAEN Digitizer DT5742 working at a sampling frequency 1 GHz and a resolution of 0.25 mV. For each registered waveform, the baseline level was determined as an average of the first 100 samples preceding the straw pulse and then was subtracted from all samples in the waveform. A distribution of the first 100 samples corrected for the baseline was fitted with a gaussian function and the standard deviation of the function was taken as the level of noise (see Fig. 34.2) represents distribution for 20 ns. The level of noise was 1.00, 0.84 and 0.80 mV, and the signal-to-noise ratio was 162, 200 and 218 respective to there stated peaking time. We conduct studies of additional shielding of the straws with extra layer of aluminized Mylar for further reduction of the noise.

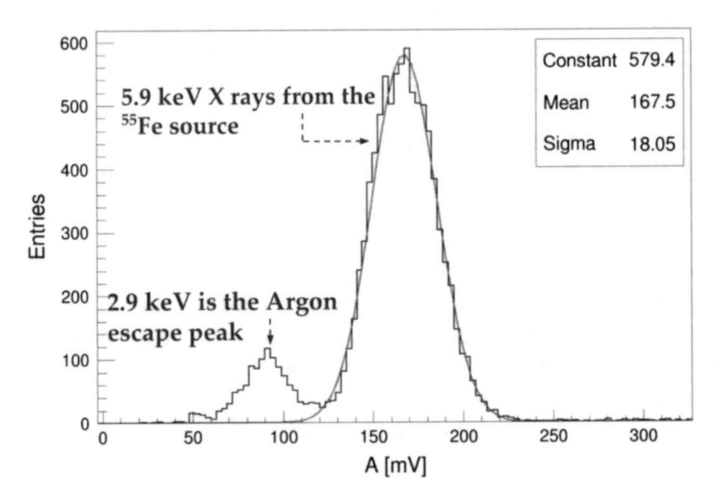

Fig. 34.1 Amplitude spectrum registered for X-rays from ^{5}5Fe source, for the anode voltage 1700 V and the s2 setting of FEE (black line)

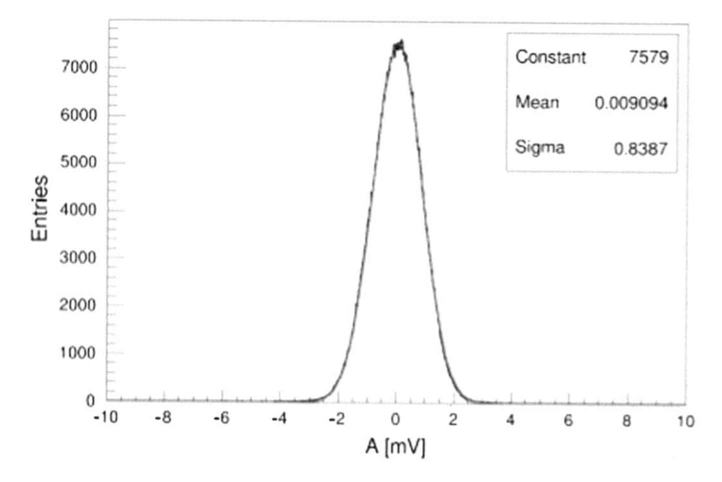

Fig. 34.2 Distribution of noise for the s2 setting of FEE (black line) together with a fitted gaussian function (red line)

Acknowledgements We acknowledge financial support by the Polish National Science Center-Centre, grant agreement no. 2016/23/P/ST2/04066 and carried out under POLONEZ programme which has received funding from the European Union's Horizon 2020 research and innovation programme under the Marie Skodowska-Curie grant agreement No 665778; by the Polish National Science Centre, grant 2017/26/M/ST2/00600 POLONEZ, as well as; and by the Ministry of Science and Higher Education grant agreement no. 7150/E-338/M/2018.

References

1. Official website: https://panda.gsi.de/
2. J. Smyrski et al., JINST **12**, C06032 (2017)
3. J. Smyrski et al., Pressure stabilized straw tube modules for the PANDA Forward Tracker, accepted in JINST

Chapter 35
Investigation of the Pygmy Dipole Resonance Across the Shell Closure in Chromium Isotopes

P. C. Ries, T. Beck, J. Beller, M. Bhike, U. Gayer, J. Isaak, B. Löher, Krishichayan, L. Mertes, H. Pai, N. Pietralla, V. Yu. Ponomarev, C. Romig, D. Savran, M. Schilling, W. Tornow, V. Werner and M. Zweidinger

Abstract For several nuclei in the vicinity of shell closures an accumulation of electric dipole strength around the particle separation threshold has been reported. This strength accumulation is often referred to as pygmy dipole resonance (PDR). As part of the systematic survey of the PDR, the dipole responses of the even-even chromium isotopes ^{50,52,54}Cr have been measured using the phenomenon of nuclear resonance fluorescence. Spin and parity quantum numbers as well as excitation energies and transition strengths were measured for numerous known and previously unknown states. The comparison to calculations within the quasi-particle phonon model and between the isotopes provide new insight about the onset of the PDR and its evolution to heavier and more neutron rich nuclei.

The electric dipole responses of numerous nuclei have been extensively studied throughout the past decades. In addition to the two-phonon states [1] and the isovector giant dipole resonance (GDR) [2] many nuclei show another common feature, an accumulation of strength around the neutron separation energy, the pygmy dipole resonance (PDR) [3]. It was shown, that the strength gathered in the PDR is connected to the so-called symmetry energy, a term of a common Taylor expansion of the nuclear equation of state (EOS) in terms of density and isospin-asymmetry, via the neutron

P. C. Ries (✉) · T. Beck · J. Beller · U. Gayer · J. Isaak · B. Löher · L. Mertes · N. Pietralla · V. Yu. Ponomarev · C. Romig · M. Schilling · V. Werner · M. Zweidinger
Institut für Kernphysik, Schlossgartenstrasse 9, Darmstadt, Germany
e-mail: pries@ikp.tu-darmstadt.de

H. Pai
Saha Institute of Nuclear Physics, Salt Lake, Block - AF Bidhan nagar, Sector 1, Kolkata 700064, West Bengal, India

M. Bhike · Krishichayan · W. Tornow
Triangle Universities Nuclear Laboratory, 101 Circuit Drive, Durham, NC 27708-0319, USA

D. Savran
GSI Helmholtzzentrum für Schwerionenforschung, Planckstraße 1, 64291 Darmstadt, Germany

J.-E. García-Ramos et al. (eds.), *Basic Concepts in Nuclear Physics: Theory, Experiments and Applications*, Springer Proceedings in Physics 225,
https://doi.org/10.1007/978-3-030-22204-8_35

skin thickness [4]. Fundamental nuclear properties depend on a reliable description of the EOS and thus determining the parameters of this series is of great interest. Part of this is the systematic investigation of the PDR.

Experiments using the nuclear resonance fluorescence method were conducted on the isotopes ^{50,52,54}Cr. The experiments were twofold: On one hand, at the Darmstadt high intensity photon setup DHIPS [5] the targets were irradiated with a continuous energy bremsstrahlung spectrum produced by the Darmstadt electron linear accelerator S-DALINAC [6]. Measuring the scattered photons with HPGe detectors yields excitation energies, spin quantum numbers and transition strengths of the excited states. Excitation energies and transition strengths are calibrated via ^{11}B measured simultaneously. On the other hand, the polarized and quasi-monoenergetic photon beam of the High Intensity γ-ray Source HIγS allows the determination of parity quantum numbers at the γ^3 setup [7]. The corresponding identification of electric dipole transitions with these observables is rather unambiguously. Calculations within the quasi-particle phonon model exclude these transition as part of the GDR and thus the onset of the PDR within the chromium isotopes was confirmed.

In order to account for transitions strength through intermediate states the HIγS measurements provide average branching ratios: For each energy setting, the summed up intensity of all ground state transitions of the states excited within the energy range of the photon beam is compared to the intensity of the decay of the first excited 2^+ state, which is only fed from the aforementioned higher-lying levels. The overall strength can be corrected by these branching ratios and expressed in percentage exhaust of the Thomas–Reiche–Kuhn sum rule [8]. The comparison of these percentages for the measured chromium isotopes shows a significant increase as a function of neutron number above the shell closure, which provides new insight in the origin and behavior of the PDR. A publication is currently prepared.

Acknowledgements The authors would like to thank the crew at the S-DALINAC, as well as the crew at HIγS for providing excellent beams. This work was supported by the Deutsche Forschungsgemeinschaft under Contract No. SFB 1245 and by HGS-HIRe.

References

1. N. Pietralla, Empirical correlation between two-phonon $E1$ transition strengths in vibrational nuclei. Phys. Rev. C **59**, 2941–2944 (1999)
2. M. Goldhaber, E. Teller, On nuclear dipole vibrations. Phys. Rev. **74**, 1046–1049 (1948)
3. D. Savran, T. Aumann, A. Zilges, Experimental studies of the pygmy dipole resonance. Prog. Part. Nucl. Phys. **59**, 2941–2944 (2013)
4. A. Brown, Neutron radii in nuclei and the neutron equation of state. Phys. Rev. Lett. **85**, 5296–5299 (2000)
5. K. Sonnabend et al., The Darmstadt high-intensity photon setup (DHIPS) at the S-DALINAC. Nucl. Instrum. Methods Phys. Res. A **640**, 6–12 (2011)
6. N. Pietralla, Nucl. Phys. News **28**, 4–11 (2018)
7. B. Löher et al., The high-efficiency γ-ray spectroscopy setup γ^3 at HIγS. Nucl. Instrum. Methods Phys. Res. A **723**, 136–142 (2013)
8. W. Kuhn, Über die Gesamtstärke der von einem Zustande ausgehenden Absorptionslinien. Z. Phys. **33**, 408–412 (1925)

Chapter 36
Modification of UO_2 Fuel Thermal Conductivity Model at High Burnup Structure

B. Roostaii, H. Kazeminejad and S. Khakshournia

Abstract The thermal conductivity of UO_2 under in-pile irradiation changes due to produced porosity. This paper has calculated evolution of fuel swelling and porosity by using a model of fuel matrix swelling due to fission gases that is valid for range of low temperature and burnups up to 120 MWd/KgU where high burnup structure (HBS) is formed. The HALDEN thermal conductivity correlation is selected for study of the fuel swelling and porosity evolution effect on irradiated UO_2 thermal conductivity. In addition, the correlation is completed with a proposed porosity factor. With considering porosity evolution by burnup, is seen a reduction about 25% in the thermal conductivity which increases the fuel temperature. Calculation results indicate a good agreement with experimental data.

36.1 Introduction

The UO_2 fuel is under irradiation-induced recrystallization at low temperatures with increasing burnup [1]. Up to now, various equations have been developed to describe the effect of porosity on the thermal conductivity. DART code [2] applies a model sequentially for modelling of dispersion fuel thermal conductivity. This model takes a unit cell of porous material represented as a cube of the solid material surrounding a gas pore and derives an analytical expression for the porosity effect on thermal conductivity. In a work, Spino et al. [3] obtained a relation for total matrix swelling, which is related to the fuel bulk density and porosity. In previous work of the authors [4] an expression was derived for total volume porosity of the fuel that was consist of two parts: volume porosity and swelling porosity. In this paper, is achieved using a model for determining porosity evolution based on the work of Spino et al. [3] as well as the Rest model [1, 2] for UO_2 swelling with progressive recrystallization. Then as a case study, the HALDEN thermal conductivity correlation is selected and combined with the porosity factor of DART code.

B. Roostaii (✉) · H. Kazeminejad · S. Khakshournia
Nuclear Science and Technology Research Institute (NSTRI), P. O. Box 14155-1339, Tehran, Iran
e-mail: broostaii@aeoi.org.ir

J.-E. García-Ramos et al. (eds.), *Basic Concepts in Nuclear Physics: Theory, Experiments and Applications*, Springer Proceedings in Physics 225,
https://doi.org/10.1007/978-3-030-22204-8_36

36.2 Models and Methods

36.2.1 Swelling and Porosity

Using the Rest method for calculation of swelling by burnup in low temperature regime for low and high burnup [1, 3] and the definition of total porosity, P, (that has two contributions: One comes from pores, P_v, including as-fabricated voids and the other named swelling porosity, P_s, stems from the irradiation induced fission gas bubbles) in our previous works [4], we have the relation for P_v and P_s by burnup.

36.2.2 HALDEN Thermal Conductivity Correlation

The HALDEN thermal conductivity correlation is as (36.1).

$$k_{95} = \frac{1}{0.1148 + 0.0035\mathrm{Bu} + \left(2.475 \times 10^{-4}\right)(1 - 0.003\mathrm{Bu})(T - 273) + 0.0132\exp(0.00188(T - 273))} \tag{36.1}$$

where k_{95} is thermal conductivity of UO_2 in 95% theoretical density (10.96 kg/cm^3), Bu is burnup in MWd/kgU and T is temperature in Kelvin (K).

36.2.3 Porosity Factor

It is assumed that the fuel has a three-phase structure consisting of the pores, P_v, and swelling porosity, P_s, dispersed in the fully dense material composed of UO_2 matrix and solid fission products. We can obtain (36.2) with using the DART code thermal conductivity model [2].

$$\kappa_P = \frac{k_{eff}}{k_0} = \left[1 - \pi\left(\frac{3}{4\pi}P_v\right)^{\frac{2}{3}}\right]\left\{1 - \left[\pi\left(\frac{3}{4\pi}P_s\right)^{\frac{2}{3}}\right]\left[1 - \frac{k_g}{2k_e^V\left(\frac{3}{4\pi}P_s\right)^{\frac{1}{3}}}\right]\right\} \tag{36.2}$$

where k_{eff} is effective thermal conductivity of porous fuel material including P_v and P_S, and k_g is thermal conductivity of the Xe [2] and k_0 is thermal conductivity of fully dense UO_2 material.

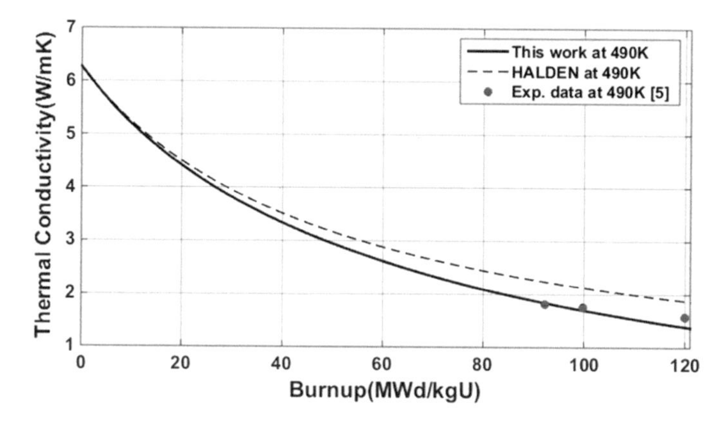

Fig. 36.1 Calculated UO_2 fuel thermal conductivity based on HALDEN correlation versus burn-up at 490 K with porosity evolution included in comparison with the case with constant porosity and experimental data [5]

36.3 Result and Discussion

36.3.1 Thermal Conductivity

Figure 36.1 compares the evolution of calculated UO_2 thermal conductivity based on the origin HALDEN correlation and its modified as a function of local burn-up. Figure 36.1 also shows that the model can predict the experimental data [5] at the temperature of 490 K.

36.4 Conclusion

In this work, we modified HALDEN thermal conductivity model with a porosity factor. It can be seen that taking into account the evolution of porosity with burn-up leads to a decrease in the thermal conductivity of about 25% at a local burn-up of 120 MWd/kgU at 490 K.

References

1. J. Rest, in *Comprehensive Nuclear Materials*, ed. by R.J.M. Konings, vol. 3 (Elsevier, Amsterdam, 2012), pp. 579–627
2. J. Rest, *The DART Dispersion Analysis Research Tool: A Mechanistic Model for Predicting Fission-Product-Induced Swelling of Aluminum Dispersion Fuels*, AN L-95/36 (1995)
3. J. Spino, J. Rest, W. Goll, C.T. Walker, Matrix swelling rate and cavity volume balance of UO_2 fuels at high burnup. J. Nucl. Mater. **346**, 131–144 (2005)

4. B. Roostaii, H. Kazeminejad, S. Khakshournia, Influence of porosity formation on irradiated UO_2 fuel thermal conductivity at high burnup. J. Nucl. Mater. **479**, 374–381 (2016)
5. C.T. Walker, D. Staicu, M. Sheindlin, D. Papaioannou, W. Goll, F. Sontheimer, On the thermal conductivity of UO_2 nuclear fuel at a high burnup of around 100 MWd/kgHM. J. Nucl. Mater. **350**, 19–39 (2006)

Chapter 37
Shell Model Calculations for Nuclei with Two Valence Nucleons Around the Doubly Magic ^{78}Ni Core

Hanane Saifi and Fatima Benrachi

Abstract In this work, we study the nuclear structure of pfg—shells in term of shell model theory, in particular, systems with two identical valence nucleons out of the ^{78}Ni core. Calculations were performed using a new effective interaction named jj45pnc. Energies of low lying states, B(E2; $J_i \rightarrow J_f$) and deformation parameter, quadrupole moment and $R_{4/2}$ ratio are evaluated for the studied nuclei. Our results are compared with available experimental data.

37.1 Introduction and Theoretical Calculations

Study of nuclei properties in the neighbouring of closed shells Z = 28, and N = 50, is an experimental and theoretical active research object. It permits testing the bases of different theoretical approaches. In this context, we have employed shell model calculation with a new realistic effective interaction jj45pnc obtained on the basis of the similarity existed between the ^{132}Sn and ^{78}Ni region; and extracted from the interaction jj56pn which already proved quite successful in describing the ^{132}Sn region. Calculations were performed using the **NuShellX@MSU** code. Single particle energies are taken from the experimental spectra of ^{79}Cu nucleus, assuming the $1f_{5/2}$ as ground state [1, 2]. Whereas, the neutron single particle energies determined realizing calculation with Hartree–Fock method employing the Skyrme potential SK19-(skxm) [3].

H. Saifi (✉) · F. Benrachi
Laboratoire de Physique Mathématique et Subatomique, Université Frères Mentouri, Constantine 1, Constantine, Algeria
e-mail: hanane.saifi25@gmail.com

F. Benrachi
e-mail: fatima.benrachi@umc.edu.dz

J.-E. García-Ramos et al. (eds.), *Basic Concepts in Nuclear Physics: Theory, Experiments and Applications*, Springer Proceedings in Physics 225,
https://doi.org/10.1007/978-3-030-22204-8_37

37.2 Results and Discussion

We study the ^{80}Zn and ^{80}Ni Nuclei with two valence protons and neutrons respectively. Structure of these nuclei has a major interest to understand the mechanism of populating the proton and neutron shells and also to test the pp and nn part of the effective interaction. The calculated level's energies are compared to the available experimental data for both nuclei and presented on Fig. 37.1.

Calculations reproduce satisfactorily the excitation energies of ^{80}Zn, and shows that low lying states are dominated by the occupation of $1f_{5/2}$ orbit for protons. Where states of ^{80}Ni have main configuration $2d_{5/2}$ orbit for neutrons with percentages range from 80 to 99%. On the other hand, we realize a fit in order to reproduce the available experimental spectroscopic properties for ^{80}Zn, which allowed us to get the best values of effective charges ($e_\pi = 1.95e$, $e_\nu = 0.5e$), when we use the standard values ($e_\pi = 1.5e$, $e_\nu = 0.5e$) for the ^{80}Ni. Results are summarized on Tables 37.1 and 37.2.

Deformation parameter β and quadrupole moment values show the existence of deformation, where the $R_{4/2}$ ratio confirm that they have shell configuration.

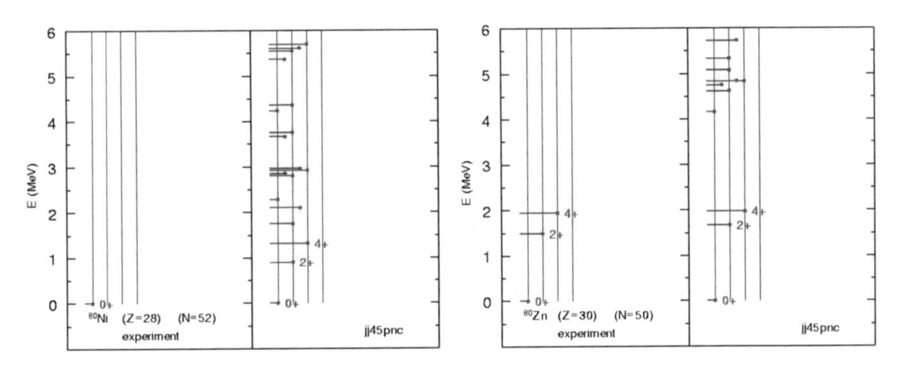

Fig. 37.1 Calculated and experimental spectra [2, 4] of ^{80}Ni (left) and ^{80}Zn (right)

Table 37.1 Spectroscopic properties of ^{80}Zn nucleus

B(E2:↑) e^2fm^4			Q(e^2fm^2)		μ(μ_N)		$R_{4/2}$		β	
	Exp	jj45pnc	J^π	jj45pnc	J^π	jj45pnc	Exp	jj45pnc	Exp	jj45pnc
$0^+_{g.s} \to 2^+_1$	730	602	2^+_1	8.36	2^+_1	0.73	1.33	1.18	0.141	0.128
$2^+_1 \to 4^+_1$	/	166.6	4^+_1	−19.2	4^+_1	1.39				

Table 37.2 Spectroscopic properties of ^{80}Ni nucleus

B(E2:↑) e^2fm^4		Q(e^2fm^2)		μ(μ_N)		$R_{4/2}$	β
	jj45pnc	J^π	jj45pnc	J^π	jj45pnc	1.467	0.053
$0^+_{g.s} \rightarrow 2^+_1$	91.11	2^+_1	−3.30	2^+_1	−1.21		
$2^+_1 \rightarrow 4^+_1$	26.41	4^+_1	−6.64	4^+_1	−2.72		

37.3 Conclusion

We have study the structure of nuclei with two identical valence nucleons out of the doubly magic ^{78}Ni in term of shell model theory. Calculations show good agreement with the experimental data, where the spectroscopic properties show that these nuclei have deformed states, on the other hand it confirms the success of the interaction jj45pnc to describe nucleus in very exotic area of ^{78}Ni.

Acknowledgements Special thanks to B. A. Brown and M. Hjorth-Jensen for their advices.

References

1. L. Oliver et al., Phy. Rev. Lett. **119**, 192501 (2017)
2. National Nuclear Data Center, www.nndc.bnl.gov
3. B.A. Brown et al., Phy. Rev. **C58**, 220 (1998)
4. Y. Shiga et al., Phy. Rev. **C93**, 024320 (2016)

Chapter 38
Investigation of the Mechanism of Proton Induced Spallation Reactions

U. Singh, I. Ciepał, B. Kamys, P. Lasko, J. Łukasik, A. Magiera, P. Pawłowski, K. Pysz, Z. Rudy and S. K. Sharma

Abstract The mechanism of proton-nucleus interactions at GeV energies is still not well understood. The agreement between data and model predictions deteriorates with increasing ejectile energy and for the forward angles. This indicates the presence of preequilibrium processes which are not taken into consideration by present theoretical models. New experimental data are needed to put constraints to any new model of the reaction mechanism. The measurements are planned to be done to study proton induced reactions at energies in the range from 70 to 230 MeV.

38.1 Introduction

The spallation reactions have a wide range of applications in many fields of science and technology, e.g. astrophysics, material science, hybrid nuclear reactors, production of rare isotopes for medical purposes. All these applications are based on the knowledge of the spallation cross sections, which frequently cannot be obtained experimentally but has to be determined from theoretical models which in turn must be tested by experimental data. Possible nonequilibrium processes are presented schematically in Fig. 38.1 together with the intranuclear cascade mechanism which is a standard part of the present day models. To proof reliability of the new theoretical models it is necessary to compare their predictions with inclusive and coincidence cross sections of proton induced reactions on several nuclear targets at various beam energies. The present project concerns such measurements to be done with the proton beam in the energy range from 70 to 230 MeV.

U. Singh (✉) · B. Kamys · A. Magiera · Z. Rudy · S. K. Sharma
M. Smoluchowski Institute of Physics, Jagiellonian University, Łojasiewicza 11, 30-348 Cracow, Poland
e-mail: udai.singh@doctoral.uj.edu.pl

I. Ciepał · P. Lasko · J. Łukasik · P. Pawłowski · K. Pysz
H. Niewodniczański Institute of Nuclear Physics PAN, Radzikowskiego 152, 31-342 Cracow, Poland

J.-E. García-Ramos et al. (eds.), *Basic Concepts in Nuclear Physics: Theory, Experiments and Applications*, Springer Proceedings in Physics 225, https://doi.org/10.1007/978-3-030-22204-8_38

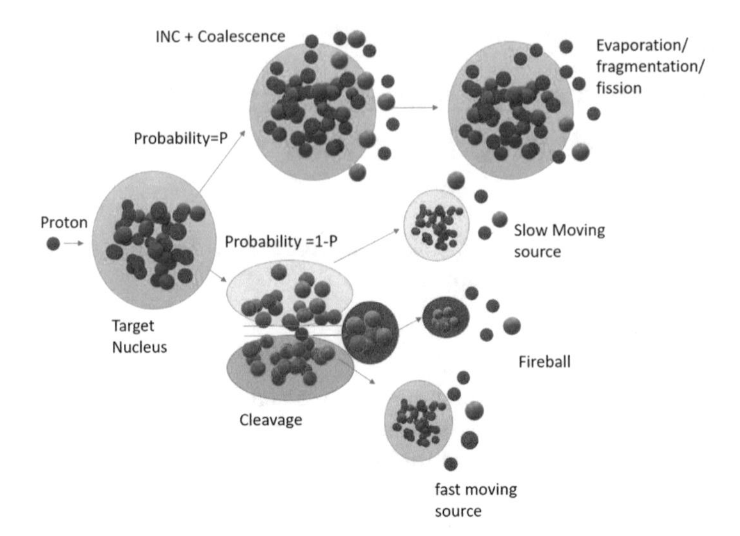

Fig. 38.1 Possible modes of the spallation reactions. The proton of the beam impinging on to the target nucleus initiates with probability P the intranuclear cascade of nucleon-nucleon and pion-nucleon collisions (present day models) or may induce with probability 1-P a cleavage of the target nucleus into three excited groups of nucleons: the smallest—fireball and two larger—the fast and slow moving sources (possible preequilibrium processes)

38.2 Investigations

It is planned using the new PROTEUS cyclotron of the Cyclotron Centre Bronowice, to measure the single spectra $d\sigma/d\Omega dE$ and coincidence spectra $d\sigma/d\Omega_1 dE_1 d\Omega_2 dE_2$ of LCP (i.e. $^{1-3}$H, ^{3,4}He) and IMF (i.e. ^{6}He, Li, Be, B, C, N, O.. ions) in proton induced reactions on various target nuclei (Al, Ni, Ag, Au). The experiment will be performed at different beam energies (70–230 MeV). The KRATTA [1] (Krakow Triple Telescope Array) detection system can be used to measure the energy, angle of emission and isotopic composition of LCP and IMF. The main goal of the present experiment is to investigate experimentally the hypothesis presented above i.e. the presence of the "fireball" contribution to the reaction mechanism.

Reference

1. J. Łukasik, P. Pawłowski, A. Budzanowski, B. Czech, I. Skwirczyńska, J. Brzychczyk, M. Adamczyk, S. Kupny, P. Lasko, Z. Sosin et al., Nuclear Instruments and Methods in Physics Research Section A: Accelerators. Spectrometers, Detectors and Associated Equipment **709**, 120–128 (2013)

Chapter 39
Challenging the Calorimeter CALIFA for FAIR Using High Energetic Photons

P. Teubig, P. Remmels, P. Klenze, H. Alvarez-Pol, E. Alves, J. M. Boillos, P. Cabanelas, R. C. da Silva, D. Cortina-Gil, J. Cruz, D. Ferreira, M. Fonseca, D. Galaviz, E. Galiana, R. Gernhäuser, D. González, A. Henriques, A. P. Jesus, H. Luís, J. Machado, L. Peralta, J. Rocha, A. M. Sánchez-Benítez, H. Silva and P. Velho

Abstract Proton induced γ-ray reactions were used as a tool to probe and characterize a CALIFA (**CAL**orimeter for the **I**n-**F**light detection of γ-rays and light charged p**A**rticles) Barrel segment consisting of 128 CsI(Tl) crystals and readout photodetectors. In this work, we present the individual crystal response to photons from inelastic and radiative capture reactions of protons on ^{27}Al, as well as a preliminary analysis of the calorimetric response of the prototype, with special focus on the energy resolution of the calorimeter.

39.1 Probing CALIFA with High Energetic Photons

The barrel-shaped section of the calorimeter CALIFA will surround the target at the upcoming R^{3}B (Reaction with Relativistic Radioactive Beam) experimental setup at FAIR (Facility for Antiproton and Ion Research, Darmstadt, Germany) [1, 2]. In this manuscript, we report on the experiment performed at the LATR (Laboratory for Accelerators and Radiation Technologies) facility at the CTN/IST [3] site (Sacavem, Portugal). A detector array consisting of 128 CALIFA crystals were exposed to high-

P. Teubig (✉) · D. Ferreira · D. Galaviz · E. Galiana · A. Henriques · L. Peralta · P. Velho
LIP, Av. Prof. Gama Pinto 2, 1649-016 Lisboa and FCUL, Portugal
e-mail: pteubig@lip.pt

P. Remmels · P. Klenze · R. Gernhäuser
TUM, Munich, Germany

H. Alvarez-Pol · J. M. Boillos · P. Cabanelas · D. Cortina-Gil · D. González
IGFAE, Univ. de Santiago de Compostela, 15706 Santiago de Compostela, Spain

E. Alves · R. C. da Silva · H. Luís, J. Rocha
IPFN, IST-UL, Lisboa, Portugal

J. Cruz · M. Fonseca · A. P. Jesus · J. Machado · H. Silva
LIBPhys-UNL, Lisboa, Portugal

A. M. Sánchez-Benítez
Dept. de Ciencias Integradas, Facultad de Ciencias Experimentales, Univ. Huelva, Huelva, Spain

J.-E. García-Ramos et al. (eds.), *Basic Concepts in Nuclear Physics: Theory, Experiments and Applications*, Springer Proceedings in Physics 225,
https://doi.org/10.1007/978-3-030-22204-8_39

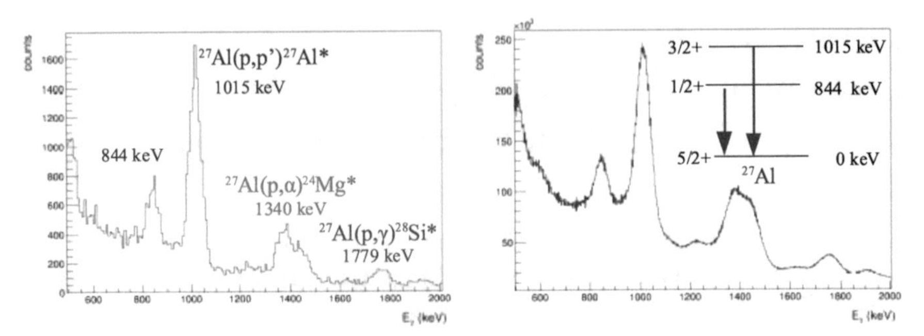

Fig. 39.1 Proton induced γ–rays from reactions on ^{27}Al (inelastic and radiative capture). The spectra measured by a single crystal (left) and by the whole assembly of 64 crystals (calorimetric, right) are shown

Table 39.1 Relative FWHM for the CALIFA prototype: single crystal versus calorimeter

Type	$E_{844\ \text{keV}}$ (%)	$E_{1015\ \text{keV}}$ (%)	$E_{1779\ \text{keV}}$ (%)
Single crystal	5.2	4.9	3.5
Calorimeter (64 crystals)	6.6	6.3	4.7

energetic γ-rays up to 14 MeV from the decay of ^{28}Si*. Several specific resonances in the reaction $^{27}Al(p,\gamma)^{28}Si$ [4] were populated to produce high-energy γ-rays, which were used to evaluate the response of the detector array.

A semi-automated calibration procedure was implemented considering photons from ^{27}Al* (844 and 1015 keV), and a ^{60}Co (1173 and 1333 keV) radioactive source. Photon spectra from different proton induced reactions on ^{27}Al are shown for a single crystal (Fig. 39.1, left) and the calorimetric response of 64 crystals (Fig. 39.1, right). Furthermore, the energy resolution for each photopeak has been evaluated (Table 39.1), pointing to a slight worsening (30%) for the calorimetric mode. The dependence of the resolution on the event multiplicity is being studied, as well as the analysis of the high-energy γ-rays observed during the experiment beyond 8 MeV.

References

1. The CALIFA Collaboration, CALIFA Barrel Technical Design Report (2012). https://fair-center.eu/fileadmin/fair/publications_exp/CALIFA_BARREL_TDR_web.pdf
2. H. Alvarez-Pol et al., NIMA **767**, 453–466 (2014)
3. http://ctn.tecnico.ulisboa.pt
4. J. Brenneisen et al., Z. Phys. A **352**, 149–159 (1995)

Chapter 40
Research and Development of a Position-Sensitive Scintillator Detector for γ- and X-Ray Imaging and Spectroscopy

Zh. Toneva, V. Bozhilov, G. Georgiev, S. Ivanov, D. Ivanova, V. Kozhuharov, S. Lalkovski and G. Vankova-Kirilova

Abstract Preliminary results of a study for a scintillator detector with a position sensitive capabilities are presented here. The first tests have been performed with a 7 mm thick plastic scintilator and 256-channel Multi-Anode Photomultiplier Tube (MAPMT). The study shows that good position resolution is in reach.

40.1 Introduction

Gamma-rays in the sub-MeV region interact with matter through photoelectric effect and Compton scattering. As such, the full photon absorption is a result of one or few point-like events, where the entire photon energy, or a portion of it, is passed to electrons from the medium. Exploiting this phenomenon, position sensitive γ-ray detectors have been used for more than 50 years in different fields, such as astronomy and space observation [1], nuclear physics [2] and medical applications [3], where γ- and X-ray imaging is required. There are different technologies existing on the market today mainly based on large scintillators or high purity germanium (HPGe) detectors. These however, have rather poor special resolution and imaging capabilities.

Research and development (R&D) study for a scintillator detector with a position sensitive capabilities is now performed at University of Sofia, as a part of the project "Novel Detectors for Gamma-Ray Astronomy". The aim of the presented project is to create a prototype of a position-sensitive detector for γ- and X-rays imaging and spectroscopy that can be utilized in space-borne nano telescopes. The detector will be used to identify lines in the range of 20 keV to 1 MeV. The prototype is envisioned to

Zh. Toneva (✉) · V. Bozhilov · G. Georgiev · S. Ivanov · V. Kozhuharov
S. Lalkovski · G. Vankova-Kirilova
Faculty of Physics, University of Sofia "St. Kliment Ohridski", Sofia, Bulgaria
e-mail: zh.h.toneva@phys.uni-sofia.bg

D. Ivanova
Military Medical Academy, Sofia, Bulgaria

J.-E. García-Ramos et al. (eds.), *Basic Concepts in Nuclear Physics: Theory, Experiments and Applications*, Springer Proceedings in Physics 225,
https://doi.org/10.1007/978-3-030-22204-8_40

be built of a single non-segmented scintillator and coupled to a position-sensitive light sensor. This approach will help to obtain the full energy of the deposited radiation while enhancing the spatial resolution of the detector.

Given that a single detector is expected to provide spectroscopic information and generate images of astronomical objects it has to have good position, time and energy resolutions. In particular, its position resolution should be better than 5 mm, which is a prerequisite for the generation of sharp images. For identification and disentanglement of different radiation sources, an energy resolution of 5% at 662 keV is required. The prototype is expected to have a time resolution of the order of 150 ps, which is crucial for the background subtraction in the telescope, while being in space mounted on a nano satellite. Further constraints on detector size, weight and power consumption are defined by the nano satellites' specifics. High efficiency is also a important factor that will reduce the observation time. Furthermore, given the constrains of the nano satillite carying the telescope, an optimal geometry has to be determined such that the telescope remains efficent for γ-rays but also to have a good position resolution.

40.2 Test Experiments

The goal of NDeGRA project is to identify and characterise possible materials, such a scintillators and light sensors, that can be used for the construction of the detector, capable to track all interaction points, unambiguously determine the first point of interaction and the initial γ-ray energy. Our approach is based on the assumption that Compton and photo-effect electrons are absorbed close to the interaction point, and where the majority of the scintillations also happen.

In the initial experimetnal setup a 7 mm thin plastic scintillator with a surface of 25 cm^2, coupled to a 256-channel MAPMT, model H9500 [4] have been used. The size of each anode is 2.8 $\times$ 2.8 mm [4]. The MAPMT was connected to one V1751 8-channel 1 GS/s digitizer with a 10-bit resolution [5]. Multiple scans with $^{90}Sr/^{90}Y$ source were performed at 5 mm steps in two orthogonal directions aligned with the detector front plane. Sampled data was recorded for different combinations of 8 arbitrary MAPMT channels.

Preliminary results from these tests are shown in Fig. 40.1, where the charge obtained from an arbitrary MAPMT channel is depicted as a function of source to anode distance. The initial tests results shows that the presently achieved position resolution is better than 8 mm. Further tests, analysis and interpretations are now ongoing.

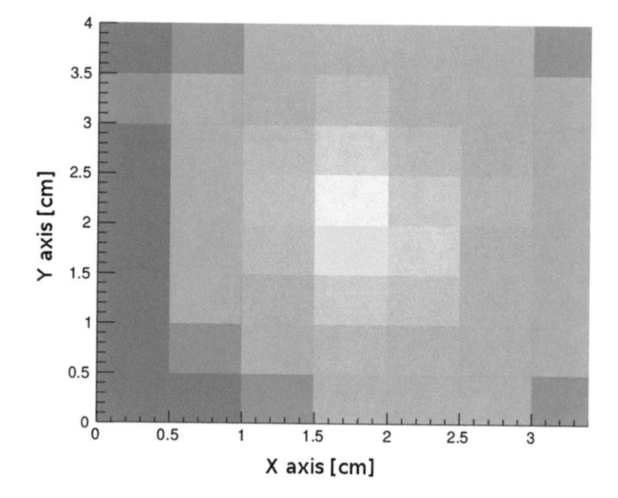

Fig. 40.1 The intensity on the image represents the charge obtained from an arbitrary MAPMT channels, when the $^{90}Sr/^{90}Y$ source is set at 1.5 cm at X axis and 2 cm at Y axis. See text

40.3 Conclusions and Outlook

Position-sensitive scintilator detector for space-borne nano telescopes, capable of detecting X- and γ-ray energies, and their interaction points in the scintillator, is now being researched under NDeGRA project. The goal is to find possible materials and to achieve detector geometries that can fit inside a nano satellite. At first, position-sensitive detector was built of plastics scintillator and MAPMT and then scanned with electron source. This approach allows for a better control on the interaction point and is considered to be a simplified version of the multi- Coulomb scattering events where the initial, and later, the scattered γ-rays would have energies less then 1 MeV. With the present setup, a position resolution better than 8 mm is achieved with $^{90}Sr/^{90}Y$ source. Next iteration will be to scan $CeBr_3$ crystals coupled to MAPMT and SiPM, with X- and γ-rays.

Acknowledgements This work is funded by the Bulgarian National Science Fund under contract DN18/17.

References

1. A. Gostojic et al., Characterization of $LaBr_3$: Ce and $CeBr_3$ calorimeter modules for 3D imaging in gamma-ray astronomy. Nucl. Instrum. Methods Phys. Res. A **832** (2016)
2. O.J. Roberts et al., A $LaBr_3$: Ce fast-timing array for DESPEC at FAIR. Nucl. Instrum. Methods Phys. Res. A **748** (2014)
3. S.L. Meo et al., Optical physics of scintillation imagers by GEANT4 simulations. Nucl. Instrum. Methods Phys. Res. A **607** (2009)
4. Hamamatsu Photonics, Flat Panel Type Multianode Photomultiplier Tube Assembly **H9500**, H9500–03 (2015)
5. CAEN Electronic Instrumentation, *User Manual UM3350* (2017)

Chapter 41
The Path Towards Low Dose CT: The Case of Breast CBCT

A. Villa-Abaunza, P. Ibáñez García, J. López Herraiz and J. M. Udías

Abstract Commercial CT systems include tools to estimate the dose received by the patient in a CT scan, but they tend to overestimate the dose for large patients and underestimate it for small/pediatric ones. In this work we consider the case of Cone Beam CT (CBCT) for breast imaging to explore the feasibility of obtaining accurate and fast estimates of the dose received during a low dose CT. The Hybrid-Ultra MC algorithm, developed in the group for fast and accurate dose estimation, has been used for the study, with the well validated MC code PenEasy as the reference. Comparing analytical and realistic projections, we estimated the improvement achievable when the later ones are used in the reconstruction, and we concluded that when a realistic projection model is used and acquisition parameters are adapted to the density and size of the breast, CBCT images obtained under the dose limits of Digital Mammography (DM) may reach diagnostic quality.

41.1 Introduction

Commercial CT systems include tools to estimate the dose received by the patient in a CT scan, but they tend to overestimate the dose for large patients, and underestimate it for small/pediatric ones, because they do not consider patient size, age, gender, or specific organs or regions [1]. The aim of this work is to explore the feasibility of obtaining accurate and fast estimates of the dose received during low dose CT. Cone Beam CT (CBCT) for breast imaging is a good test case. If the dose is adequately managed while keeping good image quality, CBCT could be used in cancer breast screening with advantages with respect to X-ray digital mammographs (DM). DM is compact and achieves high resolution, but requires compression of the breast, and it has difficulties to identify tumors in dense breasts [2]. CBCT has the potential to pro-

A. Villa-Abaunza (✉) · P. Ibáñez García · J. López Herraiz · J. M. Udías
Grupo de Física Nuclear and Iparcos, Facultad de Ciencias Físicas, Universidad Complutense de Madrid, CEI Moncloa, 28040 Madrid, Spain
e-mail: amavil01@ucm.es

J.-E. García-Ramos et al. (eds.), *Basic Concepts in Nuclear Physics: Theory, Experiments and Applications*, Springer Proceedings in Physics 225,
https://doi.org/10.1007/978-3-030-22204-8_41

vide high quality 3D images without breast compression and with dose comparable to DM.

41.2 Methods

In order to obtain accurate dose estimates in very short times, the Hybrid Ultra-Monte Carlo (HUMC) dose calculation algorithm, developed in the group [3], was used. PenEasy [4], a slow but well validated MC code, was used to obtain reference dose distributions. The image reconstruction of the simulated CBCT breast projections was performed using TIGRE [5] a MATLAB GPU-based toolkit. TIGRE was adapted to be able to incorporate projections produced with the HUMC. Comparing the standard TIGRE projections with the HUMC ones, we estimated the improvement achievable when realistic projections are used in the reconstruction. Several detector geometries have been tested in order to optimize the acquisition and reconstruction parameters.

41.3 Results

Dose estimates and projections obtained with the HUMC are similar enough to the ground truth (penEasy) both for the integral dose estimates (within 10%) and for the reconstructed images. HUMC is 500 times faster than penEasy under similar computing power (a single thread on a CPU), which enables real time dose estimates with common 10-core PCs. Furthermore, CBCT images obtained under the dose limits of DM (4 mSv) may reach diagnostic quality, especially when a realistic projection model is used, and acquisition parameters are adapted to the density and the size of the breast (Fig. 41.1).

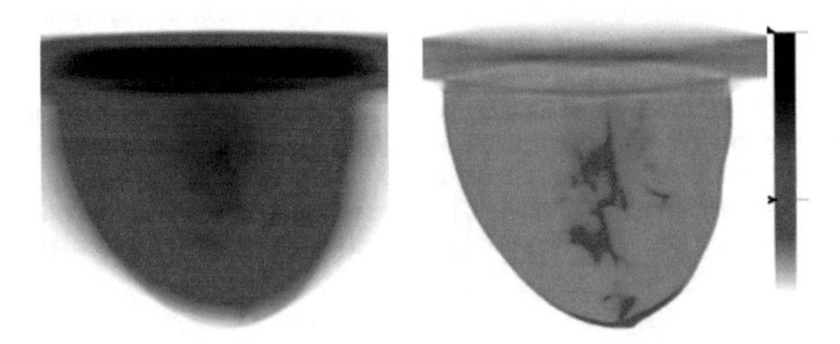

Fig. 41.1 Reconstruction of a simulated breast phantom, assuming monochromatic 35 keV X-ray beam, using the FDK analytical algorithm (left) and the iterative MLEM algorithm with 15 iterations (right)

41.4 Conclusion

This work shows that personalized CT acquisitions, together with ultra-fast Monte Carlo and reconstruction methods may provide diagnostic good quality image with reduced noise.

Acknowledgements This work was supported by Comunidad de Madrid (S2013/MIT-3024 TOPUS-CM), Spanish Ministry of Science and Innovation (Spanish Government) XIORT (IPT-2012-0401-300000).

References

1. K. Ono et al., SpringerPlus **21**, 393 (2013)
2. J.M. Boone et al., Radiology **221**(3), 657–667 (2001)
3. M. Vidal et al., Radiother. Oncol. **111**(S1), 117–118 (2014)
4. J. Sempau, A. Badal, L. Brualla, Med. Phys. **38**(11), 5887–5895 (2011)
5. A. Biguri et al., Biomed. Phys. Eng. Express **2**(5), 055010 (2016)

Chapter 42
Electron Capture of ^{8}B into Highly Excited States of ^{8}Be

S. Viñals

Abstract The experiment (IS633) was performed at ISOLDE facility by the MAG-ISOL collaboration. Our interest lies in determining the branching ratios of the 2^+ doublet at 16.6 and 16.9 MeV populated via β^+ and electron capture (EC) respectively, and also the so far unobserved EC-delayed proton emission via the 17.6 MeV state. The 2^+ doublet is interesting due to the high isospin mixing (von Brentano, Phys Rep, 264:57, 1996, [1]), leading to dominant configurations as ^{7}Li+p and ^{7}Be+n respectively. The feeding to the 17.6 MeV state is especially interesting due to the proton-halo character of ^{8}B as a core of ^{7}Be+p. Considering the halo proton as a spectator, the decay branch can then estimated from the ^{7}Be core to be in the order of 2.3×10^{-8} (Borge et al, J Phys G, 40:035109, 2013, [2]), which is to be verified by this experiment.

42.1 Experiment and Preliminary Results

The first part of the experiment, focused on the 2^+ doublet, was performed using a setup of 4 ΔE-E telescopes composed by a front Double-sided Si strip detector (DSSD) as ΔE backed by Si-PAD E-detector in a diamond configuration and an extra thick DSSD on the bottom. The setup was optimized for the detection of the fragmentation of the ^{8}Be into 2α particles. A total geometrical efficiency of 28% leads to an efficiency of 56% to detect any one of the two αs. Further, having done so, to detect the second α in coincidence (at 180°) is almost 100%. An angular resolution of 3 degrees is obtained due to the high pixilation of the DSSD-detectors.

In the left side of the Fig. 42.1 is shown the high energy part of the summed 2α spectrum; the two well-defined peaks are the 2^+ doublet at 16.6 and 16.9 MeV. The energies are in good agreement with the literature [3], and we have obtained enough

Viñals—IS633 collaboration ISOLDE CERN.

S. Viñals (✉)
IEM - CSIC, Serrano 113, 28006 Madrid, Spain
e-mail: s.vinals@csic.es

J.-E. García-Ramos et al. (eds.), *Basic Concepts in Nuclear Physics: Theory, Experiments and Applications*, Springer Proceedings in Physics 225,
https://doi.org/10.1007/978-3-030-22204-8_42

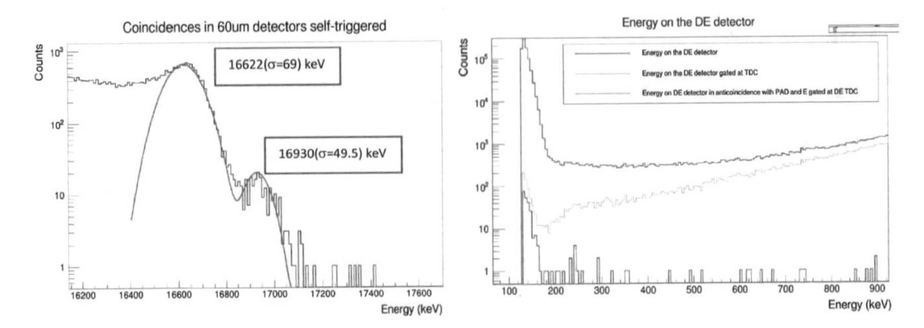

Fig. 42.1 To the **left**, peaks of the 2^+ doublet obtained from the α-α coincidences. To the **right**, the resulting single α low-energy spectrum obtained in the ΔE detector after applying different filters to remove contaminating noise and β signals in the detector. In red is displayed the raw low-energy part of the α-spectrum. In green the spectrum after the data is gated by the TDC i.e. self-triggered data. In purple what is left over in the region of interest after applying both: the TDC cut and an anti-coincidence with the other detectors in order to remove the β response

statistics ($\sim$450 counts) in the 16.9 MeV peak in order to determine a branching ratio within a 5% error, but for that the ongoing R-matrix analysis of the data is needed.

The second part of the experiment was optimised to set an upper limit on the branching ratio to the 17.6 level and thus the emission of a 330 keV proton. This proton, fed in EC, is only followed by a non-detectable low-energy X-ray, i.e. should have no other signal in coincidence. To optimize for such detection, we choose a thin ΔE detector of 30 μm with negligible β-response. This detector was fronted by a big Si-PAD covering 3 times the solid angle of ΔE to ensure that all the α-α events are in coincidence in both detectors. Another big Si-PAD detector placed behind the ΔE to VETO all β that might have passed the ΔE detector. All this to ensure as background free spectrum as possible in the region from 200 to 400 keV. After an anti-coincidence analysis, a preliminary experimental upper limit of the emission of the delayed-proton can be stablished to 4.4×10^{-6}. The right side of the Fig. 42.1 shows the resulting spectrum after the application of different filters used to clean the data in the region of interest.

References

1. P. von Brentano, Phys. Rep. **264**, 57 (1996)
2. M.J.G. Borge et al., J. Phys. G **40**, 035109 (2013)
3. Energy levels of light nuclei A = 8, TUNL (2012)

Chapter 43
Examining the Helium Cluster Decays of the ^{12}Be Excited States by Triton Transfer to the ^{9}Li Beam

N. Vukman, N. Soić, P. Čolović, M. Uroić, M. Freer, T. Davinson, A. Di Pietro, M. Alcorta, D. Connolly, A. Lennarz, C. Ruiz, A. Shotter, M. Williams and A. Psaltis

Abstract We present the first results of the experiment: "Examining the helium cluster decays of the ^{12}Be excited states by triton transfer to the ^{9}Li beam" (spokespersons: N. Soić, M. Freer), done at TRIUMF, Vancouver, CA, with the main goal of providing precise experimental data on the internal structure of the ^{12}Be excited states.

43.1 Motivation and the Experimental Method

Light nuclei, due to the small number of relevant degrees of freedom, present excellent framework in which to study the basic principles of nuclear interactions and structure: from single-particle dynamics to the appearance of clustering in the nuclei [1]. Structure of the ^{12}Be excited states decaying to the helium isotopes has been studied in several experiments so far, with scarce and contradictory results [2]. In recent years, considerable theoretical effort and advances have been made, indicating the existence of exotic clustering in ^{12}Be, namely molecular α-4n-α structure [3]. Due to it's importance for understanding the structural changes in neutron-rich light nuclei, precise experimental data is needed. Prior to the experiment MC simulations

N. Vukman (✉) · N. Soić · P. Čolović · M. Uroić
Ruđer Bošković Institute, Zagreb, Croatia
e-mail: nvukman@irb.hr

M. Freer
University of Birningham, Birningham, UK

T. Davinson
University of Edinburgh, Edinburgh, UK

A. Di Pietro
INFN-Laboratori Nazionali del Sud, Catania, Italy

M. Alcorta · D. Connolly · A. Lennarz · C. Ruiz · A. Shotter · M. Williams
TRIUMF, Vancouver, BC, Canada

A. Psaltis
McMaster University, Hamilton, ON, Canada

J.-E. García-Ramos et al. (eds.), *Basic Concepts in Nuclear Physics: Theory, Experiments and Applications*, Springer Proceedings in Physics 225,
https://doi.org/10.1007/978-3-030-22204-8_43

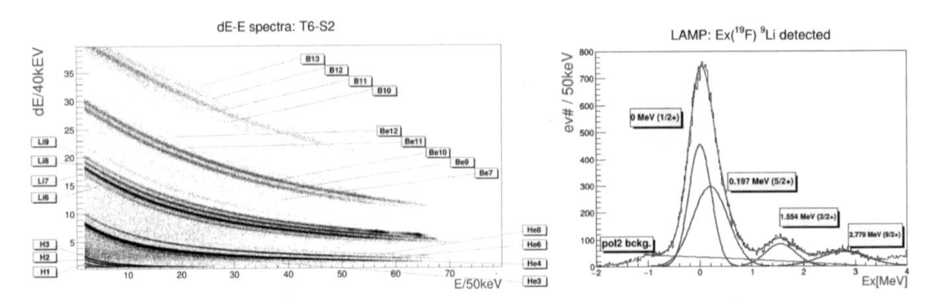

Fig. 43.1 ΔE-E (telescope 6, strip 2) and excitation spectra of ^{19}F from ^{19}F(^{9}Li, ^{9}Li)^{19}F reaction

were carried out in AUSA code, developed by Aarhus University, DK, to optimize the setup and efficiency for kinematically complete measurement. The ^{9}Li beam (75 MeV), provided by ISAC-II accelerator facility at TRIUMF, CA, was hitting the ^{7}LiF target (1 mg/cm^{2}). Two set of YY1 wedge detectors of thickness 65 μm and 1500 μm, arranged as ΔE-E telescopes, were positioned in the LAMP configuration inside the TUDA chamber, covering θ range from 16° to 48°.

43.2 Calibration Process and Preliminary Results

We obtained the energy calibration using an α source and ^{9}Li elastic scattering on the Au target, taking into account the relevant energy losses. This calibration is then used for the fine tuning of the measured geometry. The 2-body reactions excitation spectra of ^{7}Li and ^{19}F (see Fig. 43.1) are obtained, with the right position of ground state and first few excited states verifying the quality of the calibration. The next step in the analysis is to study the coincident helium isotopes detected from the break-up of the ^{12}Be* (^{12}Be* $\rightarrow$ ^{6}He+^{6}He, ^{12}Be* $\rightarrow$ ^{8}He+^{4}He) and produced predominantly via reactions on ^{7}Li (^{4}He recoil, Q > 0) or less frequently on ^{19}F (^{16}O recoil, Q < 0). Due to the large Q-value differences those two cases can be separated in the analysis.

References

1. M. Freer et al., Rev. Mod. Phys. **90**, 035004 (2018)
2. M. Freer et al., Phys. Lett. B **775**, 58 (2017); Z.H. Yang et al., Phys. Rev. Lett. **112**, 162501 (2014); Z.H. Yang et al., Phys. Rev. C **91**, 024304 (2015); A. Saito et al., Nucl. Phys. A **738**, 337 (2004); R.J. Charity et al., Phys. Rev. C **76**, 064313 (2007); M. Freer et al., Phys. Rev. Lett. **82**, 295 (1999)
3. P. Maris, M.A. Caprio, J.P. Vary, Phys. Rev. C **91**, 014310 (2015); M. Ito, K. Ikeda, Rep. Prog. Phys. **77**, 096301 (2014); M. Ito et al., Phys. Rev. Lett. **100**, 182502 (2008); Y. Kanada-En'yo, H. Horiuchi, Phys. Rev. C **68**, 014319 (2003)

Chapter 44
Fast-Timing Lifetime Measurement of ^{174,176,178,180}Hf

J. Wiederhold, V. Werner, R. Kern, N. Pietralla, D. Bucurescu, R. Carroll, N. Cooper, T. Daniel, D. Filipescu, N. Florea, R.-B. Gerst, D. Ghita, L. Gurgi, J. Jolie, R. Ilieva, R. Lica, N. Marginean, R. Marginean, C. Mihai, I. O. Mitu, F. Naqvi, C. Nita, M. Rudigier, S. Stegemann, S. Pascu and P. H. Regan

Abstract Lifetimes of yrast band states of ^{174,176,178,180}Hf have been measured using fast-electronic scintillation timing or Coulomb excitation. The lifetimes of the 2_1^+, 4_1^+ states and apart from ^{176}Hf also of the 6_1^+ states have been determined, using the slope and the centroid shift methods. By using the same setup for all isotopes of interest systematic uncertainties were reduced.

The method of fast electronic scintillation timing (FEST) [1] makes it possible to determine lifetimes down to a few 10 ps, especially with the recent development of $LaBr_3$ detectors. The typical time resolution of the $LaBr_3$ detectors is 200–300 ps and their energy resolution is about 3%. Critical for the determination of short lifetimes in the ps range a good calibration of the energy-dependent time walk has to be performed (see [2, 3] for more details).

Recent measurements of lifetimes $\tau(2_1^+)$ in the rare-earth region around the mass number A = 170 [4–6] showed discrepancies of up to 20% to literature data. In the present work we show results from a FEST experiment, that has been performed at

J. Wiederhold (✉) · V. Werner · R. Kern · N. Pietralla
Institut für Kernphysik, TU Darmstadt, Schlossgartenstraße 9, 64289 Darmstadt, Germany
e-mail: jwiederhold@ikp.tu-darmstadt.de

V. Werner · N. Cooper · R. Ilieva · F. Naqvi
Wright Nuclear Structure Laboratory, Yale University, New Haven, CT 06520, USA

D. Bucurescu · D. Filipescu · N. Florea · D. Ghita · R. Lica · N. Marginean · R. Marginean · C. Mihai · I. O. Mitu · C. Nita · S. Pascu
"Horia Hulubei" NIPNE, 077125 Bucharest-Magurele, Romania

R. Carroll · T. Daniel · L. Gurgi · R. Ilieva · P. H. Regan
Department of Physics, University of Surrey, Guildford, Surrey GU2 7XH, UK

R.-B. Gerst · J. Jolie · M. Rudigier · S. Stegemann
Institut für Kernphysik, Universität zu Köln, 50937 Cologne, Germany

M. Rudigier
Present address: University of Surrey, Guildford, Surrey GU2 7XH, UK

P. H. Regan
National Physical Laboratory, Teddington, Middlesex TW11 0LW, UK

J.-E. García-Ramos et al. (eds.), *Basic Concepts in Nuclear Physics: Theory, Experiments and Applications*, Springer Proceedings in Physics 225,
https://doi.org/10.1007/978-3-030-22204-8_44

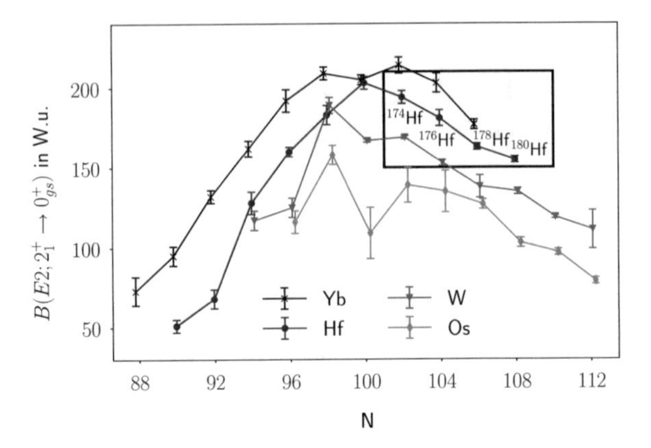

Fig. 44.1 $B(E2; 2_1^+ \to 0_{gs}^+)$ transition strength shown over the neutron number N for the isotopic chains of Yb, Hf, W and Os. The determined values of ^{174,176,178,180}Hf are marked within the black rectangle. The collectivity of the Hf isotopes maximizes at N = 100 and not at neutron mid-shell N = 104. Data taken from this work and [8]

the 9 MV FN Tandem accelerator of the IFIN-HH near Bucharest with their γ-ray detector array ROSPHERE [7]. It consisted of 11 $LaBr_3$(Ce) and 14 HPGe detectors. Excited states of the investigated isotopes were populated via fusion-evaporation reactions or Coulomb excitation. Lifetimes of the $2_1^+, 4_1^+$ states of ^{174,176,178,180}Hf and of the 6_1^+ state of ^{174,178,180}Hf have been measured. The lifetime $\tau(4_1^+)$ of ^{178}Hf has been determined for the first time and the other measured lifetimes are in agreement with the recently reported lifetimes from [5]. Figure 44.1 shows the $B(E2; 2_1^+ \to 0_{gs}^+)$ strengths of the Yb, Hf, W and Os isotopic chains around neutron mid-shell N = 104. Values determined in this work are highlighted in the black rectangle. The maximum of collectivity, i.e. the $B(E2, 2_1^+ \to 0_{gs}^+)$ strength, is not at neutron mid-shell (N = 104) as would be naively expected, but shifted towards lower neutron number for the Hf isotopic chains. A similar trend can be seen in the neighboring isotopic chains. To date there is no microscopic explanation for the early drop of collectivity. Further results and discussion can be found in [9].

Acknowledgements This work was supported by the Deutsche Forschungsgemeinschaft under the Grant No. SFB 1245 and JO 391/16-2, the BMBF under the grants 05P(15/18)RDFN1 and 05P(15/18)RDFN9 within the collaboration 05P15 NuSTAR R&D, by the U.S.DOE under Grant No. DE-FG02-91ER40609 and by the Helmholtz Graduate School for Hadron and Ion Research.

References

1. H. Mach et al., Nucl. Instrum. Methods Phys. Res. Sect. A **280**, 49 (1989)
2. J.-M. Regis et al., Phys. Res. Sect. A **726**, 191 (2013)
3. J. Wiederhold et al., Phys. Rev. C **94**, 044302 (2016)
4. A. Costin et al., Phys. Rev. C **74**, 067301 (2006)
5. M. Rudigier et al., Phys. Rev. C **91**, 044301 (2015)
6. V. Werner et al., Phys. Rev. C **93**, 034323 (2016)
7. D. Bucurescu et al., Nucl. Instrum. Methods Phys. Res. Sect. A **837**, 1 (2016)
8. http://www.nndc.bnl.gov/ensdf. Accessed 22 Aug 2018
9. J. Wiederhold et al., Phys. Rev. C **99**, 024316 (2019)

Chapter 45
Searching for Halo Nuclear Excited States Using Sub-Coulomb Transfer Reactions

J. Yang, Pierre Capel and Alexandre Obertelli

Abstract Through this study we propose to use sub-Coulomb transfer to investigate the one-neutron halo phenomenon within nuclear excited states since this method could naturally guarantee the peripherality of the transfer. Zero-range ADWA calculations are performed with the final nucleus bound under different conditions. It can be observed that there is a clear enhancement of the interaction cross sections when the nucleus is loosely bound within an s orbital.

A common feature of halo nuclei is a large spatial extension caused by the valence nucleon(s) loosely bound to a nuclear core [1]. It is mostly found in the neutron-rich nuclei with small one or two neutron separation energy. The weak binding gives rise to an extended tail in the single-particle wave function of the valence part. One of the open questions we would like to explore is the possible halo feature in the nuclear excited states since the binding energy could become very small in those situations. To study the single-particle structure of nuclei, transfer reaction has been widely employed during the past decades. In particular, they are ideal to study the one-neutron halo structure. Even though the corresponding cross sections would be reduced due to the Coulomb repulsion, performing such reaction below the Coulomb

J. Yang (✉) · P. Capel
Physique Nucléaire et Physique Quantique (CP 229), Université libre de Bruxelles, 1050 Brussels, Belgium
e-mail: jiecyang@ulb.ac.be

J. Yang
Afdeling Kern- en Stralingsfysica, KU Leuven, Celestijnenlaan 200d - bus 2418, 3001 Leuven, Belgium

P. Capel
Institut für Kernphysik, Johannes Gutenberg-Universität Mainz, 55099 Mainz, Germany
e-mail: pcapel@uni-mainz.de

A. Obertelli
Institut für Kernphysik, Technische Universität Darmstadt, 64289 Darmstadt, Germany
e-mail: aobertelli@ikp.tu-darmstadt.de

J.-E. García-Ramos et al. (eds.), *Basic Concepts in Nuclear Physics: Theory, Experiments and Applications*, Springer Proceedings in Physics 225,
https://doi.org/10.1007/978-3-030-22204-8_45

barrier would bring several advantages including: (1) Any observed event would most likely be a direct and peripheral transfer; (2) Other channels are negligible, e.g. compound-nucleus formation; (3) The distortion of the elastic Coulomb waves by the nuclear potential is small.

We consider the three-body approximation ($A + n + p$) within the ADWA method to simulate the sub-Coulomb $A(d, p)B$ transfer. A zero-range version of the adiabatic potential developed by Johnson and Soper [2] is used to include the break-up effect of deuteron for the entrance channel. The nucleon-nucleus optical potentials are obtained from the global parametrization CH89 [3] without including the spin-orbit term. For the p-n interaction, the Reid soft-core interaction is chosen to get the appropriate wave function of the deuteron. The nucleus B is described in a two-body model ($A + n$). A and n are bound together by a Woods–Saxon potential, whose depth is adjusted to reproduce the binding energy (BE). The calculations are performed with FRESCO [4].

A ^{90}Zr-like nucleus is selected as an example for testing our idea. Several assumed binding energies and states of the generated nucleus are considered (see Table 45.1). Through comparison in Fig. 45.1, we find that loosely bound state in an s orbital, which would be a halo candidate, clearly increases the transfer probability at backward angles.

Table 45.1 Several parameters of the corresponding transfer reaction. U_{coul} is the Coulomb barrier while E_d and BE represent the incident energy of the deuteron and the binding energy of the product respectively. *State* means which orbital the transferred neutron populates at

Reaction	U_{coul}/MeV	E_d/MeV	State	BE/MeV
^{90}Zr(d, p)^{91}Zr like	~8	4, 6, 8, 10	3s1/2, 2d5/2, 1g7/2	0.5, 2, 7, 10

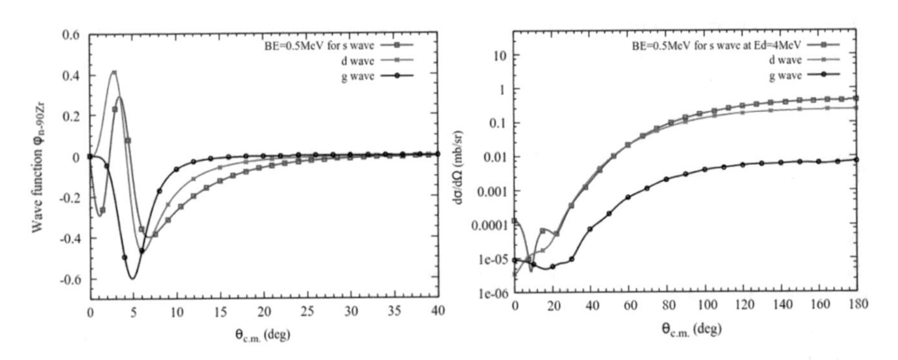

Fig. 45.1 Left side: radial wave functions obtained with the Woods–Saxon potentials for the valence neutron bound with 0.5 MeV at different states inside the ^{91}Zr-like nucleus; Right side: ZR-ADWA calculations of ^{90}Zr(d, p)^{91}Zr-like transfer probability at $E_d = 4$ MeV

References

1. K. Riisager, Phys. Scr. **T152**, 014001 (2013)
2. R.C. Johnson, P.J.R. Soper, Phys. Rev. C **1**, 976 (1970)
3. R. Varner, W. Thompson, T. McAbee, E. Ludwig, T. Clegg, Phys. Rep. **201**, 57 (1991)
4. I.J. Thompson, Comput. Phys. Rep. **7**, 167 (1988)

Printed in the United States
By Bookmasters